P9-DDZ-857

Solving Management Problems

Wiley Series on Systems Engineering and Analysis
HAROLD CHESTNUT, Editor

Chestnut
Systems Engineering Tools

Hahn and Shapiro
Statistical Models in Engineering

Chestnut
Systems Engineering Methods

Rudwick
Systems Analysis for Effective Planning: Principles and Cases

Wilson and Wilson
From Idea to Working Model

Rodgers
Introduction to System Safety Engineering

Reitman
Computer Simulation Applications

Miles
Systems Concepts

Chase
Management of Systems Engineering

Weinberg
An Introduction to General Systems Thinking

Dorny
A Vector Space Approach to Models and Optimization

Warfield
Societal Systems: Planning, Policy, and Complexity

Gibson
Designing the New City: A Systemic Approach

Coutinho
Advanced Systems Development Management

House and McLeod
Large-Scale Models for Policy Evaluation

Rowe
An Anatomy of Risk

Fishman
Principles of Discrete Event Simulation

Rudwick
Solving Management Problems:
A Systems Approach to Planning and Control

Solving Management Problems

A SYSTEMS APPROACH
TO PLANNING AND CONTROL

Bernard H. Rudwick

Planning Research Corporation
McLean, Virginia

A WILEY-INTERSCIENCE PUBLICATION
JOHN WILEY & SONS, New York • Chichester • Brisbane • Toronto

Library of Congress Cataloging in Publication Data:

Rudwick, Bernard H.
Solving management problems.

(Wiley series on systems engineering and analysis)
"A Wiley-Interscience publication."
Includes index.
1. Management—Case studies. 2. System analysis—Case studies. 3. Management informations systems—Case studies. 4. Problem solving—Case studies. I. Title.

HD31.R796 658.4'032 78-23266
ISBN 0-471-04246-3

Printed in the United States of America

10 9 8 7 6 5 4 3 2 1

To

REGINA

SYSTEMS ENGINEERING AND ANALYSIS SERIES

In a society which is producing more people, more materials, more things, and more information than ever before, systems engineering is indispensable in meeting the challenge of complexity. This series of books is an attempt to bring together in a complementary as well as unified fashion the many specialties of the subject, such as modeling and simulation, computing, control, probability and statistics, optimization, reliability, and economics, and to emphasize the interrelationship among them.

The aim is to make the series as comprehensive as possible without dwelling on the myriad details of each specialty and at the same time to provide a broad basic framework on which to build these details. The design of these books will be fundamental in nature to meet the needs of students and engineers and to insure they remain of lasting interest and importance.

Preface

This book describes and illustrates key principles and methods for anticipating and solving management problems. These techniques can be applied to a wide variety of activities, ranging from the control of manufacturing operations to the making of policy decisions. A detailed overview of the book and its objectives is given in Chapter 1.

This book documents parts of courses that I have presented at the State University of New York at Albany and at the Planning Research Corporation in systems planning, management information systems, and operations management. My students tended to be professional systems planners, managers, graduate students in business administration, and senior level students in management. Portions of the material have also been presented at seminars on resource management and systems engineering.

Most of the cases used here to illustrate the principles of planning and control are based on actual problem situations that I have had responsibility for solving during my professional career. Hence I would like to acknowledge with appreciation the following colleagues and organizations who worked with me on these problems.

Case 2 is based on the design of a management information system for the Friction Materials Division of the Bendix Corporation in Green Island, N.Y., as originally documented in "Final Report for Labor Reporting System Implementation." The work was done by a faculty-student team from the State University of New York at Albany. William Palmer, Russell Peter, Charles Friday, and Robert McMahan were the MBA students. Faculty members of the team were Donald Bishko, Harry Smith, Michael Cerullo, and myself. Appreciation for their aid on the project is also given to Charles Menz, former Bendix Division General Manager; William Messier, Controller; and Ken Nasholds, former Manager—Manufacturing.

Case 3 stems from work done for the Naval Telecommunications Command and is adapted from the technical reports prepared by the Center for Naval Analyses under Contract No. N 000 14-68-A-0091-0018: CRC 286, "Manpower Planning Handbook," Volumes I–IV, January 1976. Working with me on this problem were Michael Melich, Diego Rocque, Betsy Nunn, Janice Kofman, and Catherine Anderson. Appreciation is also due to Vice Admiral Jon Boyes, Rear Admiral Kenneth Haynes, and Rear Admiral Gordon Nagler for their sponsorship and encouragement.

Case 4 is adapted from the technical reports prepared by the Planning Research Corporation under Contract No. MDA 903 76 C 0256: TAEG Report No. 44, "Computer Managed Instruction at Remote Sites by Satellite: Phase I. A Feasibility Study," December 1976, and TAEG Report No. 49, "Phase II, A Demonstration Design," December 1977. The contract was jointly sponsored by the Cybernetics Technology Office; Dr. Harold F. O'Neill, Jr., Defense Advanced Research Projects Agency (ARPA); and the Research and Program Development Office, Chief of Naval Education and Training (CNET), Dr. Worth Scanland. CNET's Training Analysis and Evaluation Group (TAEG) served as the Project Officer for CNET and as the Contracting Officer's Technical Representative (COTR). Dr. G. S. Micheli served as Project Officer and Dr. A. F. Smode, Director of TAEG, served as the COTR. Working with me on the problems described were Ken Polcyn, Lawrence Brekka, Bernhard Keiser, and Della Kennelly.

While I have drawn on my experience at the Planning Research Corporation and the Center for Naval Analyses, the contents of this book reflect my own views and not necessarily the official views or policies of either of the organizations or their employees.

Many others have been of help to me in this effort. In particular, I thank three longtime friends who served as my "editorial board" and offered constructive comments on the manuscript. Michael Melich, an astute colleague on a number of Navy projects, including Case 3, has always added that "touch of spice" to our work by "wrestling with me" on the finer points of systems planning. Harold Chestnut's advice and encouragement as series editor motivated me to document my professional experiences for this book as well as my first one. Donald Heany, a pioneer in strategic planning and management science, my mentor, colleague, counselor and friend for 18 years, has helped immeasurably over the years in shaping and crystallizing my thoughts on planning. His discussions and painstaking review of the manuscript were of great value to me.

I also wish to give a special thanks to my many students who struggled through the cases, asked the right questions, and motivated and challenged me in the development of the materials from which the book was written.

To all these donors I express my appreciation for whatever is worthwhile in the book; the shortcomings are my responsibility.

BERNARD H. RUDWICK

Arlington, Virginia
January 1979

Contents

Solving Management Problems

Part I

FOUNDATIONS OF SYSTEMS PLANNING

1

Introduction and Overview

1.1 THE PHILOSOPHY UNDERLYING THIS BOOK

For most of the past 25 years, I have earned my living as a professional planner in a variety of problem areas. The clients for whom I have worked have been high level managers in business as well as government. Their interests have included research and development planning, information systems of many types, defense systems, health care systems, training systems, and business systems. In most cases the projects on which I worked involved finding ways to improve their organizations, which included improving the organization's products or services, reducing costs, increasing profitability, or improving management policies or procedures. These assignments have generally started with an ill-structured problem or "felt need," which was assigned to me. My task was to find a solution to the problem or to provide some analytical contribution to the problem-solving process.

During this time I became intrigued with gaining a better understanding of this planning or problem-solving process, sometimes called front-end planning of ill-defined, unstructured problems. For over 15 of these years I have been developing and teaching a variety of courses involving planning. These had various titles, including systems planning, systems analysis, management information systems, operations management, management planning and control, creativity, and engineering economics. Presenting these courses enabled me to reflect on my experiences in doing this work and to codify the approach that my fellow planners and I were using. During these courses key principles of systems planning were presented, using various case problems or situations from work that I had been involved with as examples to illustrate these principles. These courses were attended by other professional planners, managers from business and civil service, military officers, and MBA and senior-year business students at Northeastern University, Boston College, and the State University of New York at Albany. Most of these participants were planners or middle- and senior-level managers of various operations who had had up to 25 years of experience in planning but had never been exposed to a formalized method of performing this function. Since the participants in many of these courses were assigned the task of recognizing a problem in their area and applying the systems planning method to their problem as a planning exercise, this teaching experience offered me an invaluable opportunity to test and refine the planning process I was using and presenting. In addition, I was able to test its clarity on a wide variety of students and on my clients. I owe these participants a great deal for assisting me in the refinement of the planning process.

1.2 OBJECTIVES OF BOOK

The primary objective of this book is to describe the key principles and methods for planning and problem solving in a systematic, analytical way and to illustrate how to apply these principles by presenting the approach to the solution of a number of management problem situations that use such principles.

I begin this book with the assumption that the reader is familiar with the various needs for doing systematic planning.[1] Basically, without effective planning an organization cannot take advantage of opportunities for improvement as they present themselves. New technology and new ideas enable organizations to reduce costs, improve their products or services, and meet competitive forces. New user needs or those that can now be served by new technology also present opportunities. In some organizations, such as large industrial organizations or the U.S. government, there are so many requests for funding that the annual generation of a long-range plan resulting from a proper selection of investment opportunities is a requirement. Thus the problem is not whether or not to plan but rather how to do better planning.

My final objective is to share the following major conclusions I have reached based on my experiences in planning:

- There is a systematic process for doing planning that, if followed, will gain acceptance from your client for your planning work.[2]
- From this planning process can be derived several generic types of problems.

[1]See B. H. Rudwick, *System Analysis for Effective Planning*, John Wiley and Sons, New York, 1969, Chap. 2, for a further discussion of the need for systematic planning.

[2]It does not follow that the client will accept your recommended conclusions, since he may not have the funds to invest, or he may not choose to accept the risks involved. Rather I am referring to his acceptance of a plan that contains all the ingredients of sound planning.

- These same types of problems keep arising continually, and if one can identify the problem type, he can more easily apply past experience.

To show the reader how these conclusions can aid him in improving his planning efforts is the major objective of this book.

1.3 REVIEW OF THE PLANNING OR PROBLEM-SOLVING APPROACH USED

In this book the terms *planning* and *problem solving* are used interchangeably, since in both cases the planner (the individual responsible for solving the problem)[3] must make some improvement and hence needs to investigate the situation in sufficient detail that he can recommend a course of action.[4] The basic steps involved in a formalized planning approach may be listed as shown in Table 1.1. Although the literature may not agree on the exact number of steps or their names, and the order that they follow may vary with each type of problem, there is general agreement that the problem-solving process should include these elements.

STEP 1. PROBLEM RECOGNITION

During the problem recognition phase the planner observes the operation, focusing on some difficulties facing the organization. In general, the phase ends successfully when his statement of the problem is confirmed by some higher-level authority, who gives approval for continuing the planning effort by going to the next step.

STEP 2. PROBLEM DEFINITION (SOMETIMES CALLED INVESTIGATE DIRECTIONS)

In its broadest sense, problem definition should be thought of as a "quick pass" through the rest of the problem-solving approach. Its primary purpose is to bound the problem by identifying the various elements (boundary conditions, constraints, and variables) that may be causing the

[3]In some organizations this individual may be called a systems planner, analyst, systems analyst, operations researcher, management scientist, or business planner, depending on the type of problem he is solving. In the cases, I shall use that term most appropriate to the case.
[4]This is the ultimate objective of the entire planning process. Planners or analysts may also be concerned with only one facet of the entire process, as when their assignment is to estimate the cost of a set of alternatives, or when they do an analysis of the needs of a user of a system to uncover deficiencies prior to the generation of alternative improvements.

Table 1.1 *Systems Planning Process: A Formalized Problem-Solving Approach*

Step 1.	Problem recognition
Step 2.	Problem definition (investigate directions) Problem objectives Boundary conditions and constraints Variables Measures Planning resources required
Step 3.	Develop alternatives (systems synthesis)
Step 4.	Evaluate alternatives (systems evaluation)
Step 5.	Select preferred alternative (system selection)
Step 6.	Convince others

symptoms observed, and to determine objectives and the various measures of performance and effectiveness that can be used in the evaluation phase. This step results in a further definition of the problem, determination of the analytical approach to be followed, and estimation of the planning resources required to complete the study.

STEP 3. DEVELOP ALTERNATIVES (SOMETIMES CALLED SYSTEMS SYNTHESIS)

Development of alternatives includes the identification and design of alternative courses of action in sufficient detail that the alternatives may be properly compared in Step 4.

STEP 4. EVALUATE ALTERNATIVES (SOMETIMES CALLED SYSTEMS EVALUATION)

Alternatives are next evaluated on the basis of their ability to meet the objectives and of the costs and risks or uncertainties involved.

STEP 5. SELECT PREFERRED ALTERNATIVE (SOMETIMES CALLED SYSTEM SELECTION)

Here the preferred system alternative is chosen from all alternatives considered on the basis of all information relevant to the evaluation process. Some planners (myself included) prefer to consider this step as the logical outcome of a total evaluation process (Step 4). Others view Step 4 as containing only the quantifiable evaluation factors and Step 5 as including all other nonquantifiable factors important to the selection process. In either case the two steps are related.

STEP 6. CONVINCE OTHERS

Since the planning effort is not complete without consideration of all the behavioral aspects involving "resistance to change" that many times prevent a good idea from gaining approval, some planners include such considerations as a separately identifiable step. Others include this step in the design of the alternatives, which includes a time-phased plan for transitioning from the present system to the proposed alternative. In either case such consideration should be included somewhere in the planning process.

The implementation phase follows acceptance of the preferred alternative (the recommended transition plan). At this time the recommended alternative is acted on in accordance with the transition plan.

We have described the major steps in the planning process. The real challenge is how to implement these steps in an effective and efficient manner. Meeting this challenge is the main objective of this book.

1.4 OVERVIEW OF BOOK

This book has been organized as a series of interrelated building blocks for describing and illustrating the key principles of planning for change, as a systematic process, while presenting the generic types of problems for which the process may be used. The remaining two chapters of Part I deal with some of the major tools of planning: the technical aspect (what types of information need to be gathered by the planner) and the behavioral aspect (what are some of the obstacles to be avoided when gathering information from others). Before even visiting a prospective client, the planner needs to have a general understanding of the type of data or information he is looking for. He also needs to have available some sort of structure for assembling such data so that relationships among the data may be expressed. Chapter 2 discusses both of these topics and presents the various characteristics and models useful for describing the composition of a system and its operational performance. The specific ways of applying these models to various problem situations constitute the subject of the rest of the book.

It has been said that data are the ingredients of the planning process, but other people's data can only be gathered through these people and with their permission. Some of these people may be reluctant to provide the data or may even be hostile. Chapter 3 describes some of these behavioral problems and some ways of avoiding them.

The remaining parts of the book consist of a number of complex, unstructured case situations that have been constructed to illustrate the two major theses forming the foundation of this book:

- There is a fairly standard method or process for systematically performing planning.
- Most (if not all) situations or problem contexts that a planner faces can be translated into a small set of generic problem situations related to the steps in the planning approach.

To illustrate the planning process, each of the remaining parts of this book has been designed in the following manner:

- The problem as given (PAG) is presented. The PAG is a cursory statement of the problem by a client, who might also be his manager.
- The problem-solving approach presented in Part I is used in solving the case situation.
- The generalization of the problem is identified so that the reader may use the same analytical approach for other problems of this type.
- At the end of each case the key principles illustrated in that case are summarized.
- The cases build on one another.

Having described some of the techniques that planners use, we next present an example of the entire planning process. Case 1 (Part II, Chapters 4, 5, 6, and 7) illustrates how to apply the tools just developed and others to our first generic problem, system evaluation, which also involves the design of system alternatives and the selection of a preferred alternative based on a comparison of these various ways of doing a certain job.

Case 1 also presents other tools and techniques useful in performing:

- A systems design.
- A resource and cost analysis.
- A method for dealing with risks and uncertainties.

This case will be of particular interest to managers who must review the work of planners, because it describes the types of questions managers should ask of their planners. It also shows the forms that the answers to these questions should take. Although the case involves a fairly simple system, it does involve equipment investments. Hence this case is useful for illustrating how to handle investment or capital budgeting problems of many types.

A word of caution about the remaining cases: Most books describing the planning process do so at a high level of abstraction. The professional planner, however, must get down to fine details if he is to achieve the accuracy required. Hence if I am to achieve my purpose, I must get down to the working level of detail. It is relatively easy to describe the ideal data for accurate planning. However, if these data are not available to the planner within the time period allotted to the planning task, he must know what "fallback position" to assume. Illustrating how to fall back is important. For this reason, the reader may have to go through the details involved. In some cases this may be tedious, but I know no other way of illustrating the practical problems involved in planning and how to cope with them.

Whereas the first case deals with the design and evaluation of a material flow system, the second case (Part III, Chapter 8) extends to the planning process for the design of an information system, specifically one to inform the various levels of management about the performance of the material flow system, thus enabling management to take any corrective actions that are necessary (the management control function). This case focuses on the following objectives:

- Extend the systems design process for material flow systems into the area of the design of an information system required for management control purposes.
- Show how an information system can be designed to meet the needs of various information users.
- Set forth the key principles of management control systems.

Case 3 (Part IV, Chapters 9 and 10) extends the design of information systems to a management planning system for the redesign of a complex telecommunications system. The major problem in this case is to gather data that systematically relate the communications services provided by this system to its manpower cost, to provide the necessary data base for planning purposes. Such a planning data base consists of (1) the set of planning factors and (2) a planning logic needed for properly designing the system, making the necessary trade-offs between communications services available and allowable system costs.

Case 3 has several objectives that relate to the systems planning process followed in the solution of the previous two cases. In Case 1 most of the performance and cost data were provided to the analyst. His primary job was to develop evaluation models that could assemble these data to evaluate the alternatives available. But how is such cost and performance data obtained? This case illustrates how to go into a complex organization and perform a functional analysis needed to gather data for evaluating various alternatives. One of the important elements of the management control system of Case 2 was the quality, time, and cost standards used to determine whether the work was being done in an acceptable manner. Case 3 describes how such control standards may be derived.

The specific example presented is the analysis of the naval telecommunications system. Because this system performs functions similar to Western Union's telegraph operations, the same approach and planning principles also apply to such commercial operations. In fact, the principles described also apply to any type of manpower analysis.

Case 4 (Part V, Chapters 11, 12, and 13) has three major objectives. First, Chapter 11 expands the reader's knowledge of the system design process by illustrating the use of functional analysis in arriving at a system design that best interfaces with other parts of the total system. Our focus is on the design of a training system, including the use of a satellite communications system that will connect an existing computer-managed instruction (CMI) training system located in the United States with sites remote from the United States so that CMI can be conducted at these sites. This part of the case, described in Chapter 11, builds on the knowledge of the communications system developed in Case 3. Chapter 12 continues the case with the second objective: a description of how to do a more complex systems evaluation than that described in Case 1. Here we show how to compare the CMI via satellite system with alternative approaches for conducting training. Finally, Chapter 13 illustrates the practical types of analytical simplifications that must be made when time and analytical resources are extremely limited. Here we show how a "best-efforts" type of analysis can be made when all the desired data are not available.

Each of the previous cases is concerned with a problem that has already been recognized and assigned to a planner for his analysis and subsequent solution. In each case it is shown that the same systems planning approach could be applied to each of these different types of problems. In Part VI (Chapters 14 and 15) these same principles of planning are extended to an even more complex situation. We describe how to make a diagnostic audit of an organization, arriving at the method of treatment in much the same way as a physician performs a physical examination of a patient.

The problem described in Part VI involves a new manager or planner who enters an organization with which he is unfamiliar and is asked to determine the organization's current condition and generate an improvement plan. Here the main emphasis is on problem recognition, because no specific problems have been identified thus far. We begin with the perception that no organization is perfect; hence improvements can always be found. The approach to be described can also be used by the planner who is a full-time member of the organization. In addition to showing a method of recognizing and analyzing problems in a new organization, Part VI has these objectives:

- To identify the key generic functions present in every organization and to show how each of these functions may be systematically designed with respect to one another to obtain the preferred design for the entire organization.
- To use this case as a means of summarizing and coalescing the principles of systems planning developed thus far and to extend these principles to all other types of systems and organizations.

Part VI differs from the previous cases in that the objective is not to provide a specific solution of the problem as given but to use it to show how potential problems can be recognized and how the various subsystems comprising the total system may be balanced against one another to arrive at a preferred total system design.

The major emphasis of Chapter 14 is on organizing the planning effort and recognizing and generating various ways of changing the manufacturing-distribution system so as to reduce costs. Chapter 15 describes two other types of improvements:

- Redesigning the product line.
- Improving the distribution system for reasons other than cost reduction.

Chapter 15 concludes by describing how the same planning principles apply to the other types of organizations. Chapter 16 then summarizes the key principles of planning discussed and illustrated throughout the book.

1.5 THE AUDIENCE ADDRESSED

In presenting this material as a course, I have found it to be of benefit to the following audience:

1. Professionals currently responsible for planning, particularly those who have never had a course in the formalized problem-solving approach.
2. Students who aspire to be planners or systems designers, particularly in such fields as management, operations research, and engineering.
3. Managers of organizations, operations, or activities who are responsible for planning and control in their organizations and hence should be interested in improving their knowledge of how the planning and control process should be performed. If they must delegate the responsibility for planning to others within their organization, they still retain responsibility for the way planning expertise is applied to the problems at hand.

If managers understood the process of planning better and knew what planners could do for them, what their own role in the planning process is, including the right questions to ask and the necessary inputs they must provide to the planning effort, they would achieve a much more acceptable product than they are now obtaining. To this end this book is dedicated.

2

Model Building: Structuring Operations

Before a planner or problem solver even enters a problem area, he must realize that he will be seeking and be confronted with voluminous data describing the problem he has been assigned. Thus he needs some structures for assembling these data, linking together related data into clusters, and putting aside nonrelevant data. This chapter describes a number of generic structures that are useful in analyzing the performance of systems of all types. An example illustrating the application of these models is given in Case 1.

2.1 DEVELOPMENT OF THE ANALYTICAL STRUCTURES

As indicated in Chapter 1, all problem-solving efforts can be related to the improvement of some activity. Hence the first thing we should do is to develop an approach for describing or "modeling" various activities. A model is nothing more than a way of describing some activity, phenomenon, or problem area. More precisely, a model may be defined as an explicit representation of some phenomenon or problem area of interest, in terms of various factors of interest and their relationships, that can be used to help the planner to predict the outcome of certain actions. Thus a model is some analog or imitation of the real world. This definition is broad enough to include both qualitative and quantitative models.

2.1.1 Types of Models

R. D. Specht states:[1]

We can classify models according to:

[1]E. S. Quade and W. I. Boucher (Eds.), *Systems Analysis and Policy Planning*, American Elsevier, New York, 1968, pp. 221–222.

1. Purpose—training, study, etc.
2. Field of application—strategic, tactical, logistic, etc.
3. Level—from national policy to base operations.
4. Static or dynamic.
5. Two-sided or one, conflict or not.
6. Degree to which mathematics is used.
7. Use of computers—how much and how.
8. Complexity—detailed or aggregated.
9. Formalization—the degree to which the interactions have been planned for and their results predetermined.

He further develops his own model classification scheme, which he says is "as unsatisfactory as any other." His scheme consists of five main categories of models:

1. Verbal.
2. People—as an integral part of the model.
3. People and computers interacting as a part of the model.
4. Computer.
5. Analytical.

Each category has two subcategories "according to whether or not an active opponent is involved and conflict is an essential part of the model."

Thus a verbal model is a word description of some occurrence, activity, or system. A competitive business game simulating actual business decisions would exemplify a people model. If decisions are made by people but a computer is used to represent part of the problem, this would be an example of a people and computer model. Specht further classifies computer and analytical models as mathematical models. Some mathematical models, such as Ohm's law, are deterministic; however, other models that describe random processes are probabilistic, which introduces three other model classes. If the results of the analytical model are expressed as a probability distribution, it is sometimes called a *probabilistic model.* If only the expected value of the result is provided, the model is sometimes called an *expected-value model.* If the probability distribution is obtained by drawing random samples from each of the probability distributions associated with the operations involved, the model is sometimes called a *Monte Carlo simulation.* Although many of the models that we use are mathematical, our definition of a model as "an analog of the real world" would also include the following examples of models:

1. A mechanical drawing.
2. A scale model of an aircraft for wind tunnel measurement.
3. A map.
4. An electrical schematic.
5. A flight simulator for training pilots.
6. A scale model of a skeleton.
7. Pictures of man's nervous system, bone structure, and blood system.

Notice that each model on the list is either a two-dimensional drawing or a three-dimensional representation of some physical object; the specific purpose of each is to describe some real object(s) occurring in nature or created by man. It is important to note that the specific model(s) used relates to the specific problem area under investigation. For example, when trying to determine the preferred route to travel by automobile from one place to another, one model of the situation would be a road map that indicates not only all the existing roads but also the characteristics of each. Since the criterion for selecting the preferred route may be shortest time, shortest distance, lowest cost, most beautiful scenery, or some combination thereof, a road map is a good model for gaining insight into the solution of this problem.

On the other hand, if one were operating an airline company and wished to convey travel information to prospective passengers, two other models would be required: (1) a map that shows what cities are connected by the airline and (2) a time schedule showing departure and arrival times of all flights in all cities served by the airline.

Finally, in planning an interurban highway, the planner would require maps with both topographical terrain features and property locations so that property values could be inserted. Thus the costs of condemnation actions could be calculated for different highway alternatives. Note that each of these three problems requires a map, but each map is of a different type, depending on the characteristics important to the problem.

Many of the models just described were descriptive and showed elements (such as type of highway) and their relationships (such as which highways interconnect given cities). Another type of model used in systems analysis involves some form of transformation between a system input and output. An example would be the transformation of a modulated radio wave, which enters a radio receiver and is converted to sound as an output. Such models are used to describe, understand, and improve a system, such as the manufacturing system described in Part II.

2.1.2 Modeling Starts with Observation

The first step in modeling is to observe[2] the particular area of interest, see what is happening in the area, and qualitatively describe what occurred using a number of appropriate key descriptors. Later we develop quantitative relationships among these descriptors. In this section we shall mainly concentrate on material flow systems such as a manufacturing system. In the next section we shall show how this discussion also applies to information systems.

One of the first things that would be observed at an area such as a manufacturing facility is that various *activities* occur, and these activities are carried out or implemented by *entities*. Activities (also called operations, functions, and processes) are action oriented and have such names as manufacturing, selling, repairing, planning, controlling, measuring, and communications. As can be seen from these names, the objective of each of these activities is *transformation* of an *input* into an *output*. Sometimes the input is a material (such as sheet metal being cut); sometimes it is nonmaterial (such as electricity flowing into a telephone communications activity, in which the output is sound).

Entities may consist of people, equipment, or supplies. In general, a given relationship among the various entities must be maintained in order to do a unit activity according to some *concept of operation*. For example, in a machine shop one operator may operate one drill press. In another area of the same shop, another workman may be responsible for operating several machines, with most of his time devoted to loading the input of each machine and monitoring its performance. One manager may supervise 12 employees in a department. Commercial airlines always have a pilot and copilot. Some have a flight engineer. In all cases there is a proper "formula" that relates entities in achieving acceptable performance at highest utilization of entities, and hence lowest cost.

2.1.3 Relationship of Systems to Activities and Entities

How does the term *system* relate to the previously discussed concepts of *activities* and *entities?*

The term *system* is usually defined as follows:

[2] Activities currently operating may be directly observed. In modeling future activities that may not currently exist, the modeler must apply his imagination by extrapolating past observations of related activities to the hypothesized new set of conditions.

An aggregation or assemblage of objects united by some form of local interaction or interdependence; a group of diverse units so combined by nature or art as to form an integral whole, and to function, operate, or move in unison and, often, in obedience to some form of control; an organic or organized whole; as, to view the universe as a system; the solar system, any telegraph system.

Unfortunately this definition is not very appropriate to the planning of systems.

In 1964 the Systems Science and Cybernetics Society of the Institute of Electrical Engineers (IEEE) defined *system* in a way that was more meaningful to the emerging field of systems science[3]:

A system is a collection of interacting diverse human and machine elements integrated to achieve a common desired objective by manipulation and control of materials, information, energy, and humans.

With this definition, it is possible to think of systems as characterized by their objectives or by their elements, or by their major technology. Thus, one has command and control systems, naval systems, or electromechanical systems.

These definitions reinforce the following major points regarding a system:

- The objects or diverse units are what we call *entities*.
- Their function or operation is what we call the *activity*.
- The objectives of the activity are achieved by some method of control that enables the activity to respond to some degree to its environment.

Hence we may use the term *system* to mean either the activity or its entities.

2.1.4 System Organization Model

One of the best ways of starting an analysis of an operation is to obtain (or construct) a model showing what entities make up the entire operation and how these entities are organized. Industrial firms and military organizations call such a model the table of organization, as illustrated in Figure 2.1. We call this the system organization model. This model provides the following information. First, it can better define the problem area by making explicit the defined boundaries of the organization under consideration and what functions occur within the operation. Figure 2.1 is a table of organization of the production activity consisting of three primary

[3]System Sciences Symposium of the IEEE, University of Pennsylvania, September 1964.

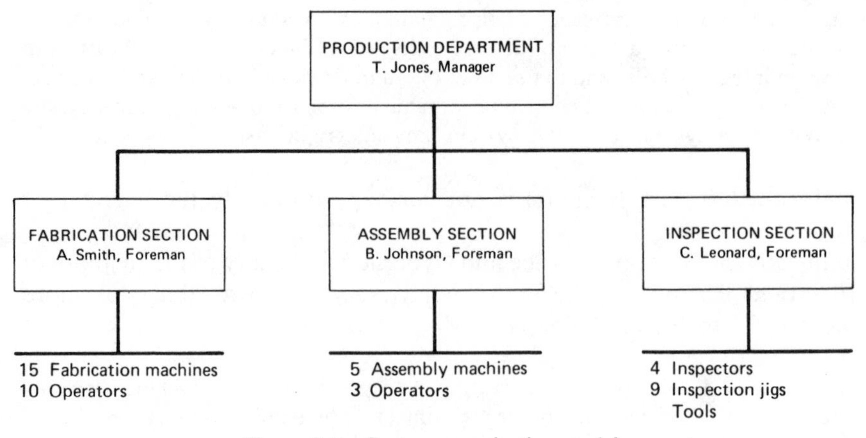

Figure 2.1. System organization model.

operating functions. Since part of the systems approach is to bound the entire problem and identify all the important elements, the analyst should obtain or construct one or more higher-level system organization models divided into two major parts, as shown in Figure 2.2. On the left is the primary system, the one that the planner was asked to analyze and improve in the initial statement of the problem given to him. Usually the client has direct control over the primary system, which might be represented by the model of Figure 2.1. Alongside are the supporting systems, which consist of all other functions required to support the primary system but found outside of this organization. Such supporting activities might include the following:

- Maintenance.
- Purchasing.
- Inventory control.
- Personnel.
- Accounting.
- Marketing.
- Plant and utilities.
- Computing center.

Obviously Figures 2.1 and 2.2 can include as much detail as is required to aid the planner in understanding the organization. The names of the

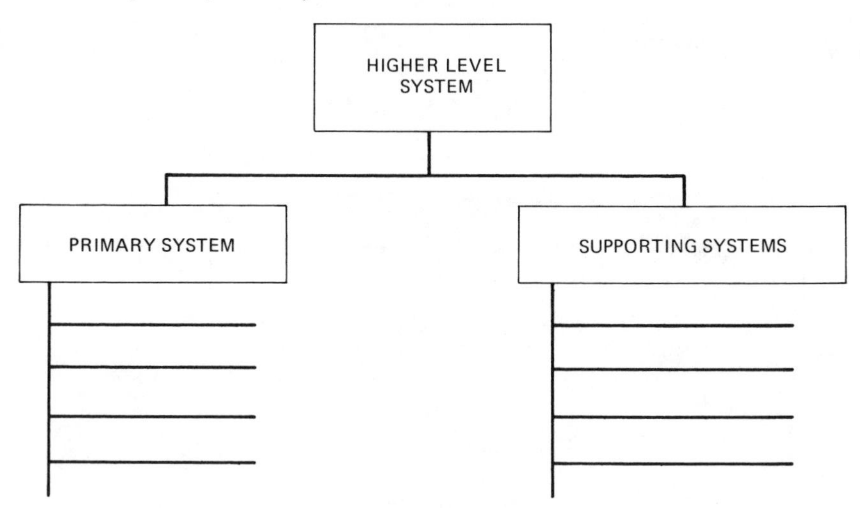

Figure 2.2. Higher-level system organization model.

various functions are helpful in visualizing the types of activities involved in the problem.[4]

System organization models also reveal the names of the people with whom the analyst must deal in gathering information, validating his approach, and determining the identities of the real decision makers. They also indicate functional responsibilities and lines of authority.

In addition, the system organization model is useful in listing what resources are required to operate, maintain, and support the operation; hence it is used for cost analysis as well as for system design.

2.1.5 Input-Output Model

Once the system entities and the activities they perform have been defined, the planner examines in more detail the activities and the various "flows" among them. The basic input-output model, shown in Figure 2.3, is a useful tool in describing the dynamics by which the system translates the inputs into outputs that meet the objective. Consider the primary system activity to be that of production. Thus the box labeled "primary system activity" can represent the factory floor. Inside the box are all the equipment and people needed to perform the specified production func-

[4]Here the analyst must avoid a possible pitfall. In some organizations the functional names may not really represent the activities which are really occurring.

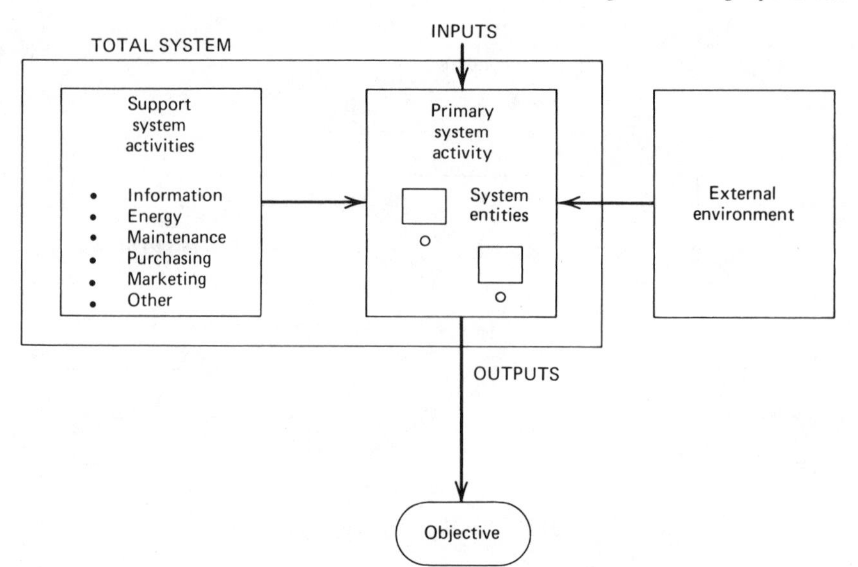

Figure 2.3. Basic input-output model.

tion of transforming the basic inputs of parts and materials into product outputs. But the primary system is also impinged on by two other influences that are required for the transformation or that affect it. The first consists of the set of support systems. One of these is an information system providing various types of information needed to ensure that the system meets its objective. For example:

- A set of *instructions* indicating the way in which people and machines are to perform their functions. Instructions may be of two types.

  Company instructions, including working hours (starting and stopping times, lunch breaks, coffee breaks), vacation periods when the organization will close, etc.

  Individual job instructions, such as (1) the operational procedure to follow in doing a particular job or (2) the production schedule to be followed in terms of how many units are to be produced over a given period of time.

  Instructions are constraints that limit the freedom of machines or people. What is not made specific may be left to the discretion of the individual. Of course, instructions for machines must be quite specific.
- A set of *standards* by which to assess the performance of the system

and the outputs it produces. In general, these standards involve the quality of the output, the time required to produce the output, and the cost required for such production.[5]

The primary system also requires energy from the support system, both for machine operation and for "people operation." Machines require electricity, heating gas, gasoline, and so on, to operate; people require wages or salaries and various "fringe benefits" that together make up the entire employee compensation package. Furthermore, machines (and even people) break down periodically and hence require maintenance (corrective as well as preventive). For this reason maintenance personnel and their tools and equipment enter the primary system activity, perform their maintenance support functions, and then leave that area.

Second, the system is also influenced or constrained by the external environment (e.g., the market or customer demand for a product, actions of competitors or adversaries, and such physical forces as weather). Sometimes the primary system exerts a direct influence on its environment, as in the discharge of pollutants from the system. Environmental regulations may be a constraining influence on the design of the system.

2.1.6 Further Observations of the System

Having specified the activity and its entities, the planner must observe the activity over its entire life and describe the movement or flow of entities through the activity. One of the first things he notices is that the physical location of some activities are fixed (e.g., a factory, hospital, retail store) and some are mobile (e.g., a radio-dispatched TV repairman, a physician making a house call, a truck making deliveries, a huckster selling his wares from the back of a truck). All systems have to be initially formed (i.e., set up). In fixed-location activities such as a factory we would note all the factory equipment and tools initially installed to construct the factory (creating the system).

The next set of events to be observed is that the system may not be operated continually. For example, people come in and leave at times associated with their work shifts. Energy, information, and wages also enter the system in varying amounts, depending on the time of day or week. All of these are associated with making the total activity support its operation.

[5]A more detailed description of the information and control system is contained in this chapter and in Chapter 8.

2.1.7 System Objectives

The primary purpose of an activity is to translate an input into an output, thereby accomplishing a given objective. The output may be a product (such as an automobile) or a service (selling, repairing). Thus one of the important relationships to construct is the input-output relationship. Examples of some input-output relationships include the following:

- Sheet metal and other materials entering a factory and converted into products, such as automobiles and washing machines.
- Ill patients entering a hospital and leaving as treated patients.
- Potential customers entering a retail store, some of whom leave with a product they have purchased.
- Students entering a school and leaving as educated graduates.

Associated with this relationship is the term *system demand function,* which is the primary driving force for the system. In general, there are two types of system demand functions:

- In some primary systems the input drives the system. For example, one objective of a TV repair shop is to repair all TV receivers whose owners contact the shop and request such a repair. Hence for a maintenance system the set of all requests of people desiring corrective maintenance is the major part of the system demand function. We must also not lose track of the preventive maintenance schedule for all equipment in the repair shop.
- In other primary systems the output drives the system. For example, the objective of a General Motors factory may be to manufacture 100,000 automobiles during the month. This level of output determines the quantities of materials needed as the input to the manufacturing activity.

2.2 QUANTIFYING THE CHARACTERISTICS OF THE ACTIVITY

The planner must now determine the quantitative relationships that exist between the input, output, objective, system entities, and performance characteristics of the activity. There are two major sets of characteristics that describe the performance of the activity:

- Those describing the quality of the output.
- Those describing how fast the output is obtained.

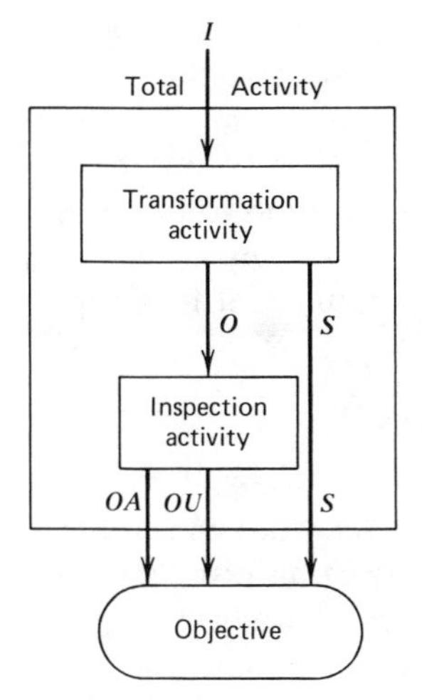

Figure 2.4. Input-output model.

The input-output model of Figure 2.4, a different version of Figure 2.3, will help the planner in developing these quantitative relationships.

2.2.1 Quantifying the Quality of the Output

This input-output model assumes an input of I units of parts or materials entering the transformation activity. On the average, the I inputs are transformed into O output units, as well as S units of scrap emanating from the process, including residues, scraps of materials, grinding dust, other waste materials, and so on, which are essentially by-products of the transformation. Following the transformation activity, it is found by subsequent inspection that of the O units produced, an average of OA output units are judged to be "acceptable," and OU output units are judged "unacceptable." The latter consists of all units with characteristics that do not, in the judgment of the inspector, meet the quality standards established for an acceptable output. This inspection activity may be accomplished by the worker who performed the activity preceding the inspection, by an inspector or foreman not directly involved in the work process himself, or by test equipment.

From such observations and measurements two performance characteristics having to do with losses within the system can be derived. The first might be called the transformation factor, TF, where

$$\text{TF} = \frac{O}{I} = \frac{O}{O + S}$$

and O, I, and S are all measured in the same units of output and input. For example, in building a frame house, we might take the ratio of the amount of linear feet of "two by four" lumber used in making the frame to the total amount of two by four lumber needed to be purchased. The transformation factor will be different for each type of material used.

A second performance characteristic describing an activity is the yield, defined as the proportion of the total output that is acceptable. Thus

$$Y = \frac{OA}{O} = \frac{OA}{OA + OU}$$

where Y is the yield of the activity.

Both of these measures give us a different measure of the losses attributed to the activity and aid us in recognizing problems by focusing on such losses. The planner should then explore ways of (1) reducing the loss (if it is economically feasible) or (2) finding some use for the scrap or unacceptable units. Logging and plywood companies look for good uses for bark and sawdust. The unacceptable units can often be reworked at an average cost that is less than the average cost of producing a new unit.

2.2.2 *Quantifying Time Required for Output*

The next performance characteristic to be quantified is the amount of time required to produce one unit of output or N units of a repetitive production operation. This same measurement is an important performance characteristic for two reasons. First, the objective of the activity may be to complete the unit within a given time. Second, "time is money," particularly in calculating manpower costs.

Times required for output may be calculated by observing the total operation over a long enough time and measuring the following performance characteristics:

- Production time or rate characteristics, excluding setup and teardown times.
- System failure characteristics.
- Maintenance characteristics.
- Production setup and teardown characteristics.

Each of these characteristics is now considered in a different order.

2.2.3 *System Failure and Maintainability*

If the planner observes the system operation over time, he will note the various operational events shown in Figure 2.5. First, there are times

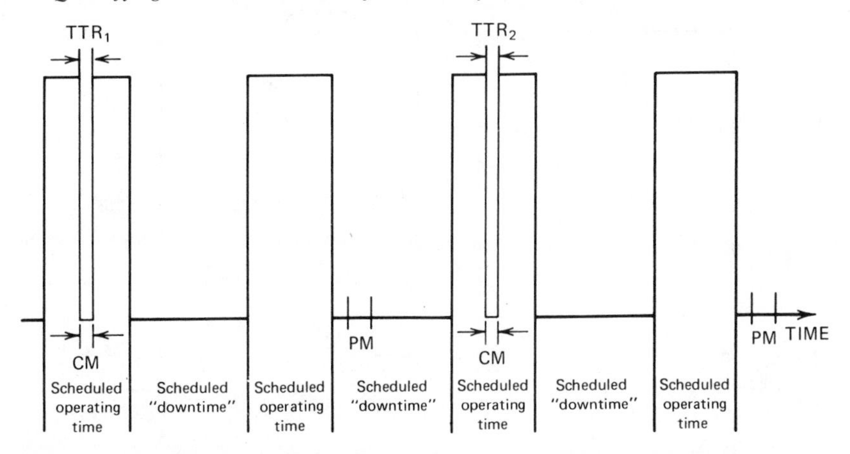

Figure 2.5. Machine operation over time (TTR = time to repair or replace; CM = corrective maintenance period; PM = preventive maintenance period).

when the system is scheduled to be operating, or "up" (as for the various shift operations shown in Figure 2.5). However, during these periods machine failures do take place, forcing the machine to be "down" for corrective maintenance (CM) until the machine is repaired.[6] This time is labeled TTR for "time to repair" or "time to replace," whichever is the case. In addition, at scheduled times the system is down for preventive maintenance (PM). Typically PM takes place during off hours, so as not to interfere with normal production.

From such observed data the following performance characteristics may be derived:

- The *mean time between failures* (MTBF). The arithmetic mean of the total operation times between equipment failures. In calculating MTBF ignore the times when the system was not scheduled to be operating.
- The *mean time to repair (or replace)*. The arithmetic mean of the times taken to repair (or replace) the equipment once it has failed.[7]

2.2.4 Other System Lost Time

Other activities may regularly occur during an operating shift and prevent the system from producing output units (Figure 2.6). These are (1) the

[6]To minimize the time that the production activity is halted, sometimes an entire subassembly is replaced and the actual repair occurs later at the repair shop.

[7]Some systems are "well-behaved" and these two arithmetic means have significance. In other systems, the actual data for each may have wide variations from the mean. In these cases a probabilistic analysis may be required.

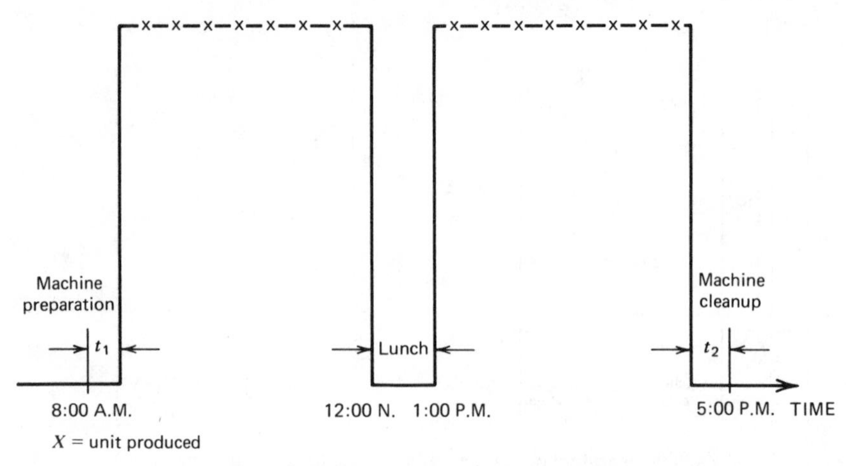

Figure 2.6. Time sequence of occurrences.

activities required to set up and tear down a machine following completion of a job or shift (including cleanup), (2) adjustment of a machine, (3) lunch and coffee breaks.[8]

2.2.5 *Calculating Total Production Time or Rate*

Output units are produced during the times between system setup and teardown (excluding maintenance disruptions). If the planner keeps track of the total units completed and the time required to produce them, he can calculate the following performance characteristics:

- *Total production time for one unit.* The objective of many systems is to produce one and only one unit (a "custom design unit"). In such cases the planner calculates the total time required for the operation (Figure 2.6). The total time is equal to the sum of all times, including

  Setup time.

  Time to produce the unit.

  Machine teardown or cleanup time.

  Any operational time required for both forms of maintenance.[9]

[8]Other activities, such as job actions and acts of God (e.g., blizzards), may interfere with production on a sporadic basis, and the planner may not wish to include these as "normal" performance characteristics. They can be included under *uncertainties*, as described in Chapter 7.

[9]Corrective maintenance time is a random variable and cannot be completely predicted. Hence the chance that a malfunction will occur and the time required to repair or replace the

- *Effective acceptable productive rate for multiple units.* In analyzing a system that produces a number of identical units, the planner calculates the average rate for producing acceptable units rather than the time to produce one unit. This production rate may be calculated by considering the following factors associated with a particular production facility:

  The *nominal production rate* (NPR). The ratio of the total number of output units produced (both acceptable and unacceptable) to the production time used when the production facility was actually operating. (No down time is included.)

  The *effective production rate* (EPR). The ratio of the total number of output units produced (both acceptable and unacceptable) to the total production time used (All down time is included.)

  EPR may be calculated from NPR as follows:

$$\text{EPR} = \frac{(\text{AOST} - \text{ASLT})\ (\text{NPR})}{(\text{AOST})}$$

where AOST = average operational shift time. For a standard work shift this is generally eight hours, with lunch assumed to be outside the scheduled operational hours

ASLT = average system lost time during the operational shift, for repair, setup, teardown, coffee break, and so on, as previously described.

NPR = nominal production rate

By taking the production yield into account, these additional performance characteristics may be calculated:

The *nominal acceptable production rate* (NAPR). The ratio of the total number of acceptable output units produced to the total production time used.

$$\text{NAPR} = Y \cdot \text{NPR}$$

where y = the yield of the system, as previously defined.

The *effective acceptable production rate* (EAPR). The ratio of the total number of acceptable output units produced to the production time used when the production facility was actually operating.

$$\text{EAPR} = Y \cdot \text{EPR}$$

---

defective equipment can both be included as probability distributions to arrive at an estimate of this time.

This characteristic (EAPR) is the one we actually use in our calculations.[10]

Figure 2.7 plots these four characteristics as cumulative production versus time. The scope of each of these functions represents the characteristics.

2.2.6 Generic Types of Activities

Each of the activities that occur in a material flow system can also be classified as being one of three primary types:

- *Transformation* (such as cutting, drilling, assembling). Here transformation time (or rate) and yield are the key performance characteristics.
- *Storage*. Here storage volume and yield (the ability to store without damage or spoiling) are the key characteristics.
- *Transportation*. Here time and yield (the ability to travel without damage or spoiling) are the key characteristics.

2.3 ANALYSIS OF INFORMATION SYSTEMS

We have described how to analyze and model a material flow system; let us now consider information systems. As we shall see, the three types of activities described earlier also apply to information systems.

2.3.1 Types of Information Systems

All the different types of information system applications within industrial companies or the government can be classified under one of three basic types:

1. *Routine data-processing systems:*
   - Payroll operations that generate payroll checks based on the number of hours worked and the wage rate.
   - Billing operations that bill customers for services rendered and credit such accounts for payments received.

[10]A numerical example illustrating how these calculations are made is given in the production planning case in Chapter 4.

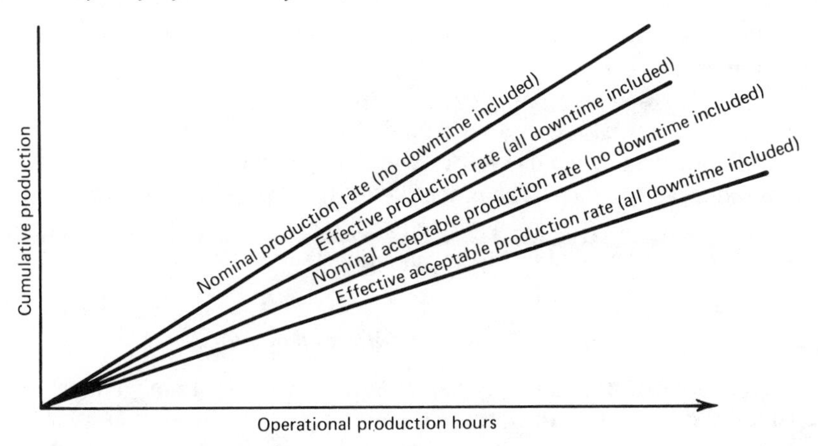

Figure 2.7. Cumulative production vs. time.

In such applications the environment is fairly stable; hence the set of procedures for transforming the necessary data into the final product can be specified rather precisely.

All routine data-processing systems may be thought of in the same way as a production system (or any material flow system): a flow of materials containing incoming data, in which the data are extracted, manipulated, or processed into some final form to meet the well-defined objective.

2. *Control systems:*
   - Management control systems in which some material flow system is to be monitored and timely information delivered to the managers of the system, comparing its actual operation or output with that which was expected. If significant deviations from this schedule are detected, corrective action can be taken to get the operation back on schedule.
   - Weapon control systems, to provide military commanders (the managers), with information regarding how their weapon systems are operating. These are the defense counterparts of industrial management control systems.
   - Industrial process control systems (such as those in sheet steel rolling plants) are just like management control systems. However, much faster reaction time generally is required. Thus much of the information is sent directly to the machines in the form of electrical control signals, avoiding human intervention.

In control systems the sole purpose of the information system is to cause a material flow system to operate in a prescribed fashion even in a changing environment.

3. *Planning systems,* such as the annual strategic planning process that many industrial firms use to prepare a longer range plan as well as the coming year's budget. The objective of this type of information system is to generate a set of plans of an acceptable nature, within given time and resources available to the planning team.

2.3.2 Basic Activity Functions of an Information System

We have defined the three types of information systems; let us now consider the sequence of activities generic to all information systems. Figure 2.8 is a functional flow diagram of these activities, which will now be explained, showing that they may also be classified as a transformation, storage, or transportation function.

1. *Environment,* though not actually part of the information system, contains certain observables of interest that are used by the system as its inputs.
2. *Data collection or gathering* derives the input data from observations, recordings, or measurements of the observables of interest. A time card, for example, records the arrival of a worker, and a stethoscope or electrocardiogram provides information about a heartbeat.
   Note that the data-gathering function can also be considered a transformation, since the measurement of these characteristics is sensed in one form and transformed into another form that is useful and compatible with the total system operation.
3. *Data communications* may be considered as the transfer (or transportation) of data. Two basic functions of data communications are the following:
   a. Transfer of the data via an appropriate transmission medium, whether via the mail, pneumatic tubes, or electronic carrier (wire or radio frequencies).
   b. Transformation of the data into suitable form so that the given transmission medium can be used (e.g., writing the message on a piece of paper for mailing or using a microphone for voice transmission using telephone or radio). On the receiving end, the data may again need to be transformed into suitable form to interface with the entity receiving the data (e.g., a telephone receiver).

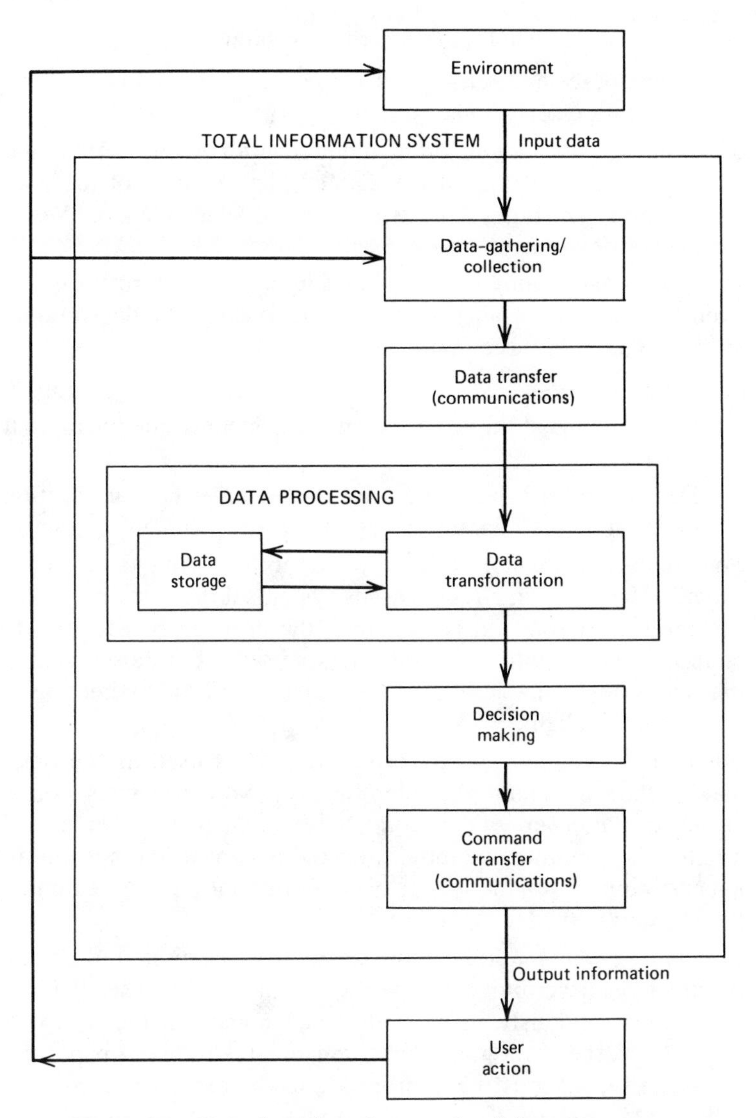

Figure 2.8. Operational flow diagram of an information system.

4. *Data processing* usually involves two subfunctions:
   a. Data storage and retrieval of data, as in a manual filing system or an airline reservations system.
   b. Data transformation, often required prior to the storage and retrieval functions, as in the manual filing records of the previous example. The types of transformation are dependent on the storage and retrieval mechanisms used.
5. *Decision making* may also be considered a transformation process similar to data processing. In general, it consists of a decision-making model containing three components:
   a. A set of all possible actions permitted or the data to generate them.
   b. Observations of the environment, which constitute the input to the model.
   c. Decision rules, which determine the preferred course of action as a function of the characteristics of the environment.

   Sometimes the decision rules are deterministic. An example of this is provided by the airline reservations system. If the customer asks for a certain flight and a seat is available, the decision rule followed is to confirm the reservation and subtract one seat from those available. If the seat is not available, the reservations agent is instructed to suggest an alternative flight.

   Sometimes inference, a deductive process, is used as the decision-making model. Thus a physician makes a patient diagnosis based on certain test or observed data collected from the patient. On the basis of a patient's symptoms he may order the patient to the hospital for an appendectomy even though the need is not totally confirmed until after the incision is made.
6. *Command communications* consist of the transmission of commands or directives describing the action to be taken to the user.[11] The same comments previously made under data communications also apply here. Now the form that the command takes (e.g., paper report, electrical signal) must be appropriate to the users involved.
7. *User action* involves the change which is made in another system based on the command generated by the information system. For example, the doctor's directive (the command) may result in the sick patient being transported to a hospital for the appendectomy (the other system's activities). Feedback loops are shown connecting a user action to the environment and to the data-gathering function. For

[11]The user may be an individual or a "machine."

example, certain actions may change the environment, which in turn changes the characteristics of the observables. In fact, part of the action may be to redirect the data-gathering function. For example, because of his preliminary observations, a physician may request a set of X-rays.

Although the information system ends with the command communications function, the user system must be considered in designing the information system, as is illustrated in Case 2.

2.3.3 Characteristics of Information

We have described the various types of information system activities; let us now consider some of the characteristics of the information entities and the information it provides. These characteristics form the basis of the system demand function, as well as the performance characteristics of the various information system activities.

2.3.3.1 Quality Aspects of Information

The information provided by the data-gathering activity depends on

1. The observables being measured.

A saw and a grinder provide different types of transformations in a production system; similarly, a thermometer and an electrocardiograph are data sensors that provide different outputs.

2. The accuracy and precision to which each observable is being measured.[12]

Here we are concerned not only with the number of significant figures in the data provided by a sensor but with certain sensors or test procedures that inherently have two types of errors associated with them. Consider, for example, the In-Home Early Pregnancy Test that has recently been announced. As the name implies, this test allows women to test themselves early for pregnancy. All such tests have a specific procedure to follow. In this test a woman waits at least nine days past the date she

[12]Some scientists differentiate between accuracy (the degree of truth in a statement) and precision (the degree of exactness in a statement). In this book the term *data accuracy* is used to denote the degree of certainty of the measurement and also includes the concept of "precision" (the exactness of the measurement).

expected her period, and puts three drops of her first morning urine into the EPT test tube containing a test solution. If a brown ring appears in the tube 2 hours later, it is an indication that the woman is pregnant. Clinical tests have shown that such a positive reading is accurate 97% of the time. Thus 3% of the time women taking the test who are not pregnant are wrongfully classified as being pregnant. This is the first type of error, a "false alarm."

On the other hand, if the test reading is negative (i.e. a brown ring does not appear), the inference is that the woman is not pregnant. However, clinical tests have shown that this inference is accurate only 80% of the time. Thus 20% of the time women who are pregnant receive an indication that they are not pregnant, which is a second type of error, a "miss."

However, the accuracy of test results increases with time. Thus if a week passes and the woman's period has still not begun, a second test should be run. This time a negative reading is accurate 91% of the time. Thus the second type of error has been reduced from 20% to 9%. It is important that the planner know the error characteristics for the sensors he is planning to use and is certain that they are sufficiently low to serve his purposes, also considering the losses which may occur because of each of these two types of errors.

2.3.3.2 Timeliness Aspects of Information

If the environment is changing, serious errors will occur in the total system unless the total time required to collect, transmit, and process data corresponding to this environment is much faster than the changes in the environment. Hence we are concerned with how long it takes for each entity to collect, transmit, or process data. This characteristic corresponds to production time or production rate for a production system.

Time lags occur for two reasons: (1) It takes a given amount of time to collect each sample of data and (2) many samples must be taken for the accuracy desired, since each sample requires a given amount of time. The time required for the entire collection process is directly proportional to the total number of samples required. A market research survey is a simple example of this.

2.4 COMPOUND ACTIVITIES IN FLOW SYSTEMS

Our discussions have thus far dealt with single activities. However, it should be apparent that most systems consist of more than one activity. These can be modeled as a series-parallel network of individual activities arranged in proper sequence, as shown in the operational flow diagram of Figure 2.9. This diagram could represent the production of a product consisting of two parts. The first part, $P1$, is fabricated through a process consisting of two basic activities, $A1$ and $A2$. Note that the acceptable outputs of $A1$ becomes the inputs to $A2$. Similarly, the second part, $P2$, is fabricated through a process consisting of three basic activities, $A3$, $A4$, and $A5$. Assembly of the two parts takes place in Activity $A6$. Obviously, compound activities such as this one can be analyzed in the following fashion:

- The proportion of acceptable units reaching the final stage will be the product of the individual proportions of each of the stages in series.
- The production rates of each activity in sequence should be reasonably "balanced", since the total production outputs can never exceed the rate of the slowest activity. The outputs of the faster activities will start to "pile up" and storage arrangements will have to be made if the process is not balanced.

Figure 2.9 can also represent a single "fluent" passing through a flow

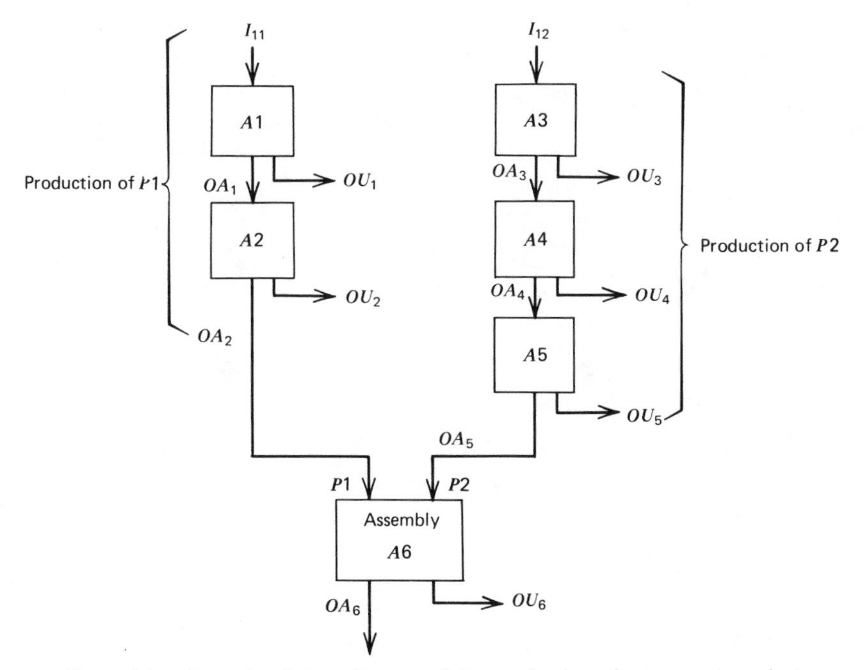

Figure 2.9. Operational flow diagram of the production of a two-part product.

system consisting of a total set of activities (e.g., a space vehicle rendezvousing with a satellite station). The assembly stage represents "docking" of the two vehicles. In this case it would be important to calculate the time required for the total operation and the probability of a successful voyage. The latter could be done by applying probability theory to the set of sequential events. The output would be the joint probability of success, based on the probability of success of each of the constituent activities. The total time required for each of the two sets of activities may be calculated as the sum of the times of the sequence of activities.

The travel of the fluent from one activity to another is represented by the line joining the activities. If we are interested in the time of travel or the "quality" of travel (e.g., aircraft arriving successfully or goods not damaged in shipment), a separate activity representing "travel" should be constructed so that these performance characteristics can be explicitly analyzed.

3

Gathering Data: Behavioral Aspects

When a planner enters a client's organization to conduct an analysis, he is invariably forced to rely on other individuals inside the organization to furnish data required for his work. Hence effective data-gathering methods are important to him. This chapter lists the key points to consider in preparing for client contacts and gathering data as part of a planning effort. Some of the behavioral problems that planners may encounter are described, along with some ways of avoiding them.

The subject of how to go into an operation and how much data to gather is not simple. The answer depends on a number of related problems, which must be considered simultaneously. In the beginning the interviewer may know little about the organization or the people in it. He has to "feel his way" until he develops further knowledge or contacts. The individuals he is to deal with may be extremely cooperative, evasive, or hostile. Hence the format of an interview and the phrasing of the questions depend on how far the planning process has advanced as well as on the reception the planner is getting.

Two aspects of data gathering are emphasized:

- What information are we looking for? (the technical aspects)
- How do we get this information? (the behavioral aspects)

The total data-gathering process can be divided into three main phases:

- *Previsit,* in which you plan your strategy and tactics for the visit.
- *Visit,* in which you execute the tactics.
- *Postvisit* analysis, in which you analyze the results and replan the subsequent events in light of what you have learned thus far.

Each of these phases is discussed in this chapter.

3.1 PREVISIT PHASE

The key principle involved in the previsit phase is *do your homework!* The planner must prepare for each interview. He must take his own medicine and consider (1) the objective of the particular interview, (2) who has the information, and (3) how to ask for this information.

The description that follows has been structured in terms of work tasks, with the behavioral aspects fitted into the description as needed.

3.1.1 Determine the Objective of the Interview

The objectives of an interview depend on where the planner is in the planning effort. Here is a list of different objectives common to every planning effort:

- Making a courtesy call to get permission from the head of an organization to interview subordinate personnel.
- Understanding the various problems that the operation has or might encounter.
- Understanding the operation as it relates to each problem.
- Gathering "hard data" or opinions useful in building qualitative or quantitative models of the operation.
- Validating the analytical work and results you have obtained thus far.
- Convincing others of the soundness of your approach and recommendations.

3.1.2 Preparation of a Work Plan

Having decided on the objective for the client visit, the planner should prepare a set of questions and notes to be used in the interview. At the end of the interview the planner can use them as a checklist.

- What questions do you intend to ask?
- What "hard data" do you hope to collect?
- Where might these be found?
- What are your "fall-back" positions if the data are not available?
- Whom do you plan to see to get the data?
- In what sequence should you make these interviews?

Make sure you have thought out how you are going to use any data you request. It requires an expenditure of resources to get data (an expenditure from both you and those you visit). Nothing destroys your credibility as a planner faster than being asked how you are going to use the data and being unable to answer such a question satisfactorily.

3.2 VISIT PHASE

Here are some points to consider, depending on the objective of the visit:

3.2.1 Initial Courtesy Visit into New Area

Once the project has been started the planner must establish satisfactory relations with personnel in the client organization having jurisdiction or actual possession of the information he needs. Several points should be considered:

- Try to start with top management. Introduce the project, its objectives, and the approach to be used. Then and only then work your way down to lower levels. This is not just a matter of courtesy. It will dispel suspicion regarding what is going on in a manager's area and will give you an opportunity to explain both the type of assistance you will need and the benefits the organization will receive.
- One source of help to the planner, particularly during the initial phase of the project, is the "friendly insider." This is an individual who is already a part of the client organization, known to be sympathetic to the project, and available for discussion. Sometimes he is the "patron," the one who initiated the planning project. The patron can tell you:

  Who the key people are in the organization and their frailties.

  Pitfalls to avoid.

  Problem priorities.

  Where the data are located and who is best qualified *and has the time* to assist you in the study.

  Ideally, the patron is the manager of the higher-level organization. Such an individual can provide you with a view of the problems facing the organization, the people to see, and the data relating to the problem. He can even make formal personnel assignments, which motivates others to contribute to your planning effort. In addition, he controls the budget and can release funds.
- Obtain the table of organization. This will aid you in understanding the

various functions performed and may indicate the personnel who can best provide the information you require. In making your contacts, consider *rank* in organization, *skill,* and *motivation* to participate. However you should be aware that sometimes the organization chart may not reflect the relative power or ability to accomplish things.

- When conducting an initial visit, be ready to introduce yourself and your project. Try to anticipate the type of questions you may be asked. Appendix 3.A is a list of questions normally asked and their answers. New planners can use it as a checklist.

It is vital to gain the confidence and cooperation of your data source. Establish rapport with the client. Try not to be seen as a threat to his authority.

3.2.2 Obtain an Understanding of the Operation

During an interview of this type explain to the person being interviewed what you are looking for

- An understanding of his operations.
- The information needed to operate satisfactorily.
- How this information is used in his operation.

Explain the importance of his help in supplying data and information needed to do your work. Write up the various functions and work elements performed in descriptive form as a time sequence of events or a scenario, with different scenarios for different environmental conditions.

3.2.3 Identify and Understand the Problems of the Organization

The most important contributions from the various people in the client organization are their views concerning the problem you (and they) are facing. Gathering these views is particularly important during the problem definition phase. Several planning contexts can be described along with the type of dialog that might go with each. If the planner enters an organization "cold" and has little knowledge of its problems, he can resort to the direct approach and ask for a list of the problems the client has recognized within the organization. If he obtains one, he can then inquire about alternative solutions. This initial phase of problem recognition is not easy. Progress can be slow for a number of reasons.

Sometimes there are behavioral obstacles. Some people are afraid to admit their organization has problems. They wonder if higher management will view any problems they cite as an indication of personal weakness and a sign that they are not doing their work well. In such cases the planner can inquire about what sort of pressures for improvement are being placed on him from above. Sometimes the client says, "*I* do not have any problems, but *my people* have many problems." This gives an opportunity to discuss "his people's problems." Another approach is to ask about the problems with an interfacing organization. For example, the manufacturing manager may complain of the poor service that maintenance is providing for his equipment. Using this approach you can lead the discussion to other parts of the organization and combine the responses given by the various groups.

Still another avenue is to take the positive approach and ask about planning efforts under way that might lead toward improving the operation. Instead of asking for deficiencies in the operation, ask what proposed improvements the person being interviewed is exploring, is recommending, or has proposed unsuccessfully in the past.

Often the responses are not in operational form. For example, the response may be, "My biggest problem is that my budget was frozen. I need more money and people." Unfortunately, this statement does not help the analyst, but it does offer him the chance of asking, "What would you do if you were to get the money you need? How would you use it?" This question is useful in getting the interviewee to indicate what improvements he would propose and what additional resources he needs. Another approach to the same question is, "If your budget were raised 10%, what would you do with the money?"

In both instances the planner follows the "cause-and-effect" logic. For each solution proposed, you ask what this solution leads to and what problems it solves. Continuing with this logic, for any solution offered, you can ask, "What gain does it provide?"

For any problem you can further ask, "What is the loss associated with it?" For example, if the planner hears the complaint, "We have poor communications," he should ask for some examples and what losses were incurred. In this way the planner can tie these kinds of problems back to an operational system and look for the losses that have resulted.

Indicate to the interviewee the benefits to him of your working with him, that any proposed improvement in his area resulting from your report will be credited to him. In addition, promise to assist him in getting his improvements "sold" to higher-level management. Remember your function is to analyze and structure the operation. You may be only the catalyst for the planning effort. Your success is tied to the success of the planning effort. Your job is to try to design systems that work, regardless of where the ideas for an improvement originate. If the organization is improved as a result of your planning efforts, the credit will be shared by all.

If the benefits approach does not work with him, *as a last resort* mention that all the other organizations you are working with are providing inputs and suggestions for improvements. Suggest that if one organization holds back, it will not be a part of the planning effort and that in due time its share of the budget may suffer. *But do not threaten.*

In summary, personnel in the organization itself constitute the richest source of ideas for change and improvement. Capturing these ideas is beneficial for a number of reasons:

1. The ideas they have may really be good ones, but the originator has not been able to convince others of their value. Here is where the analyst's knowledge of cost-effectiveness analysis and systems planning may be useful.
2. The basic idea may be sound but flawed in ways that a redesign may overcome. If the planner can be helpful in this, so much the better.
3. Starting the analysis with a set of proposals is good because the analyst can work back from these to get a better understanding of the *problems* that they are supposed to overcome.
4. From a behavioral point of view, show the interviewer that you are sympathetic to his specific ideas and anxious to obtain his support during your analysis.

3.2.4 Conducting a Data-Gathering Interview

For complex situations a two-person team is much more effective than a single individual. One person leads the interview and the other records, observes, and adds to the interview if something is being left out or not understood properly.

One last set of pointers may also be helpful. Always remember that in addition to gethering data, you are teaching, explaining, and allaying the fears of the respondent. Be careful of his time. Be sympathetic to his problems. Most operating personnel are busy and subject to pressures. Your project probably causes him additional work; so be appreciative of this fact.

Before terminating the interview use your work plan as a checklist to make certain you have covered everything. Then leave the interviewee with an assignment covering the data he has agreed to supply you by a certain time.

3.3 POSTVISIT ANALYSIS PHASE

Immediately following a data-gathering interview, the planner must analyze and relate the data obtained from that interview to other data available or to be obtained later. The various models described in Chapter 2 are useful for this purpose, and can also show what data are still missing. The planner must determine how and where he is going to get these data. Use the interviewer's experience to set (and reset) priorities for future work efforts. For example, an interview may provide information that convinces the planner that the problem priorities should be changed.

Finally, prepare a trip report for the record. Appendix 3.B offers a recommended format. Such reports

- Prevent the loss of data, particularly qualitative data.
- Are useful in determining the missing parts of the problem.
- Validate the planner's current understanding of the problem.
- Aid the creative process.
- Force the planner to look ahead to the next set of activities.

3.4 VALIDATING THE FINDINGS

A planning project can and should be looked on as a series of "thrusts," in which information and data are being gathered and structured so as to implement the steps of the planning approach. One type of information the planner should be gathering is confirmation that his model of the problem is reasonably consistent with his client's views of the problem. Periodic reviews of the planning effort with the client offers the way of keeping "on track."

Completion of each step in the systems planning approach provides a natural milestone when client feedback should be obtained (Table 1.1):

- *Problem definition.* The output of this step consists of a structured definition or formulation of the problem. It includes a definition of the job to be done, alternative solutions recognized by that time, the analytical approach to be followed, and an estimate of the time and the planning resources required to complete the planning process.
- *Creating system alternatives.* The output of this step includes a more detailed design of each alternative solution in sufficient detail to provide an estimate of (1) each system's performance characteristics and (2) its resources required over the entire defined system life to do the

specified job. All system alternatives are generally to be designed to do the same job. Hence it is most important that the decision makers validate the models of the objectives and the concept of operations[1] before the system design process begins.

- *Evaluating system alternatives*. Once the objectives, performance, and effectiveness considerations have been validated, the next step is to validate the resource and cost models to be used.
- *Final results*. This step consists of inserting the gathered data into the appropriate models and performing the required calculations. Thus the review should include (1) the descriptions of the preferred system, (2) other alternatives analyzed that were found to be inferior, (3) the rationale by which such judgments were made, (4) all uncertainties considered, and (5) the impact of such uncertainties on the selection of the preferred system.

APPENDIX 3.A. INTRODUCING YOURSELF TO THE CLIENT

When the planner initially enters an organization seeking data, he is generally asked a number of questions about what is going on. A series of questions normally asked and possible answers to these questions follow. It was prepared for students taking courses in management information systems and operations management who were assigned as planners on projects in these fields.

1. *Who are you?*

   I'm ____________________ from ____________________. I'm a member of the management information systems project now being initiated at your company.

2. *What are you doing?*

   We're studying (*your operation*) so that we can design an effective management information system (MIS) for your division.

3. *What's an MIS?*

   It's a system for collecting, processing, and distributing information in such a way as to help the general manager and the other managers of this division better manage their organization.

[1]The concept of operations is defined as the description of the sequence of activities to be followed by the system in accomplishing its objectives. The operational flow diagram can be constructed from this.

4. *How does it work?*

   Generally this involves setting objectives and goals in some way that can be measured and then recording the outputs of your organization and projecting trends so that you can better "steer" the organization toward the desired objectives.

5. *How will this help me?*

   Usually these approaches make it easier to communicate with others, especially when one wants to show how effort and resources have been used. Where it is used, an MIS system helps deserving supervisors better justify their need for more manpower or to use their people more effectively or get a fairer share of the budget.

6. *Will you guarantee that this will happen?*

   No. All we can do is help set up the system so that it can work. The results depend on how you as a manager use it in your organization.

7. *What are your qualifications?*

   I have completed my first year at the graduate school taking courses that relate to this assignment. I have had experience in ____________________. Also, before seeing you I've discussed some of the problems you face with some of your management as well as my own management.

8. *How shall I introduce you to others in my organization?*

   I think that if you say I'm from the . . . Graduate School of Business and a member of the company's management information systems project, and that it would be useful all the way around for us to explore the operation. That should do it.

9. *Will you review your findings with me?*

   Absolutely. We will then make them part of the overall study effort that we will review periodically with you and higher-level management.

10. *How do you expect that to happen when I haven't got enough people now?*

    What we're going to propose won't be that complicated. We hope to use existing reports as much as possible and, in many cases, simplify your reporting system. In any event, what is developed should be a management tool that will help you *cope* with a shortage of people.

11. *How much time will you require of me and my people?*

    I've been assigned to your organization to review your operations and procedures and document the type of information you need to do your job effectively and efficiently. To do this I would like a little of your time to start off with so that I can accurately understand your operation and some of the problems you are facing. I would then appreciate your introducing me to one of your people who can work with me in providing me with the data I need to do my work. Later, I shall try to describe your operation quantitatively and will need to gather data to do this. Your man will be extremely helpful in showing me where the data may be. I will be meeting with your man on an average of approximately (____) hours per week to gather data as well as having him review my findings.

"Openers"

1. *Describe your organization.*
   a. Table of organization.
   b. Sequence of activities performed.
2. *How many people are in your area and how are they organized?*
3. *What equipment is available?*
4. *What is your annual budget (more or less)?*
   a. Percentage salaries.
   b. Other costs.
5. *What is the mission of your organization?*
   a. In general.
   b. In detail (functions).
   c. Rank by resource expenditures.
6. *What reports and data do you regularly keep?*
   a. Internally.
   b. Transmit to "outside."
   c. Use of reports you transmit to others.
   d. Percent of your effort spent on outside formal reports.
7. *What reports and other information do you currently receive?*
   a. How often sent to you.
   b. Origin of report.
   c. Your use of report.
8. *What reports and other information would you like to receive periodically?*
   a. How often.

b. Origin of report.
c. Your use of report.

9. *What improvements would you like to see in your operation?*
   a. Type of quality improvement.
   b. Any cost savings.
   c. How to implement these improvements.
      (1) Resources.
      (2) Time.
      (3) Changes in other operations.

APPENDIX 3.B. DATA-GATHERING INTERVIEWS

A. *Planning for the interview*

Plan your data-gathering visit considering the following:

1. People to see, in what order.
2. Questions to ask.
3. Answers expected.
4. "Hard data" expected to be collected.
5. Source of data.
6. How to establish good relationships with interviewee.

B. *After the interview*

Make a trip report including the following information:

1. Date and time of visit.
2. Objective of visit.
3. Person visited.
4. Information gathered.
5. Behavioral aspects of visit.
   a. His attitude.
   b. Your response.
6. Problems encountered.
7. Analysis of visit.
   a. Actions you are taking.
   b. Actions others should take.
   c. New potential projects uncovered.

Part II

SYSTEMS EVALUATION AND THE DESIGN OF MATERIAL FLOW SYSTEMS

CASE 1: THE PRODUCTION PLANNING CASE

Having described some of the tools and techniques planners use, we now present the entire planning process, illustrating how to apply the tools just developed and others to be used in our first generic problem, namely, system evaluation. This involves the design of a number of alternatives and the selection of a preferred alternative based on a comparison of these various ways of doing a certain job.

In the course of doing this, we also present other tools and techniques useful in performing

- A systems design.
- A resource and cost analysis.

This case is of particular interest to managers who must review the work of planners, since it describes the types of questions managers should ask of their planners. It also shows the forms the information that answers these questions should take. Although the case involves a fairly simple type of system, it does involve investment in equipment. Hence this case is useful for illustrating how to handle investment or capital budgeting problems of many types.

4

Analyzing System Performance of Alternatives

4.1 PROBLEM AS GIVEN

You are a staff planner with the Acme Plastic Products Corporation. John Williams, the Manager of Manufacturing, calls you into his office to ask you to consider the following problem. He says,

Over the past three years we have been producing a plastic container that holds six half-gallon cartons of milk. The product wears out after about one year and needs replacement. Our sales over this time have been increasing at a rate of about 10,000 units per year, as shown in the first three entries of Figure 4.1.

The company has recently received an order of 50,000 units from a new customer. Marketing feels they can retain this customer's business and has just projected a new sales forecast (Figure 4.1).

My problem is that up to now we have been doing all of our production in a one-shift operation with some overtime. The projected sales for next year will far exceed the total first-shift capacity of the equipment we now own. I'm now faced with the problem of having to change our current method of operations in some way. I've discussed this problem with my staff and the following possibilities seem to be open to us:

ALTERNATIVE A1

We could continue to use the same production equipment but go into additional 8-hour shifts of operations as needed. This entails a premium labor cost of 15%. If a machine is not needed for an entire shift, we might handle the volume with planned overtime at time and a half. We do not anticipate much difficulty in hiring and training new workers, but there is a training cost associated with this. Also, we would have to keep a limited part of the plant open for the extra time.

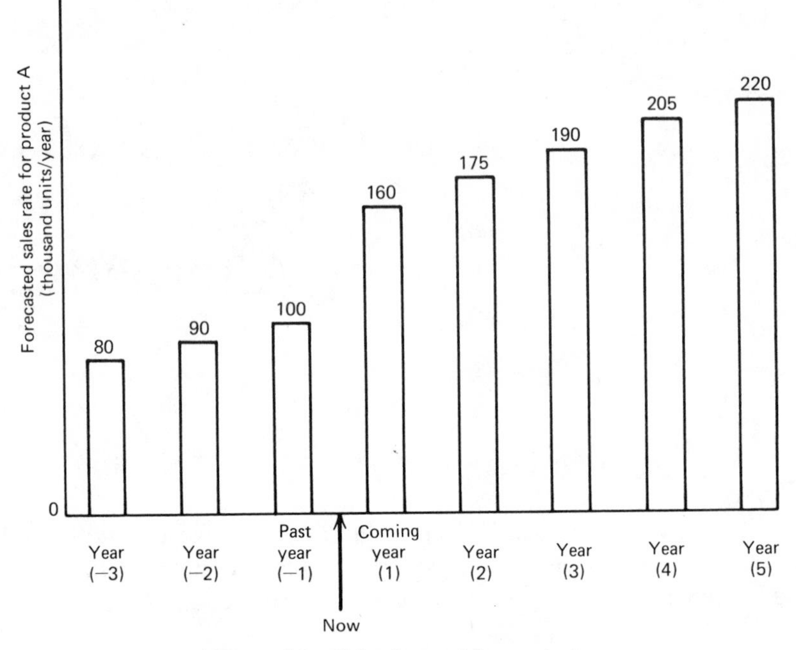

Figure 4.1. Sales forecast for product.

ALTERNATIVE A2

We could buy additional plastic casting equipment identical to that currently installed as sales build. Tables 4.1 and 4.2 summarize the price of this equipment, its performance characteristics, and other cost considerations. Note that 5 years ago the same equipment cost $50,000.

ALTERNATIVE A3

We could take delivery on additional semiautomated casting equipment with improved performance characteristics (Tables 4.1 and 4.2).

ALTERNATIVE A4

In another year an automated casting unit with performance characteristics superior to the semiautomated equipment will become available (Tables 4.1 and 4.2). This could be purchased then or later as needed.

To give you a better understanding of the current operations, Mr. Wil-

Table 4.1 *Performance Characteristics of Various Alternatives*

	Alternative A1 Semiautomated	Alternative A2 Semiautomated	Alternative A3 Improved Semiautomated	Alternative A4 Automated
Unit purchase price	$50,000 (5 years ago)	$60,000	$100,000	$300,000
Allowable depreciation	10 years	10 years	10 years	10 years
Production rate	5 units/hour	5 units/hour	15 units/hour	50 units/hour
Number of machines currently available	10 machines	Available to buy	Available to buy	Available to buy next year
Production yield	0.8	0.8	0.9	0.95
Rework characteristics				
Unacceptables reworked acceptably at or less than cost of new unit	60%	70%	80%	80%
Nonreworkable	40%	30%	20%	20%
Average failure rate	During past year each machine failed average two times per week	Same as Al	Each machine fails average one time per week	Each machine fails average one time per week
Average downtime	15 minutes	15 minutes	15 minutes	30 minutes
Scheduled maintenance required	One hour per week per machine	¾ hour per week per machine	One hour per week per machine	Two hours per week per machine
Operator getting machine ready for operation	10 minutes, beginning of shift	10 minutes, beginning of shift	15 minutes at beginning of shift	0 (automatic preparation)
Operator cleaning machine at end of shift	15 minutes at end of shift	15 minutes at end of shift	0 (self-cleaning)	0 (self-cleaning)

Table 4.2 *Production Planning Case: Cost Data*

	Alternative A1 Semiautomated	Alternative A2 Semiautomated	Alternative A3 Improved Semiautomated	Alternative A4 Automated
Purchase price (each unit)	$50,000	$60,000	$100,000	$300,000
Number of operators required per machine	1	1	1	2
Operator wages/hour	$4	$4	$4	$5
Cost of materials used per product unit molded	$1	$1	$1	$1
Sales price per product unit	$10	$10	$10	$10
Current total indirect costs	90% of direct labor costs (during past year)	Assume the same		
Maintenance costs:[a]				
Number of men required per repair	2 men	2 men	2 men	2 men
Hourly wage	$5	$5	$5	$6
Average cost of materials per repair	$10	$10	$20	$50
Rework costs (part of direct costs)				
Number of work operators required	During past year, average of 1 man required half-time	Same costs as A1		
Hourly wage	$5	$5	$5	$5
Cost of rework materials this past year	$10,000			

[a]All maintenance costs are included as part of indirect costs.

liams then walks you through the manufacturing area and describes the current production process:

- Plastic granules constitute the input to the production process. Each day these are carried by material handlers from the factory store to the input raw materials inventory hopper serving each of the 10 plastic machines. This hopper is never allowed to be empty.
- At the beginning of each shift, each operator readies his machine, preparing it for acceptance of the plastic granules. He then feeds the casting machine, applying heat and pressure. The container product is removed after a given baking time.
- The operator then inspects it, separating the acceptable products from those below specs. Some units must be scrapped. Some units contain burrs and sharp edges that can be ground down. Others have small gouges that can be filled with epoxy by the rework operator, who services all 10 operators. These units are placed in a rework bin awaiting service by the rework operator.
- The rework operator also inspects each unit selected for rework prior to beginning his work. If he estimates that more than 5 minutes would be required the unit would be scrapped rather than reworked. Otherwise rework costs would exceed the cost of fabricating a new unit.

4.2 ANALYSIS TO BE PERFORMED

The planner is to construct an analysis that will aid Mr. Williams evaluate each of his alternatives. Specifically, you are to

1. Describe the approach you would use in comparing these alternatives. Specify the evaluation procedure and all decision-making criteria.
2. Identify additional factors you feel should be considered in analyzing this problem.
3. Indicate where you might get such information pertaining to each of these additional factors.
4. Describe how you would evaluate Alternative A1. Include the set of possible alternative assumptions that could be made in this analysis and the degree of accuracy of the result.
5. Derive the set of parametric equations and logic statements that can be used to evaluate the other alternatives. This set of equations should be in sufficient detail for programming into a computer.

6. Discuss any additional factors you might include in the analysis if you had access to a computer and time to make such calculations.
7. Where could you obtain the data needed to quantify these factors?
8. How would you include these additional factors into the analysis? If Monte Carlo Simulation is to be used, describe the logic used in the analysis in detail sufficient enough to serve as an input to a computer programmer.

4.3 DEVELOPING THE ANALYTICAL APPROACH

After one is given a problem (problem recognition) the next step is to develop an analytical approach to its solution (problem definition). Although we shall describe a rational, straightforward approach for solving a problem, experience has shown that the initial stage of formulating and defining the problem and then developing an approach to its solution is really similar to a creative process. That is, a series of "thrusts" are made to gain a greater understanding of the problem to arrive at an initial structuring of the approach. The main thing is to work through the problem-solving approach at least once, following as a guide the steps in systems planning, as listed in the summary of the process given at the end of these four chapters. Consider and record all you understand about the problem.

As new information is gained (and some gestation takes place in the planner's mind) further refinements in the approach are made. Thus the planning process can be thought of as a series of iterations. At least three iterations through the planning process should be planned for. As a first step consider "emptying" your mind. Put down on a large sheet of paper all the information that you know about the problem. The steps in the process serve as an excellent checklist for doing this. The generic models described in Chapter 2 also help and further serve as a structure for both attacking the particular performance characteristics that are available and for indicating the type of data still to be gathered.

Later, this understanding of the problem, the assumptions that have been made, and the approach to be used can be validated with the client. Do not worry about the exact order in which you consider each factor. Using the steps in the process as a checklist helps make certain that you have not overlooked any of the elements of the problem.

4.3.1 Understanding the Problem

The following statements characterize a current understanding of this problem:

- There exists an assumed sales demand over the next 5-year period for the product under consideration.[1]
- It is desired that the production system provide the number of finished units on a schedule in accordance with this sales demand.
- The current production system can no longer meet the production schedule without going into extensive overtime or even beyond a second shift of operation, thereby requiring some premium in labor cost.
- A number of alternative system concepts can be designed to meet this required production, but at a lower premium labor cost, by adding additional units of production equipment and operators.
- Any new machines added will require an additional investment of money.
- Which is the preferred way of changing the production system:

  Buying additional men and equipment, and of what type?

  Increasing the production time beyond the first shift?

  Both of these?

4.3.2 *Identifying the Type of Problem*

Another planning aid is identifying the type of problem, that is, associating this problem with a problem type with which you are familiar.

By classifying all past problems on which you have worked and have experienced successful solutions into problem types, you can relate each new problem you encounter to your entire set of experiences with this problem type. In this way you are rarely starting a new problem from scratch. Each now becomes a new application of a particular problem type. Of course, with each new experience you must update your problem-solving *process* for that problem type.

This new tool you have (your own "corporate memory" or file of problem types) can save you much time. Not only can you set out your analytical approach more rapidly; you can also avoid the pitfalls you have encountered in the past and avoid making the same mistakes. By keeping past reports in your "corporate memory," you can use these as structures or guides for the new reports you must furnish. Finally, if you encounter difficulty in finding a good approach to a problem you have not worked with before, you can use analogies. Ask, "In what way is this problem like others with which I have worked?" Sometimes this technique pro-

[1]An important problem we shall later have to deal with is the uncertainty of this market estimate, as well as the uncertainties associated with the rest of the data provided. For now we assume the data are exact. Treatment of these uncertainties is discussed in Chapter 7.

duces the strangest pairings of problems, but it invariably works in getting you started on the new problem.

Comparing this problem to the traditional problem-solving steps listed in Table 1.1 shows that a problem has already been recognized (Step 1); further problem definition will be required (Step 2); but several alternative *approaches* to the problem solution (operational concepts) have been offered (Step 3), although these alternative approaches still have to be *designed* (i.e., converted into systems that will exactly meet the objective) (Step 3). The primary task will be to evaluate these designed alternatives and to select the preferred approach (Steps 4 and 5). Thus we classify this type of problem as systems evaluation. Sometimes it is also called capital budgeting, since it is concerned with choosing a particular course of action involving capital expenditures for investments (defined as expenditures for equipment or facilities whose useful life is greater than one year). Sometimes this type of problem is also called an economic analysis, systems analysis, cost-effective analysis, or cost-benefit analysis, depending on the background of the planner. However, in all cases the problem consists of evaluating alternative courses of action.

4.3.3 Approach to a Systems Evaluation Problem

Having identified the type of problem, here is a summary of the key steps to be followed in performing this systems evaluation[2]:

1. Determine the primary objective of the activity, construct the hierarchy of objectives, and choose a specific level of effectiveness for the job to be done as the basis of the systems design.

The primary objective of the new production system is to meet the forecasted sales demand over time. However, this is only a means to an end, as illustrated by the model called the *hierarchy of objectives,* shown in Figure 4.2. This shows that we would like to produce enough units to meet the entire system demand (the job to be done), because this would meet our forecasted production demand. If the problem involved a nonprofit organization, we could define our key objective as meeting the production demand at lowest total cost. However, since we are dealing with a profit-making organization, we must consider two other factors of importance, gross profits and depreciation of equipment, since both of these influence the net cash return available to the organization after

[2]It should be emphasized that Steps 1, 2, and 3 of this description are related and are essentially done in parallel through a series of iterations.

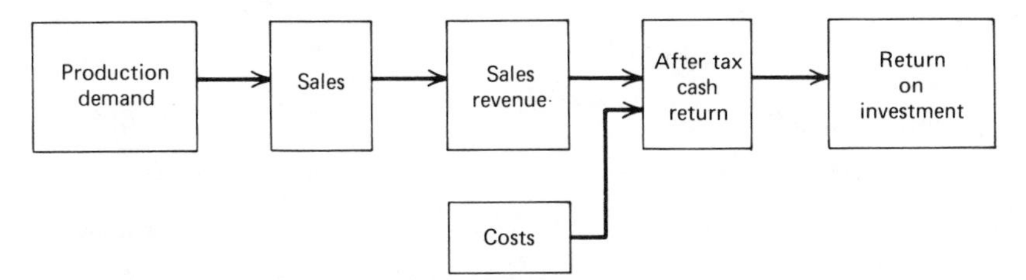

Figure 4.2. Hierarchy of objectives in a profit-making organization.

taxes. All of these factors contribute to the return on investment, an important higher-level measure if additional investment is to be made. All of these concepts are described in greater detail in Chapter 6.

One other point should also be considered at this time. The investment costs for new equipment required to produce the additional units may be so high compared to the additional revenue received that the return on investment is not satisfactory. Hence, in addition to the objective of meeting the forecasted sales demand, we should probably also consider meeting other lower levels of effectiveness, that is, meeting lower production levels, if these involve a lower level of investment. In other words, we do not want to be forced to produce more units if our total profits are reduced by doing so.

2. Identify the various alternatives available.

In addition to the current system, three alternative equipment types have also been identified thus far, and more may be identified as the project progresses. However, the units of equipment required for each type and the operating schedule of each type have not been determined as yet. These determinations need to be made during the system design phase so that the costs of each alternative can be calculated.

3. Identify the differences in the characteristics of performance and cost among each of the alternatives.

A systems evaluation can best be thought of as a comparison of alternatives. The performance characteristics that differ among the alternative equipment types are listed in Table 4.2 and include

- Production rate.
- Production yield.

- Ability to rework defects.
- Scrap rate.
- Equipment failure rate.
- Time to repair.
- Scheduled maintenance time.

Note that each performance characteristic affects the amount of man-hours required to achieve a given amount of production. Furthermore, man-hours can be directly translated into operating costs, which affects profits. The investment cost of each alternative also varies.

Another type of difference concerns the risks and uncertainties in being able to achieve the expected performance and cost characteristics. All alternatives have such uncertainties, but this is particularly important in the case of A4, since only prototypes of this equipment have been made thus far, and the final performance and cost characteristics, as well as delivery time, may differ from the current estimates.

4. Design all system alternatives to meet the same level of effectiveness selected in Step 1.[3]

This design process interrelates most, if not all, of the system performance characteristics. The final specifications of the design must be given in sufficient detail so that its total cost can be determined. In our case, at the very least we need to specify:

- An operational concept to be employed, out of which can be constructed an operational flow diagram showing the functions to be performed.
- How many entities (units of each type of equipment and personnel) are required over time, since these may affect investment required.
- The time schedule over which these units will be operating, since these will determine cost of operations, maintenance, and support.

This procedure (Steps 1 to 4) for synthesizing design alternatives is described in this chapter.

[3]This design approach is called *pivoting on constant effectiveness*. Sometimes all system alternatives are designed to cost the same amount. This approach is called *pivoting on constant cost*.

5. Determine the total system life cost for each system alternative.[4]

Here the main costs are

- Preoperations cost (investment and installation costs for the particular units of equipment to be acquired and training costs for all new men hired).
- Operations and maintenance cost (the cost of all manpower, materials, utilities, etc., needed to produce the units).
- Postoperations costs (the residual value of any equipment no longer needed).

6. Formulate a quantitative decision-making model that enables the alternative systems to be compared on the basis of the incremental benefits provided and the incremental costs required.

A number of decision-making models, based on the hierarchy of objectives described in Step 1, are available for this purpose.[5]

7. Identify any differentiating characteristics among the system alternatives that may influence the comparison but have not been included in the quantitative decision-making model. Determine the economic impact of such characteristics.

Quite often risks and uncertainties may not have been included in the main analysis. This step affords an opportunity to consider these as well as any other factors not included in the quantitative analysis. At the very least, such factors should be considered on a qualitative basis.

8. Select the preferred system alternative on the basis of both the quantitative decision-making model used and any other information provided that was left out of the quantitative decision-making model.

The lowest-cost alternative for meeting the objectives, or the one yielding the highest estimated profit, is not always the alternative selected. Risks and uncertainties and the desire of the decision makers to assume them are also important. In addition, any other factors that were not included in the decision-making model must also be considered at least qualitatively.[6]

[4]The procedure for determining total cost of a system alternative is described in Chapter 5.
[5]These are described in Chapter 6.
[6]Steps 7 to 9 are described in Chapter 7.

4.4 DESIGNING THE SYSTEM ALTERNATIVES

Having described the overall analytical approach to be used, we now proceed to the first step: designing each of the system alternatives to meet exactly the assumed sales demand function[7] and determining how many equipment-hours per year are required by each design alternative. We follow the same approach described in Chapter 2, starting with the design of the current system.

4.4.1 Designing the Current System to Meet Future Demand

The planner cannot avoid analyzing and attempting to redesign the current system to do the job. In general, the current system requires the least investment, and a good decision maker will always find out why the current system cannot be made to work. The planner also needs the current system design and cost as a basis of comparison with the other system alternatives, particularly in comparing profits and return on investment. In addition, analyzing the current system first aids in creating improved system alternatives by showing the perceived deficiencies in this system as well as the operational concept used and the functions required, a good starting point in doing functional analysis as part of the systems design task. Finally, let us not overlook the human element: the boss may have designed the current system.

4.4.1.1 Design Procedure

The basic procedure used to design a "production-type" system, such as the one under consideration, is as follows:

- Choose the type of machine under consideration for the operation under consideration (in this case 10 A1 machines).
- Using the following formula, determine the total number of equipment-hours (EH) required to meet the production output demand.

$$\mathrm{EH} = \frac{OA}{\mathrm{EAPR}}$$

[7]This assumes that profits and return on investment are maximized when the maximum value is produced and sold. Sometimes (and perhaps in this case) a large expenditure is required to reach the maximum volume and profits are reduced radically. If so, the smaller production volume before the large expenditure should also be analyzed. In other words, we should design the system for the volume that maximizes profits.

where OA = number of acceptable output units required in the period of time under consideration (say 160,000 units in the first year of production)

EAPR = effective acceptable production rate (in acceptable units per hour)

- Determine the amount of hours required. Here the designer has a design option since he can trade off the number of machines required (N) against the amount of time required (T):

$$T = \frac{\text{EH}}{N}$$

This trade-off permits the designer to reduce the total time required by increasing the number of machines used. Such reduction in time may be required if production demand is to be met or if premium time is to be reduced.

4.4.1.2 Calculation of Total Production Equipment-Hours Required

The next step is to use the information accumulated to determine how many machine-hours of production are required by the current system to meet the required production schedule. Figure 4.3 illustrates the type of analysis conducted to find this unknown (MH).

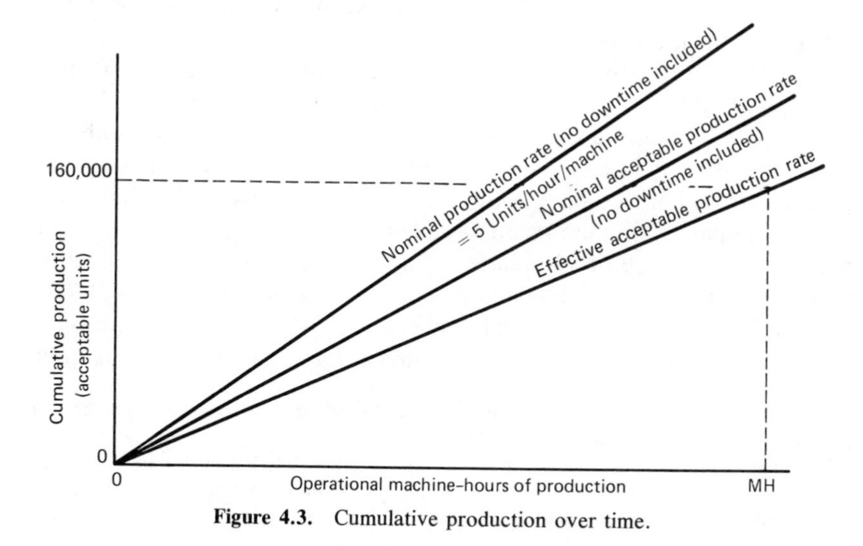

Figure 4.3. Cumulative production over time.

The nominal production rate of all output units (acceptable and unacceptable), assuming no time lost because of maintenance and setup and cleanup, equals 5 units per hour for each of the 10 machines. The nominal acceptable production rate takes into account only the acceptable units produced, and is equal to the product of the nominal production rate and the yield.

As described in Chapter 2, the yield may be found from the *operational flow model,* which in turn is constructed from the *scenario*, a word description of the operation under study that contains an "operational concept" or step-by-step procedure by which it operates and the environment under which it operates. The scenario (as described in the problem as given) is obtained by observing the complete cycle of operation. From the scenario can be constructed the operational flow model of the current primary system, shown in Figure 4.4.

The operational flow model enables us to illustrate the "flow" of units as they proceed through the various activities of the total production process. The probabilities associated with the units being "acceptable," "scrap," or "reworkable" at each stage in the process are shown.

In the current system, after the machine operator completes a machine cycle, he inspects the output units to see if they meet the quality standards. Only 80% do ($p_{oa1} = 0.8$). The operator then distributes the remaining 20% of the output either in the rework bin for later action by the rework operator or in the scrap bin, depending on their condition ($p_{ou1} = 0.2$). Periodically, the units in the rework bin are collected by the rework operator, who is able to repair 60% of the 20% within an acceptable time period and cost[8] ($p_{oa2} = 0.12$). As can be seen from the flow of units in Figure 4.4, the total yield of acceptable units of the current system (p_{oat}) is 92%, with 8% (p_{out}) being scrap. Thus the nominal acceptable production rate is (0.92)(5.0) = 4.60 units/machine-hour.

Finally, the effective acceptable production rate takes into account all downtime. EAPR is calculated by calculating all time lost from actual production, as indicated in the time sequence of operations of Figure 4.5. During the past year each machine failed on an average of twice a week, requiring an average of 15 minutes of unscheduled or corrective maintenance.[9] Thus last year's total maintenance time loss was (30)(50)(10)/60 = 250 machine-hours per year.[10] However, this loss is for the total production of 100,000 units, and it can be assumed that higher production would

[8]This means that 40% of these are so damaged that it would be less expensive to scrap them and produce new units than to rework them.

[9]Note that the scheduled maintenance can be accommodated during times when the machines are not operating, and hence does not contribute to lost production time.

[10]It is assumed that the plant is shut down for 2 weeks per year.

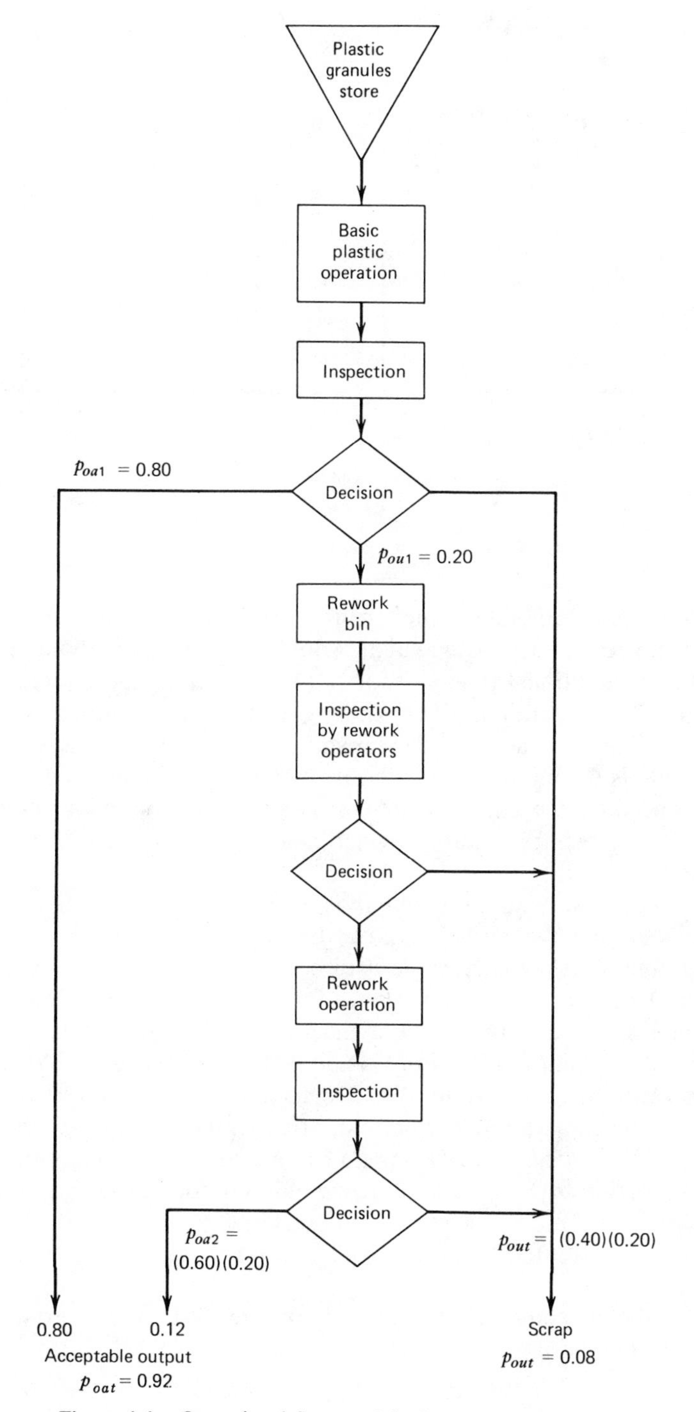

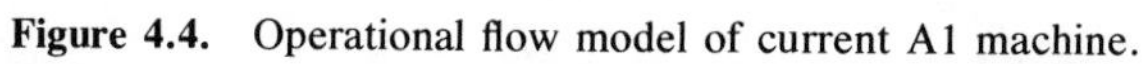
Figure 4.4. Operational flow model of current A1 machine.

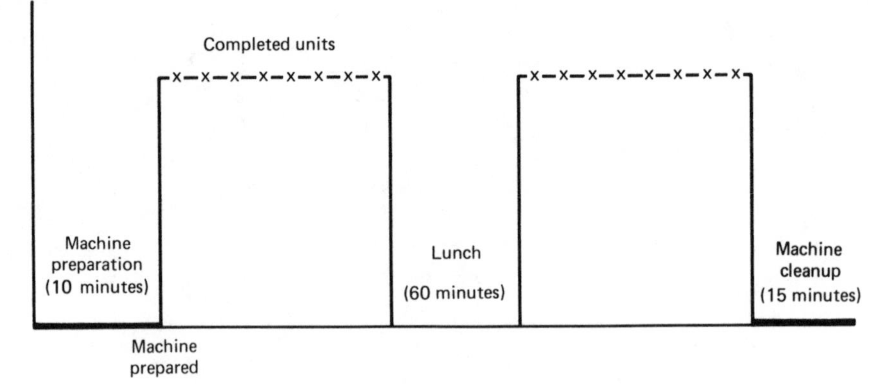

Figure 4.5. Time sequence of operations.

result in proportionately higher unscheduled maintenance. Thus the maintenance loss for 160,000 units would be 400 machine-hours per year.

There is additional time lost for setup and cleanup time (preparing the machine for operation at the beginning of the shift, requiring 10 minutes, and cleaning it at the end of each shift, requiring 15 minutes). Thus this time loss is 25 minutes per 8-hour shift per machine. This presents a problem, since we cannot calculate this time loss without knowing the total machine shifts required, which is a function of EAPR, the primary unknown.

One way of solving this problem is by calculating EAPR by a trial-and-error solution. We already know that EAPR must be less than 4.60 acceptable units/hour. Assume it to be 4.50 acceptable units/hour. Thus the total number of machine-hours for year 1 is 160,000/4.50 = 35,556 machine-hours, or 4444.5 8-hour machine shifts. Thus the time loss for setup and cleaning is (25 min)(4444.5 machine shifts)/60 = 1852 machine-hours. And the total time loss is 2252 machine-hours. For the assumption of 35,556 machine-hours of operation, this represents a downtime proportion of 2252/35,556 = 6.3% (or an uptime of 93.7%). Thus the second approximation to the effective acceptable production rate is

$$\text{EAPR} = (5)(0.937)(0.92) = 4.31 \text{ acceptable units/machine-hour}$$

Thus the machine-hours required for the first year's production of 160,000 units is

$$T = \frac{160{,}000 \text{ units}}{4.31 \text{ units/machine-hour}}$$
$$= 37{,}123 \text{ machine-hours, or } 4640 \text{ machine shifts}$$

rather than 35,556 machine-hours.

Using these figures the next iteration can be made. Now the setup and cleanup time loss is (25)(4640)/60 = 1933 machine-hours. Since the unscheduled maintenance time loss is the same (400 machine-hours), the total loss is 2333 machine-hours. This results in a proportional loss of 2333/37,123, or 6.3%. Thus this iteration of EAPR is

$$\text{EAPR} = (5)(0.937)(0.92) = 4.31 \text{ acceptable units/hour/machine}$$

the same as the previous iteration, and our final answer. Thus a total of 37,123 machine-hours are required, as previously calculated.

Since the first shift of 10 machines provides 20,000 machine-hours, an additional 17,123 machine-hours of production time are also required during the first year. Since this is too large a requirement for planned overtime, a second shift would have to be scheduled for Alternative A1, requiring an average of 8.56 machines operating on a 2000-hour second shift during the first year. Obviously, the need for a fraction of a machine during a shift is met by hiring the additional machine operator at the appropriate time of the year—after (1 − 0.56)(365 days) = 161 days in the example given where a ninth machine is needed for only 0.56 of the second shift.

Similar calculations can be made for the other 4 years of operation. Under this "zero investment policy" all 10 available A1 machines would be operated up to full capacity (i.e., three shifts or even Saturdays, Sundays, and holidays) to meet the desired production schedule. It may be found that the current system of 10 A1 machines does not have sufficient production capacity to produce the desired number of units in the later years. In this case the maximum obtainable production capacity per year will be calculated.

Having analyzed the key part of the total system (plastic production operations), the remaining parts of the system must also be analyzed. For this the system organization model, shown in Figure 4.6 and containing all the functions of the total system, must be constructed. As shown in this model, the remaining part of the primary system is the rework operation. Further analysis of the data in Table 4.2 indicates that last year the rework operator worked one-half time for only one shift, or 1000 man-hours. Since this corresponded to last year's production of 100,000 units, suc-

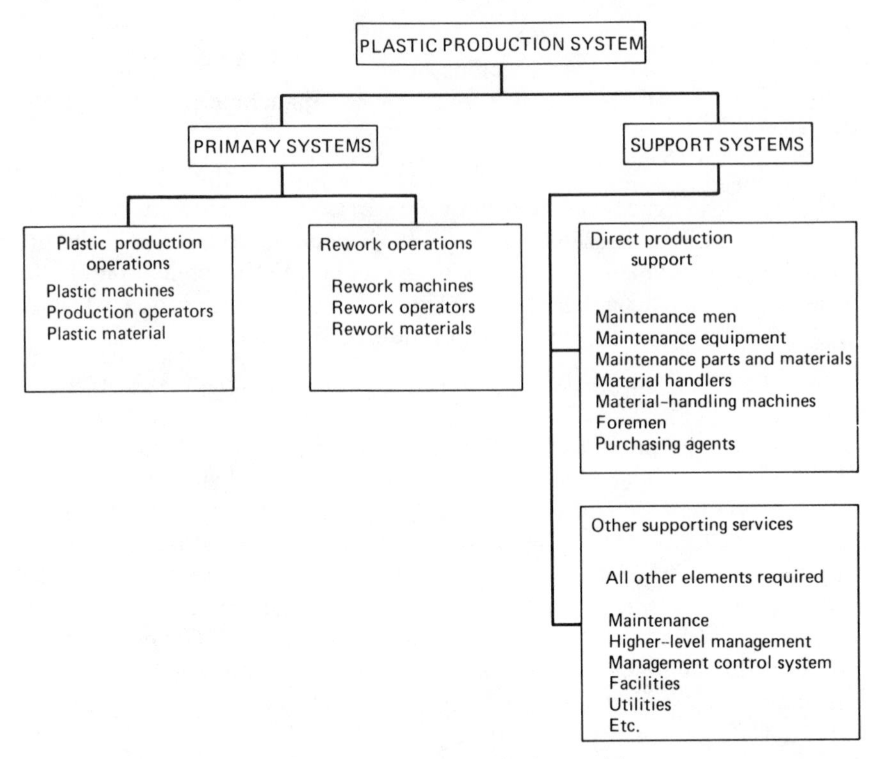

Figure 4.6. System organization model.

ceeding years will require proportionately more man-hours (1600 this coming year, for example).

Finally, the current support systems must be analyzed. As shown in the system organization model of Figure 4.6, the support systems include all those system entities outside of the primary system (required to actually produce the units) but still required for the primary system to function. It is necessary to identify all of these support system entities since they must be included in the resource and cost analysis, described in Chapter 5.

This completes the design of the current system, which is defined as the *base case,* since it is the basis of the cost comparison with the other systems, as described in Chapters 5 and 6. This design consists of the complete description of operations, including the number of machines required and the number of machine-hours of operation required each year to meet each year's production objective.

4.4.2 Designing Alternative Systems

The same design procedure can be used for a number of alternative systems. Since the operational concept of these other alternatives will be the same as that of the current system, the same operational flow diagram can be used. The only difference will be in the performance characteristics of the other system entities used (Table 4.1). These new characteristics can also be applied as new parameters to the equation for the effective acceptable production rate, which must be calculated for each new alternative. Our systems design problem then really becomes a combinational problem, in determining what different types of entities might satisfy the production demand and how many of each type of machine should be used in each alternative system. The generation of these alternatives can be approached systematically by characterizing these alternatives as follows: The first set of alternatives can be called the *minimum investment* strategy. This consists of using the 10 A1 machines plus X A2, or Y A3, or Z A4 machines. Additional machines are added each year as the demand increases. A second set of alternatives can be called the *minimum labor cost* strategy. Since the total labor hours are constant for a given number of units and type of equipment used, labor costs can only be minimized by operating only one production shift, thereby eliminating premium pay.[11] Unfortunately, this results in a maximum investment cost for the new equipment required. Since the current system produces a total first shift production of P = (43.1)(2000 hours/year) = 86,200 units/year, there is a need of 73,800 additional units to be produced during the first year. Thus with this strategy the design problem is to calculate how many A2 machines are required so that these 73,800 units can be produced in the first shift. By placing the performance characteristics of each A2 machine in the generic operational flow model, as shown in Figure 4.7, the production yield of A2 can be calculated as before. It is 0.94. Since the total production time losses for A2 machines are the same as for A1 (i.e., 6.3%), EAPR = (5)(0.94)(0.937) = 4.40 acceptable units per hour per machine. Now the additional production machine-hours required are: T = 73,800/4.40 = 16,773 machine-hours.

For a single shift operation (2000 hours), this means that an average of 8.39 additional A2 machines are required, with the ninth machine to be procured when needed later in the year. As an additional option, you could buy eight machines in the first year and put in some planned overtime. Under both alternatives, additional A2 machines would also have to be purchased in subsequent years. The same approach should also

[11] Utility costs are also reduced under this option.

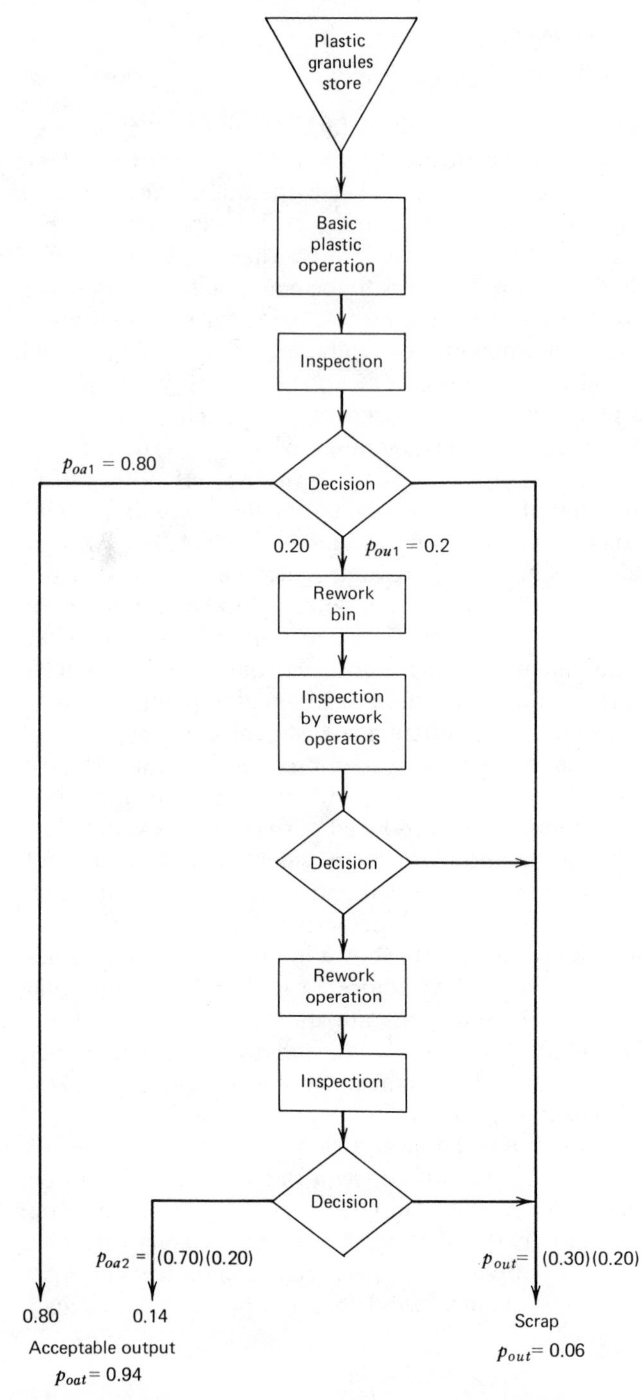

Figure 4.7. Operational flow model of Alternative A2 machines.

be followed using A3 or A4 machines instead of A2 machines, so that the cost of these alternatives can be compared.

A third set of alternatives might be called the *replacement strategy*. This consists of trading in (or selling) the 10 A1 machines for a number of higher-performance machines, such as one or more A4 machines, if the calculations show this to be a less costly way of meeting the demand.

There are also a number of "in-between" alternatives consisting of different combinations of equipment types and numbers, all designed to meet the demand. Since this demand is increasing over time, some of these alternatives include buying a new unit of equipment during the five-year period, perhaps even during the fifth year. How the investment cost is treated for systems in which the full 5 years of use is not obtained is discussed in Chapter 6.

4.5 SUMMARY OF KEY PRINCIPLES

This chapter described the first two steps of the systems planning process (problem definition and development of system alternatives) following the recognition and assignment of a specific problem.

4.5.1 Problem Definition

Given that a problem has been recognized and assigned to you, the first step is problem definition, which consists of a quick, preliminary analysis of the entire problem and contains the following elements:

1. Summarize all available data, including your understanding of the current system and its objectives and deficiencies in meeting the objectives. Also include suggested alternatives and operational constraints. Models of the current system helpful in identifying and structuring the data required are
   - The scenario(s) describing the operations.
   - The input-output model yielding the system objective and the system demand function.
   - The "means to an end" model showing the hierarchy of objectives.
   - The time sequence of events and the operational flow model. These models show the operational concept or sequence of functional activities being used by the primary system in providing the sys-

tems output. In addition, these models provide the key performance characteristics of the system:

The former indicates all unproductive time obtained during the operation of the system.

The latter provides the system yield or acceptability rate of the system based on the acceptability proportions of the various system activities.

Combining these data provides the effective production rate of the system.

2. Classify the generic type of problem in terms of the steps in the planning process. This helps the planner by relating the specific problem to past examples of this type of problem that he has solved or whose solution he is familiar with. This case represents the generic problem of systems evaluation, in which a problem has been recognized and a set of alternative courses of action are presented in fairly general form:
   - Operating the current system for a larger number of operational hours per week. It still has to be confirmed that it is possible to meet the maximum production demand (as measured in acceptable units per hour) without increasing the units of production equipment. The major deficiency of this alternative is the premium cost of labor when working beyond the first shift.
   - Increasing the production rate or capacity by buying additional equipment of various types, which have already been identified. The major problem here is the additional investment cost required.
   - It is the task of the planner to identify any other alternatives to be considered, complete the design of each of these alternatives, evaluate each of the alternatives, and decide which one is preferred. This type of problem is also called systems analysis, cost-effectiveness analysis, economic analysis, and capital budgeting.

Problem definition provides the analytical approach to be followed. This is the set of eight key steps described in section 4.3.3, "Approach to a Systems Evaluation Problem," which we plan to follow.

4.5.2 Systems Design

Since the alternative courses of action were specified only in general form, each alternative must now be more rigorously specified or designed. The systems design process is accomplished as follows.

1. Rigorously identify each alternative to be considered in terms of a set of system elements to be used and an operational concept describing how the system would operate, including the sequence of activities involved and the work schedule to be followed in producing the desired output.
2. Redesign the current system to best meet the system demand function and consider this a base case. This step is important because it furnishes a baseline of comparison for the other system alternatives. In addition, an understanding of the current system aids in synthesizing other system alternatives, since it identifies the current operational concept and functions (serving as a good starting point for an intensive functional analysis). Begin this process with the design of the primary system, that part of the total system that actually performs, rather than supports, the primary objective of the system. Construct the operational flow model of the primary system, consisting of the operational flow diagram showing the sequence of activities and the set of system performance characteristics of each system entity (e.g., each production equipment) used in each activity. Calculate the effective production rate of each activity. Based on the production schedule assumed for this system alternative (i.e., work as many hours as needed to meet the system demand function), determine how many machine-hours are required to meet the assumed production demand.[12]
3. Redesign the support systems of the current system using the same design steps as for the primary system. In this case, the system demand function for each support system is determined by the performance characteristics of the primary system (e.g., maintenance requirements are related to the frequency of failure).
4. Design each of the other alternatives to be considered in the same way as described above. Design concepts may be formulated in a systematic fashion by considering that a system consists of a combination of the following factors:

   - An operational concept, leading to an operational flow diagram. A good starting point in modeling these for the other system alternatives is to consider using the same operational concept as that of the current system (as was done for this case), modifying the operational flow diagram to accommodate the differences in the new system elements introduced. Many times these new elements permit the elimination or restructuring of certain activities, thus re-

[12]If a single output were produced, the production time for each activity would be calculated (rather than the rate) and the system designed to meet the required completion time.

sulting in a system whose operating costs are less than those of the base case.

- System elements to be used (e.g., A1 machines, A2 machines, A3 machines, A4 machines, or various combinations of each type).
- An investment strategy that affects the work schedule permitted, for example:

  Zero investment strategy. Use only the 10 A1 machines available, even though this may only meet a lower system demand function.

  Minimum investment strategy (System A1A). Get maximum production out of the 10 A1 machines. Then add a minimum number of A2 or other type machines as needed.

  Zero overtime labor cost strategy (System A1B). Add enough machines so as to operate only the first shift of production.

  Intermediate strategies (System A1N).
- The output of the design process for each of these alternatives consists of the number of machines needed, the performance characteristics of each machine, and their operating schedule to meet the objective (providing the required number of units in accordance with the production schedule).

5

Resource and Cost Analysis

5.1 KEY PRINCIPLES OF COST ANALYSIS

Performance or effectiveness is only half of the consideration in a systems evaluation. The other half involves cost. We now describe the method used to calculate all the costs that would be incurred for each of the alternatives identified in this case. Since we are interested more in the reader understanding the approach used than in a numerical solution, we concentrate on describing the cost elements to be considered in constructing the cost model and showing how the numerical results would vary for the various alternatives.

The key principles of cost analysis as they apply to the production planning case may be summarized as follows[1]:

1. Choose a common starting time when operations are to begin for all system alternatives.

This starting time is determined by the system demand function. Certain entities of a particular system alternative may not be available at that time, since they may not yet have been developed. Alternative A4, for example, will not be available for 1 year. Hence in considering A4 as a system option, it must be understood that the current system (A1) would have to be operated until that time. Sometimes it is possible to speed up the development and availability of new system elements, but this generally involves overtime and other additional costs that must be reflected in the costing of the alternative. The construction of the Alaska oil pipeline is a good example of this.

[1]Also see B. H. Rudwick, *Systems Analysis for Effective Planning*, John Wiley and Sons, New York, 1969, chap. 9 "Resource Analysis," pp. 219–251. For a more elaborate treatment see G. Fisher, *Cost Considerations in Systems Analysis*, American Elsevier, New York, 1970.

2. Define a common operating life to be considered.

One of the problems in comparing the various alternatives is that the useful life of the various alternatives may differ; for example, A1 has 5 years remaining, whereas the other alternatives have a predicted life of 10 years. However, the true *economic* life of a new production system may be only 5 years when one considers various uncertainties in sales demand, technological obsolescence of the system, and so on. Hence, to aid in reducing the effect of uncertainty in predicting future conditions, many organizations consider only a conservative estimate of operating life. For this reason, we shall consider only 5 years of operation. This may initially seem unfair when considering equipment that could last (and bring in revenues) beyond the 5-year period. It obviously penalizes new investments as compared with the current system. However, there are two other parts of the analysis that compensate for this conservative approach. First, the use of a present value analysis reduces the residual value of all systems beyond the useful life defined in the analysis. Second, in a profit-making organization, if the equipment does become obsolete prematurely (5 years instead of 10), the organization can take a tax write-off of the remaining book value. Hence the residual value can be included in the analysis in this way. Both of these concepts are discussed in greater detail under equipment depreciation in Chapter 6.

3. Include all net, "out-of-pocket" system costs incurred over its useful lifetime.

It is necessary to include all the net costs of all system resources required (for both the primary and support systems) for all activities performed over the entire system life being assumed. These costs include

- *Preoperations costs,* including:

  Any research and development that may be performed. In general, this is purchased separately for equipment unique to only one customer, such as when military aircraft are developed for the government. When commercial goods are sold, R&D costs are generally prorated over the entire production run, and hence are included in the purchase price.

  Procurement or investment costs (i.e., buying equipment initially, including its delivery and installation). Any trade-in allowance for existing equipment must be subtracted from the purchase price of the new equipment.

  Initial recruiting and training of operating personnel.

- *Operations and maintenance costs,* over the entire useful life of the system, including the cost of any additional manpower required.
- *Postoperations costs,* including the cost of dismantling and removing the system after its useful life is completed. However, the equipment may have a net residual value and hence may be a "negative cost."

Note that only "out-of-pocket" costs are considered. Any resources (such as in the current system) made available for use in the improved system are considered "free." These are called *sunk costs,* since they have already been purchased and no additional cost may be necessary to obtain their use. This is true even if the equipment is purchased on the installment plan or through a loan from a bank since this debt must still be paid back whether the equipment is used or not. On the other hand, if any parts of the current system are sold or used as a trade-in, thereby reducing the net cost of an improved system alternative, only the net cost is attributed to the new system.

4. Choose the time periods for making the calculations.

A cost calculation should be made for a particular period of operating time; this could be yearly, monthly, or even daily. Our calculations are made on a yearly basis.

5. Treatment of manpower savings.

The analyst must determine if the manpower savings that occur because of the purchase of the higher production rate equipment (A3 and A4) really result in true savings. Sometimes strong union pressures may prevent layoffs or reassignment of workers to other areas of work. In this analysis we assume that all displaced personnel are absorbed by normal attrition or are assigned to other shifts or to other jobs elsewhere in the division, particularly with the assumed production growth.

6. Treatment of inflation.

Although sometimes the analysis projects future costs (and revenues) at an assumed rate of inflation, we assume no future inflation. Two reasons justify this assumption. First, if inflation occurs, it will probably be compensated by a corresponding inflation in sales price, and hence sales revenue. But more important, inflation will affect *all* the alternatives, and as a nondifferentiating factor, it can be ignored without affecting the *comparative* results very much.

This is a very important point to be considered in the modeling process.

Because of the various risks and uncertainties involved it is rarely possible to get a precise *absolute* solution. The more factors considered, the more analytical resources and time are required to obtain the solution. By focusing on the key *differentiating* factors, a good comparative analysis can be made.

5.2 CALCULATING SYSTEM COSTS

To obtain all of the system-life costs associated with each alternative, the analyst must find answers to four questions:

- What entities are involved in each system?
- What cost elements are involved?
- What resources are associated with each entity over the entire system life?
- What costs are required for these resources?

There are two models or structures that the planner uses in tandem to answer these questions systematically. The system organization model, shown in Figure 4.6, is useful to identify all the entities relevant to the production process. This model, as shown, divides all the entities required into two major subsystems, including those entities that are directly involved in the production operation as well as those providing required support. It is not critical just how the entities are divided in the model. The main objective is to make certain that all entities are included under some descriptor.

The second structure used is the resource analysis matrix, illustrated in Table 5.1 for the primary production system. A similar table showing all resources required for the support subsystems, for such functions as maintenance, purchasing facilities, and so on, is required. This structure is used to gather and store systematically the major resource data required for each alternative being analyzed. If the operating resources required change for each year (as in our case), additional listings of column 4 or separate tables will be required for each of the 5 years. The following steps will be followed in accumulating the data required:

1. Classify all entities which comprise the system alternative.

Column 1 of Table 5.1 lists the various entities comprising the system (as obtained from the system organization model) as well as the type of resource required for R&D, operations, and maintenance.

2. Determine how many units of each type of entity (equipment and people) listed in column 1 are required to meet the production schedule for that year and list in column 2 of Table 5.1 on the appropriate row.

This data is obtained from the system design analysis.

3. Determine all resources required prior to the beginning of operations for such things as equipment development, acquisition, installation, and training, and list each in column 3 of Table 5.1 on the appropriate row. Note that some rows will be empty (i.e., not requiring preoperations resources).
4. Determine all resources required after operations begin for all functions of operations, maintenance, and support and list each in column 4 on the row of the entity with which it is associated.

Having determined which resources are required, the analyst determines the cost of these resources as well as the time required to complete the function. Each resource requirement should be related to the number of system entities required.

Three measures of these requirements are listed in each column. "Time" represents the total time required to complete an activity such as development, procuring equipment, installing, or training. "Dollars" are used to represent the out-of-pocket costs of this activity. Man-hours represent manpower resources required.

The planner should also provide some indication of the range of uncertainty in each estimate by providing three numbers—the expected value, the optimistic value, and the pessimistic value—if considerable uncertainty exists.

5. Determine the net resources required to disassemble the system at the end of the assumed 5 years of operations, and list each in column 5 of Table 5.1 in the row of the entity it is associated with. Again, some rows will be empty.

These net resources should include expenditures for disassembly and residual value (if any) of the equipment.

A new matrix must be constructed for each alternative analyzed.

5.3 DERIVING THE COST EQUATIONS

The cost equations can now be derived from the data previously assembled as well as from the cost data related to the production process, as

Table 5.1 *Systems Resource Analysis Matrix*

SYSTEM DESIGN ALTERNATIVE NO. ________
SYSTEM DESIGNATION ________________

		Resources Required								
(1)	(2)	(3)			(4)			(5)		
	Number of Entities Required	Preoperations			Operations Maintenance and Support (year ___)			Postoperations (Residual Value)		
System Entities		Time	Man-hours	Dollars	Time	Man-hours	Dollars	Time	Man-hours	Dollars
1.0 Production System										
1.1 Plastic molding machines										
1.1.1 Development										
1.1.2 Acquisition										
1.1.3 Installation										
1.1.4 Operations										
1.1.5 Preventive maintenance										
1.1.6 Corrective maintenance										
1.1.7 Depreciation										
1.1.8 Residual value										

1.2 Machine operator 1.2.1 Training 1.2.2 Operations										
1.3 Plastic materials 1.3.1 Quantities used										
1.4 Rework machines 1.4.1 Operations 1.4.2 Preventive maintenance 1.4.3 Corrective maintenance										
1.5 Rework operators 1.5.1 Operations										
1.6 Rework materials 1.6.1 Quantities used										

obtained from cost accounting and shown in Table 4.2. They may be obtained by tracking each entity as it moves across the time phases represented by columns 3, 4, and 5. They could also be obtained by focusing on each time phase and then considering each entity in turn. We took the latter course.

5.3.1 Preoperations Costs

- *Research and development costs.* In our case no explicit costs are to be borne by the company.
- *Equipment investment costs.* Cost of A2, A3, and A4 equipment depends on the number of machines bought. Net investment cost must also include "trade-in" or sale of any A1 machines.
- *Equipment installation costs.* This may be included as part of the sales price (investment cost) or it may be an additional cost. If any of the company personnel are involved in the installation, their costs must be included as the product of the man-hours required and the wage rates.
- *Training of new operators.* This is based on the number of student and instructor man-hours required. Other instructional costs, such as cost of training materials, may be required.
- *Rework equipment costs.* Since the existing rework machine is currently used only half of the time, it is assumed that it offers adequate capacity for future production demand.

5.3.2 Annual Costs

- *Plastic machine operator costs.* The annual cost of the production operators on each shift (POC) is calculated as follows:

$$\text{POC} = (\text{PMH})(\text{OW})$$

where PMH = the sum of production man-hours per year associated with each production shift

OW = the operator wage per hour on each shift. Note that this cost may be straight time, premium time, or overtime, depending on the system design

POC must be calculated for each shift and yields the total production operator cost for each year. Hence for the first year of operations of the current system:

$$\text{POC (first shift)} = (20{,}000 \text{ machine-hours})(\$4/\text{hour}) = \$80{,}000$$

$$\text{POC (second shift)} = (17{,}123 \text{ machine-hours})(\$4/\text{hour})(1.15) = \$78{,}766$$

Thus the first year's production operators' costs are $158,766. It is further assumed that the production schedule is formulated so that each operator is productive during his entire shift. This can be accomplished by stockpiling finished goods, doing other work required, and so on. If this is not done, additional man-hours of unproductive time will have to be added to the productive hours required.

- *Material costs.* With the current A1 system for each 100 units of material we obtain a total of 92 acceptable units. Thus the total direct material costs (CM) for the first year of production will be

$$\text{CM} = \frac{160{,}000(\$1)(100)}{92} = \$173{,}913^{2}$$

The cost of materials for the other system alternatives is found by using the appropriate yield and rework characteristics for that system.

- *Rework operator cost.* Since we know that one man is *currently* required half-time for this past year's production rate of 100,000 units, we may use this information to estimate the future year's rework load. Thus the rework labor cost (RLC) for this year's production of 160,000 units using the yield obtained from the current system is

$$\text{RLC} = \left(\frac{160{,}000}{100{,}000}\right)(\tfrac{1}{2}\text{ man})(2000\text{ hours})(\$5/\text{hour}) = \$8000$$

However, when other types of machines are used, such direct proportionality no longer holds. Now the rework material costs are the product of the number of units reworked and the cost of reworking each unit. Using the operational flow model of the current system (Figure 4.2), we can illustrate the past year's flow in the following way, as shown in Figure 5.1:

For the 100,000 acceptable units produced, 100,000/0.92 = 108,696 input units of plastic granules are required.

Of these 108,696, the number of reworkable units is (108,696)(0.12) = 13,043.

Thus the rework labor cost per unit is (½)(2000)($5)/13,043 = $0.38 per unit. This factor can be used for the rework labor cost of all other systems.

In the case of the other system alternatives it is assumed that the amount of rework labor is the product of the rework labor cost ($0.38) and the number of units requiring rework. It is also important to

[2]Note that if the scrap units can be ground and mixed with the regular plastic granules, the yield factor of 0.92 is irrelevant and the material cost becomes $160,000.

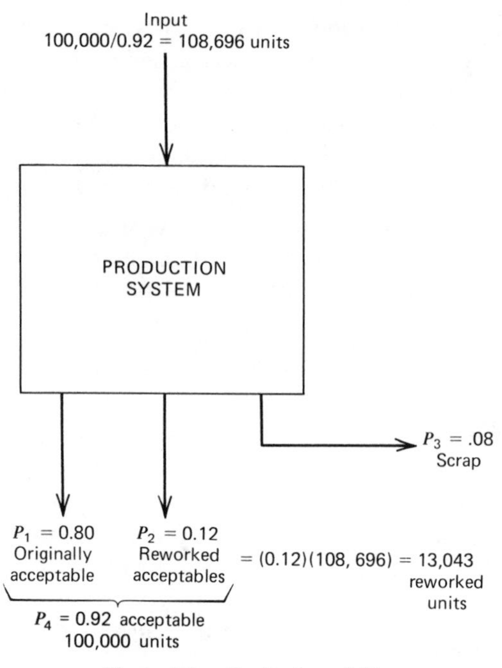

Figure 5.1. Production yield.

determine how the remaining time of the rework operator is being used and to allocate these costs accordingly.

- *Rework material costs*. During the past year the cost of rework materials for the current system has been $10,000. Thus if the current system is used, next year's rework cost for 160,000 units is readily calculated as ($10,000)(160,000/100,000) = $16,000. It is assumed that this factor will be the same in calculating the rework material costs for the different systems. However, to calculate the rework material costs for the other systems, we use the same type of analysis as above. The rework material cost for each of the 13,043 units reworked last year equals $10,000/13,043 = $0.77 per unit.
- *Indirect costs*. In the cost accounting system used by this company all other costs associated with the product (except depreciation) are defined as indirect costs. These include

  The fringe benefits associated with employee costs, such as holidays, vacation, illness, pension contributions, and all other personnel benefits.

  Workmen's Compensation and payroll tax.

All factory support such as maintenance, production control, purchasing, general tools, and supplies not directly attributable to the product (plastic material).

All management, clerical, and marketing personnel.

All costs related to facilities and utilities.

Such costs can be included in the total costs in different ways by different cost accounting groups. In our company all these costs for the past year were summed and divided by the total direct labor cost of the entire factory for the past year, and an overhead rate of 90% was obtained.[3] Thus each product's pro rata share of the total indirect cost is 90% of its direct labor cost.

Although the 90% overhead figure may be a convenient way of calculating these costs for the current production system, this figure is based on both the current *systems organization model* as well as the past year's *volume* of production. As the volume is to increase over these next 5 years, how will the system organization change and how will the ratio of indirect costs to direct labor costs change? The planner must find out what cost elements have changed (or will change) since the last overhead calculation was made. For example, as the production volume increases each year, certain fixed costs, such as rent, may remain constant, but if production goes into the other shifts, utility, security, and other support costs will increase. The new indirect costs for future years and for different system alternatives may be calculated as follows:

First, calculate the base indirect costs, which are 90% of the total direct labor cost for the current system for 100,000 units of production.

Second, calculate what changes will occur in each of the resources in the direct cost category, based on the new volume and for each new system alternative. In the case of personnel express the resource change in man-hours. For other resources, express the changes in dollar costs. One example of such savings in resources would be the reduced maintenance man-hours required for improved machines such as A4. These can be claimed as "equivalent" cost savings due to the new machine and can be calculated as the product of the man-hours saved and the hourly wage.

Third, calculate the indirect labor requirements for each year by considering the changes in work elements for the entire facility, the ways

[3]Sometimes these indirect costs are divided into several categories, such as employee benefits, factory overhead, engineering overhead, and general administrative and selling costs. These individual costs may be allocated differently among the different products.

in which related work elements may be clustered into a single job for a worker, and the number of hours (straight time and overtime) that each type of worker will work. In this way we can see whether reductions in indirect labor hours will result in actual savings or merely result in some unapplied or unproductive time. Even if the use of these manpower savings is not identified, the equivalent savings are still a good indicator, since creative planners can generally find a way of using these hours saved in a cost-effective manner (e.g., by using saved maintenance hours to do the additional rework needed for increased production).

Fourth, calculate all other incremental indirect costs, such as nonpersonnel costs.

Thus the new indirect cost for each year is the sum of the original base indirect cost plus all incremental changes in indirect labor cost caused by the new production rate and the new system machinery.

5.3.3 Other Cost Elements

Equipment depreciation and the residual value of each unit of equipment are two other cost elements that must be considered. However, these are more appropriately introduced and described at the end of Chapter 6.

This completes our summary of a method of calculating each of the out-of-pocket costs for each of the 5 years of life for each system alternative, including a numerical example of how it is done for the current system. These calculations are made for each system alternative designated in Chapter 4. Next we describe various methods for comparing the costs with the benefits of each alternative (the essence of cost-benefit analysis, or cost-effectiveness analysis) and show how to choose the preferred alternative on this basis.

5.4 SUMMARY OF KEY PRINCIPLES

This chapter began the systems evaluation step of planning. Having designed each of the system alternatives to meet the same objective, we begin the systems evaluation step by determining all the incremental costs required for each system alternative. This is done as follows:

1. Determine all the incremental resources required (as compared with the current system as the base case) over the entire system life (both

inside the system and outside if it imposes additional costs on an interfacing system). These resources are of the following types:

- Research and development.
- Investments for such things as equipment, material, or training, whose use is obtained beyond one year.
- Operation, maintenance, and support resources (man-hours and materials required).

2. Convert these resources required to dollar costs (or credits for residual values at the end of the operating period) as follows:
   - Investment costs.
   - Labor costs (product of man-hours and wage rates).
   - Material costs of all materials used.
   - Residual value of capital equipment at end of operations.

The system organization model and the resource analysis matrix are helpful in this process of determining the resources required, particularly in making certain that no required resource is omitted.

6

Analyzing Investment Possibilities

Having calculated the cost stream (costs over time) for each system alternative under consideration, we must now compare these costs to the benefits obtained for each alternative and among alternatives. Benefits may be treated in either of two ways, depending on whether the organization we are analyzing is a profit center (an organization held accountable for the profits it achieves) or a cost center (generally a service organization held accountable only for the costs expended in rendering that service). Since our organization is a profit center, we first illustrate how to do the evaluation for profit centers. Later in the chapter we describe how to do evaluations for cost centers.

The purpose of this discussion is

- To present a review of present value analysis (sometimes also called present worth analysis) for those unfamiliar with the concept.
- To apply present value analysis to investment problems, assuming no risks or uncertainties (treatment of uncertainties is discussed in Chapter 7).
- To describe the various ''decision-making'' models used in analyzing investments for both profit centers and cost centers.
- To describe various methods of treating equipment depreciation and residual value.

We now detour from the main case and discuss a number of tools applicable to all investment problems.

6.1 EVALUATING INVESTMENTS FOR PROFIT CENTERS

In a profit center the benefits obtained can always be expressed as a payback of money in return for the investment expenditures. The key

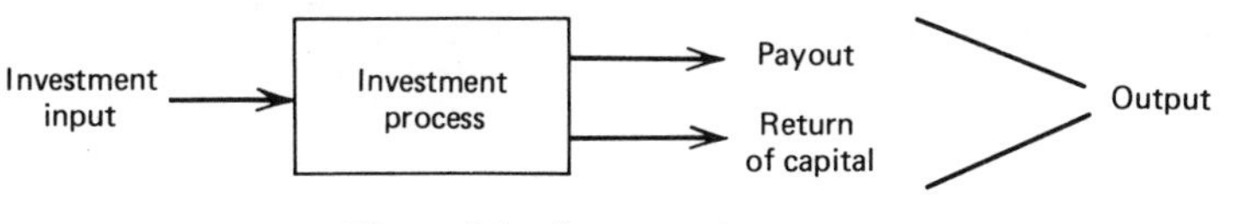

Figure 6.1. Investment process.

principle in analyzing an investment is to consider an investment as an input-output transformation. As illustrated in Figure 6.1, an investment consists of money paid *in* at some initial time that is converted to resources and used by an organization. At some *later* time the organization periodically pays money back to the investor in the form of interest or dividends. At some even later time an additional lump sum may also be recovered, such as when a bond matures or shares of a business are sold to other investors. At this time the investment ceases (as far as the original investor is concerned).

Consider the following examples, which illustrate how to analyze this input-output process.

Example 1.
You are considering purchasing a number of high-quality $1000 bonds selling for $800 each, using funds you have on deposit in a 5½% savings account. These bonds have 5 years to go for redemption and yield 6% coupon interest (on a semiannual basis). Bonds of this type normally provide a rate of return of 8%.

a. Does this bond exceed this rate of return?
b. What should the selling price of the bond be to yield 8%?

To present the concept of present value discount analysis, this investment alternative is now analyzed.

6.1.1 Analysis of Bond Investment

The first step in analyzing the bond is to construct the cash flow diagram of Figure 6.2 showing costs and income as they occur over time. Here the initial purchase price of $800 is represented as a negative value since it represents an expense to the investor, whereas the payments to the investor are represented as positive values. These consist of the coupon payment of $30 every 6 months for the remaining 5-year period until maturity and the bond redemption of $1000 at the end. To analyze this investment the analyst must compare the positive values (the payouts) with the negative values (the investment). Since the payouts occur over different time periods, they must be accumulated using present value analysis, as is now illustrated.

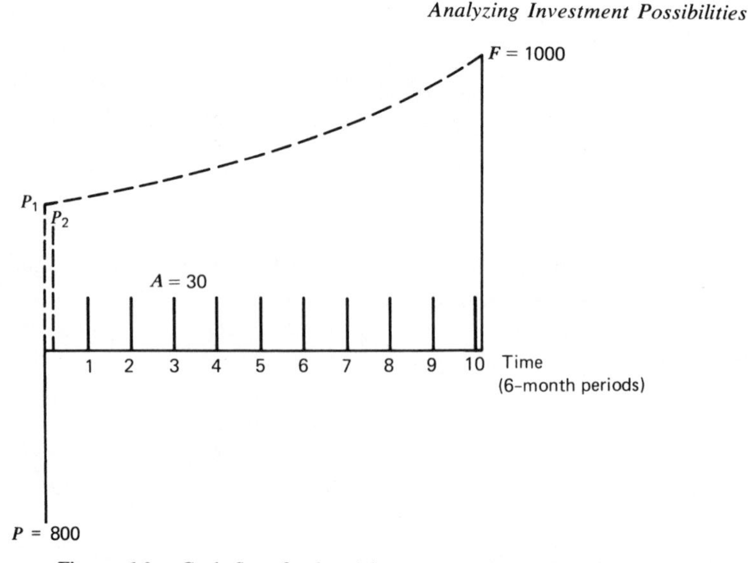

Figure 6.2. Cash flow for bond investment (Example 1).

The basic premise made in present value analysis is that a decision maker would prefer to have a sum of money (say \$100) available to him today rather than the same sum at some future time. Saying this another way, for the decision maker to be indifferent to the two alternatives, if the first alternative is \$100 received today, the second alternative would have to be greater than \$100 received at some future time. This relationship is best illustrated by the interest compounding obtained at a savings bank. If a saver is willing to take 8% interest compounded semiannually in return for the bank's use of his money, the future value (F) of his present savings of \$100 (P) is illustrated in Figure 6.3. In general, the future value (F) of a given investment (P) is

$$F = P(1 + i)^n = P(\text{SPCA})$$

where i = the interest earned in that period (4% each 6-month period)
n = the number of (6-month) periods over which the investment is held
SPCA = single payment compound amount factor, sometimes also called the amount at compound interest factor, which is equal to $(1 + i)^n$

We introduce this factor since these values are usually obtained directly

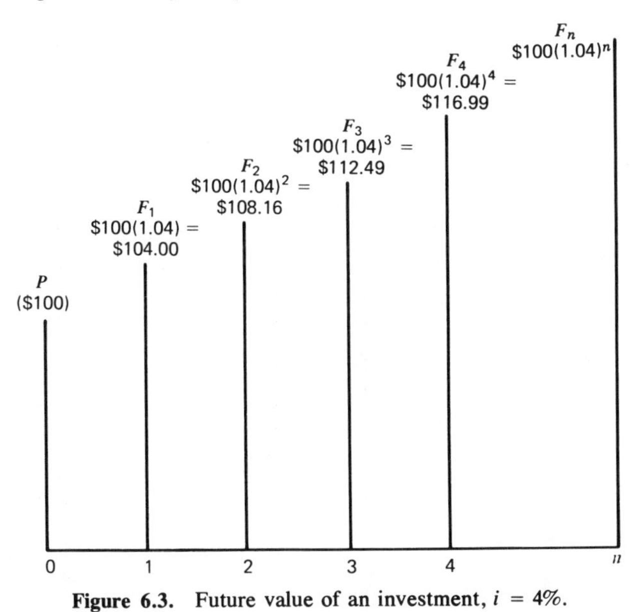

Figure 6.3. Future value of an investment, $i = 4\%$.

from interest tables rather than by calculating them directly.[1] Conversely, the present value (P) of a given future value (F) may be obtained from the same formula:

$$P = F(1 + i)^{-n} = F(\text{SPPW})$$

where SPPW = single payment present worth factor, sometimes called the amount of annuity factor, equal to $(1 + i)^{-n}$

Thus these two formulas provide a way of translating P to its equivalent F, or F to its equivalent P, as illustrated in Figure 6.3.

In a similar fashion, if a sum of money (say $100) were deposited in the bank *at the end of each 6-month period,* it would be worth the amount shown in Figure 6.4 by the dotted lines at the end of each period. Again this assumes semiannual compounding at the rate of 4% per period. Note that this amount is always more than the sum of the payments since the

[1]See E. L. Grant, W. G. Ireson, and R. S. Leavenworth, *Principles of Engineering Economy*, Ronald Press, New York, 1976; or R. C. Weast (Editor-in-Chief) and S. M. Selby (Editor-in-Chief of Mathematics), *Handbook of Tables for Mathematics*, Chemical Rubber Co., Cleveland, Ohio, 1967.

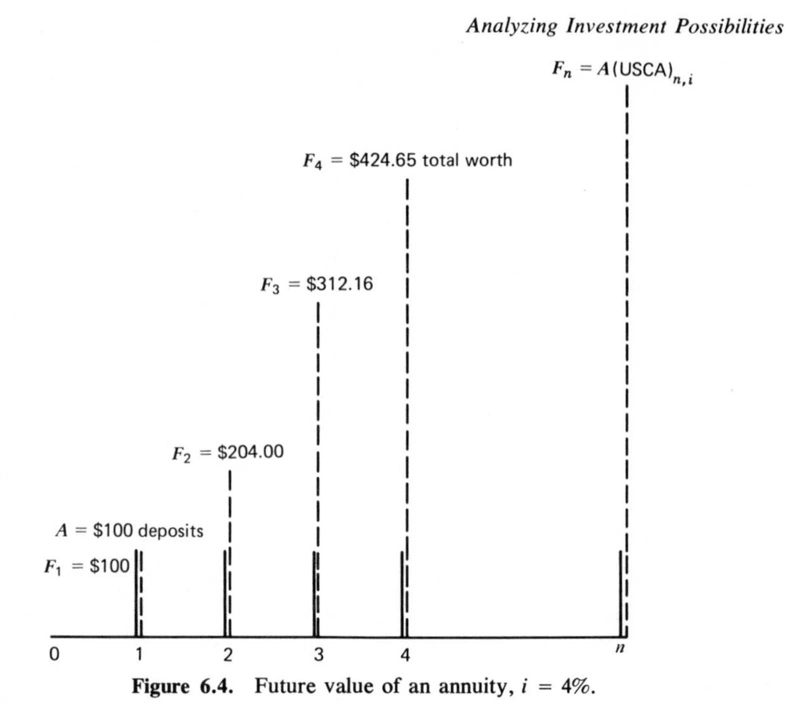

Figure 6.4. Future value of an annuity, $i = 4\%$.

money earns interest while on deposit. In this case the general formula that relates S and A is

$$F = A(\text{USCA}) = \frac{A\,[(1 + i)^n - 1]}{i}$$

where USCA = uniform series compound amount factor, sometimes called the amount of annuity factor, since it is used in calculating annuities that consist of periodic savings of this type

Next we can calculate the present value (P) of a uniform series of end-of-period payments (A), as shown in Figure 6.5.

$$P = A(\text{USPW}) = \frac{A\,[1 - (1 + i)^{-n}]}{i}$$

where USPW = uniform series present worth factor, sometimes called the present value of annuity factor, since this formula describes how much money would have to be deposited (P) to obtain a periodic annuity payment (A) over n periods of time.

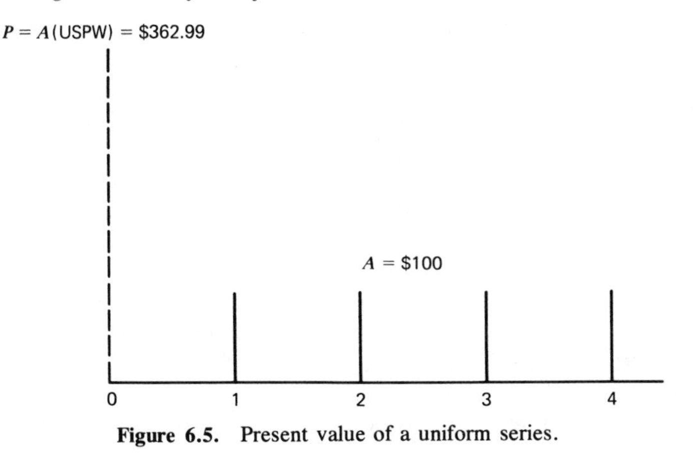

Figure 6.5. Present value of a uniform series.

Thus the present value (P) of four \$100 end-of-period payments and an interest rate of 4% per period is

$$P = (\$100)(3.6299) = \$362.99$$

Finally, the last relationship in the interest tables is that which calculates A from P:

$$A = P(\text{USCR}) = P \frac{i}{1 - (1 + i)^{-n}}$$

where USCR = uniform series capital recovery factor, sometimes called the annuity whose present value is 1 factor, and represents the mortgage payments (A) to be made in paying off a given mortgage (P) over n periods.

Having described the main factors available from the interest tables, let us return to the analysis of the bond alternative and the analysis of its cash flow of Figure 6.2. We want to determine the total value of all the payments received. Since they occur at different times, one way to do so is to take the sum of the present value of each payment. In our example the present value (P_1) of the final payment (F = \$1000) may be calculated as follows:

$$P_1 = F(\text{SSPW})_{i,n}$$

where P_1 = present value
F = future value
i = interest rate for the period
n = number of periods in the future
SSPW = single payment present worth factor

In this case, since the interest is paid semiannually, $n = 10$ periods and $i = 4\%$ per period. Thus SSPW = 0.6756 and $P_1 = 1000(0.6756) = \$675.60$.

Next we calculate the present value (P_2) of the 10 interest payments (each of \$30). Since this is a uniform series of end-of-period payments, its present value (P_2) may be found as follows:

$$P_2 = A(\text{USPW})_{\substack{i = 4\% \\ n = 10}}$$

Thus $P_2 = 30(8.1109) = \$243.33$.

Hence the present value of the total payout is the sum of its two components, or \$918.93. Since this is greater than the \$800 selling price of the bond, the effective rate of return must be higher than the 8% used in the calculations. In fact, assuming an 8% rate of return, the selling price of the bond should be \$918.93.

Example 2.
Your broker also has available some municipal bonds which are exempt from both federal and state income taxes. These \$1000 bonds sell for \$950 each, pay 5% coupon interest, and mature in 10 years. Your state income tax bracket is 10% and your federal income tax bracket is 40%. Assuming that these bonds as well as the previous ones are of high quality and risk is not a factor, which bond is the better investment?

6.1.2 Analysis of Second Bond Investment

Let us now analyze the second bond investment, whose cash flow diagram is shown in Figure 6.6. Since the income from this bond is exempt from both federal and state income taxes, to compare it with the first bond, an *aftertax* cash flow analysis of both investment alternatives must be made. The aftertax cash flow for each bond is also shown in Figures 6.6 and 6.7. Note that federal and state taxes (50%) must be paid on the income of the first bond, but no taxes are paid on the income of the second bond. In both cases it is assumed that federal and state capital gains tax (totaling one-half of 50%) must be paid on the capital gains obtained when the bond matures.[2]

Another change to be made in the aftertax analysis is the interest rate to be used. Now the worth of money to the investor (after taxes) is half of what it was before, or 4%.

Since a capital gain of \$50 was obtained, a capital gains tax of

[2]For simplicity, this analysis has ignored the fact that any state taxes paid can be deducted from next year's federal taxes. This factor could be inserted for a more accurate analysis.

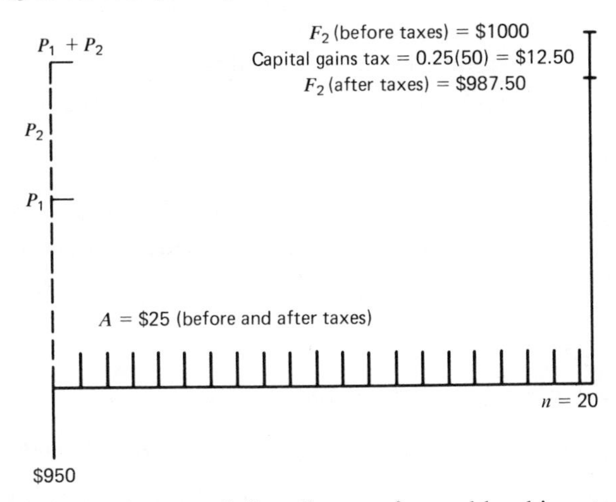

Figure 6.6. Aftertax cash flow diagram of second bond investment.

(½)(0.50)(\$50) = \$12.50 is due when the second bond matures. Thus the present value of the \$987.50 aftertax payment received 20 periods from now is

$$P_1 = F(\text{SPPW})_{\substack{i = 2\% \\ n = 20}} = \$987.50(0.67297) = \$664.56$$

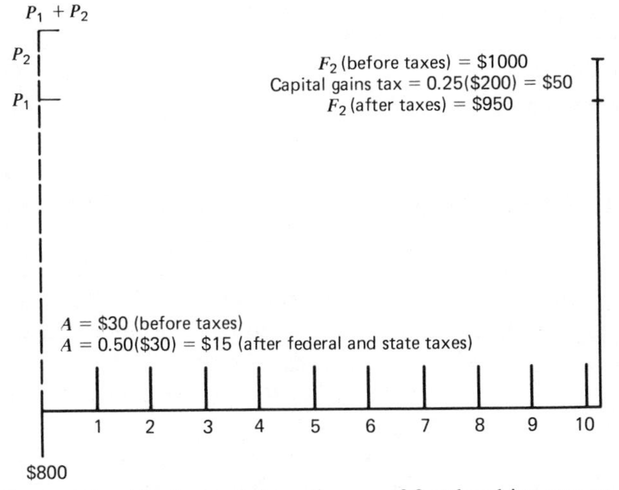

Figure 6.7. Aftertax cash flow diagram of first bond investment.

Note that the aftertax discount rate of 2% per period is used. The present value of the interest payments (after taxes) is

$$P_2 = A(\text{USPW})_{\substack{i = 2\% \\ n = 20}} = \$25(16.351) = \$408.78$$

Thus $P_1 + P_2 = \$1073.34$, which when compared to the original cost of \$950 means that the second bond has an aftertax return on investment greater than the 4% per year (compounded semiannually) used in the calculations.

6.1.3 *Aftertax Analysis of First Bond*

The present value (P_1) of the final payout of \$950 after taxes is

$$P_1 = F(\text{SPPW})_{\substack{i = 2\% \\ n = 10}} = \$950(0.82035) = \$779.33$$

The present value (P_2) of the interest is

$$P_2 = A(\text{USPW})_{\substack{i = 2\% \\ n = 10}} = \$15(8.9829) = \$134.74$$

Thus $P_1 + P_2 = \$914.08$. Since this is greater than the original investment of \$800, this bond also has an aftertax return greater than 4% compounded semiannually.

The analysis shows that

- Both investment alternatives achieve an aftertax return on investment greater than 4%.
- We do not know which is the better alternative.

6.2 TYPES OF ECONOMIC EVALUATION MODELS

A number of different economic evaluation models (shown in Table 6.1) are used by different profit and cost center organizations in analyzing their investment alternatives. These models all involve the calculation of system costs and revenues received as they occur over time. They differ in how they combine the time flow of costs and revenues and the decision criterion used to select the preferred alternative.

Using the two bond alternatives as illustrative examples, each of these models is described in terms of these characteristics:

Table 6.1 *Types of Economic Evaluation Models*

For profit centers
Conventional present value of net returns model
Incremental present value of net returns model
Internal rate of return model (marginal efficiency)
Profitability index method
Payback period model
For cost centers
Present value of costs model
Savings/investment ratio (SIR) method

- What data are needed as inputs to each model.
- How the model interrelates this data.
- The criterion of choice used in selecting the preferred alternative on the basis of the model.
- The relationship between the accuracy of the results obtained from the model and the analytical resources required to implement the analysis, including problems inherent in its use.
- The advantages and disadvantages of using each method and under what circumstances each should be employed.

6.3 MODELS FOR EVALUATING PROFIT CENTER INVESTMENTS

6.3.1 Conventional Present Value of Net Return Model

The conventional present value method was used in the previous two examples. A given discount rate is first applied to the stream of costs (negative entries in the cash flow diagram) and then to the stream of revenues (positive entries). If the present value of the revenues exceeds the present value of costs, this means the rate of return (ROR), or return on investment (ROI), exceeds the discount rate. If the present value of revenues is less, the ROR is less than the discount rate.

This method is straightforward and requires that only one calculation be made. Yet there are two problems associated with this method:

- What interest rate should be used as the standard in making the present value calculations? Generally, a company chooses a minimum acceptable rate based on the historical rate of return achieved on its past investments.

- For each investment alternative analyzed, this method only tells us whether or not the ROR exceeds the chosen discount rate. It does not enable us to select the best alternative from among those that do exceed the discount rate chosen.

6.3.2 Incremental Present Value of Net Returns Method

A variation of the conventional present value method previously described is the incremental present value method. In this approach two alternatives are compared by calculating the *incremental* cash flows between the two alternatives (say, cash flow 2 minus cash flow 1). If the resulting present value of the incremental net return obtained is positive, alternative 2 is superior to alternative 1, and vice versa.

Let us apply this approach to the example of the two bonds being compared. First, choose bond 1 as the base case of comparison (since this requires the least investment). Next construct the cash flow diagram of the difference in cash flow between the two bonds (i.e., the cash flow of bond 2 minus the cash flow of bond 1). This is illustrated in Figure 6.8. The next step is to calculate the present value of the negative stream (P_1 and F_1).

$$P_1 = -150$$
$$P_2 = F_1(\text{SPPW})_{\substack{i = 2\% \\ n = 10}} = -950(0.82035) = -779.33$$

Thus the present value of the negative stream is $P_1 + P_2 = -150 - 779.33 = -929.33$.

In a similar fashion we find the present value of the positive stream (R_1, A_2, and F_2).

$$P_3 = A_1(\text{USPW})_{\substack{i = 2\% \\ n = 10}} = 10(8.9829) = 89.83$$

P_4 equals the present value of the A_2 stream. This is calculated in two steps. First, reflect the A_2 stream as a present value at $n = 10$. Call this P_5. Then find P_4, the present value of P_5 at $n = 10$.

$$P_5 \text{ (at } n = 10) = A_2(\text{USPW})_{\substack{i = 2\% \\ n = 10}} = 25(8.9829) = 224.57$$

$$P_4 \text{ (at } n = 10) = P_5(\text{SPPW})_{\substack{i = 2\% \\ n = 10}} = 224.57\ (0.82035) = 184.23$$

Let P_6 be the present value of F_2.

$$P_6 = F_2(\text{SPPW})_{\substack{i = 2\% \\ n = 10}} = (987.50)(0.67297) = 664.56$$

Thus the present value of the positive stream $= P_3 + P_4 + P_6 = 89.83 + 184.23 + 664.56 = 938.62$. Since the present value of the positive stream is greater than the present value of the negative stream, the base case is the inferior investment, and bond 2 is the preferred alternative.

Although this method enables a comparison of two alternatives, it still requires considerable calculation in selecting the best alternative from a number of alternatives.

6.3.3 *Internal Rate of Return Method*

This method, sometimes called the marginal efficiency method, is the most accurate way of comparing a large number of different investment alternatives, particularly those requiring different-sized investments. Essentially, it consists of applying the conventional present value model to each investment, using different values of the discount rate on a trial-and-error basis to find the value at which the present value of the net cash flow equals zero (the present value of the payouts equals the present value of the investments). When this condition is satisfied, the rate of return of the investment is equal to the discount rate. Table 6.2 shows the results of such present value calculations for the positive cash flow of bond 1, using different values of the discount rate. They indicate that the ROR in which the positive cash flow equals the negative cash flow of \$800 is slightly less

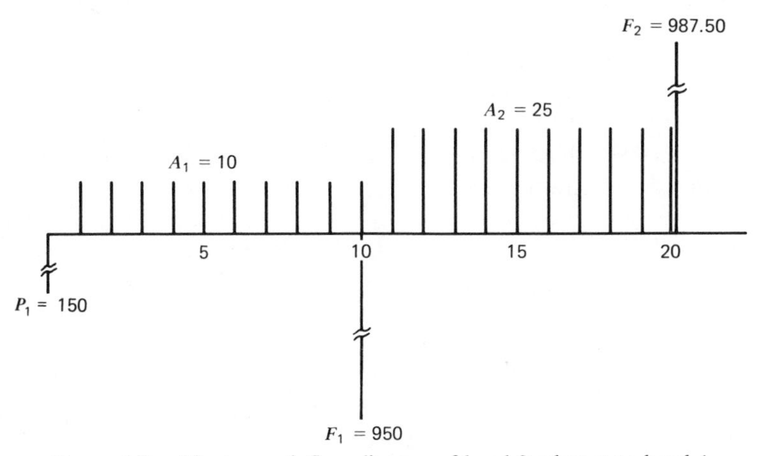

Figure 6.8. Aftertax cash flow diagram of bond 2 relevant to bond 1.

Table 6.2 *Present Value Calculations to Determine the ROR for Bond 1*

	Discount Rate			
	2%	2½%	3%	3½%
$(\text{SPPW})_{i}{}_{n = 10}$	0.82035	0.78120	0.74409	0.70892
$P_1 = 950(\text{SPPW})$	779.33	742.14	706.86	673.47
$(\text{USPW})_{i}{}_{n = 10}$	8.9829	8.7521	8.5302	8.3166
$P_2 = 15(\text{USPW})$	134.74	130.82	127.95	124.75
$P_1 + P_2$	914.08	872.96	834.81	798.22

than 3½% per 6-month period. The same approach is then applied to bond 2, and the alternative yielding the largest rate of return can then be chosen.

Although the rate-of-return method offers the most accurate way of comparing a large number of different alternatives, its major disadvantage is the added calculations required for the trial-and-error solution. In addition, the use of any ratio as a means of comparing alternatives presents certain pitfalls. Two examples illustrate these pitfalls.

First, consider the four investment alternatives whose present value of investment cost and calculated ROR are shown in Table 6.3. Assume that these alternatives are completely independent of one another so that all combinations of these three investments are possible (unlike the alternatives in our production planning case where only one alternative can be chosen). In Table 6.3 we have ranked these alternatives in the order of highest to lowest ROR. The fourth alternative is to leave our investment funds in the bank at 5% return.

We can define our investment problem as a combinatorial problem: Choose the combination of available alternatives that maximizes our total return, taking into account the limitation in investment funds. Let us show how the decision process would operate, treating the investment funds available as a variable. It really is a variable because it consists of two basic parts: the funds available in the treasury that exceed what is needed for day-to-day operation of the business and the funds that can be borrowed from outside sources at a rate of interest that depends on a number of factors, including how much debt is already outstanding. Thus the task of investment analysis may be thought of as comparing the ROR of

Table 6.3 *Ranking of Investments*

(1) Investment Alternative	(2) Investment Cost	(3) ROR	(4) (SPCA) Single Payment Compound Amount Factor (n = 5 Years)	(5) Total Return from Investment (Future Worth in 5 Years)
A1	$ 7,000	30%	3.7129	$25,990
A2	$ 5,000	15%	2.0114	$10,057
A3	$10,000	10%	1.6105	$16,105
A4 (savings bank)	Any amount	5%	1.2763	$ 1,276 per $1,000 investment

available investment alternatives with the cost of money, also taking into account the premium that should be obtained because of uncertainties in the investment outcome.[3]

One way to use these data is to calculate the future worth obtainable at the end of the planning horizon[4] as the total size of the investment funds available increases. For example, as shown in Figure 6.9, the data in Table 6.3 indicate that if less than $5000 is available for investment, it should be kept in the bank (A4), since all higher-earning investments require $5000 or more. Thus amounts less than $5000 earn interest at the rate of 5%. This will produce a future worth (F) at the end of the 5-year period of $1276 per $1000 investment, or $6380 for a $5000 investment.

However, if $5000 is available, A2 is the better investment, yielding $10,057 at its 15% ROR. Between $5000 and up to $7000, A2 plus A4 is the better alternative to only A4. If $7000 is available, A1 is chosen. At $10,000, A3 is available, but this gives a return lower than A1 plus assigning the remaining $3000 to A4. Hence A3 is rejected. Figure 6.9 shows all the other investment alternatives available, including the inferior alternatives not chosen (A2 + A3 at $15,000 and A1 + A3 at $17,000).

The planner must be even more careful when alternatives that are not independent of one another are being evaluated, such as in the production

[3]Uncertainties are described in Chapter 7.

[4]A planning horizon of 5 years is assumed in this example.

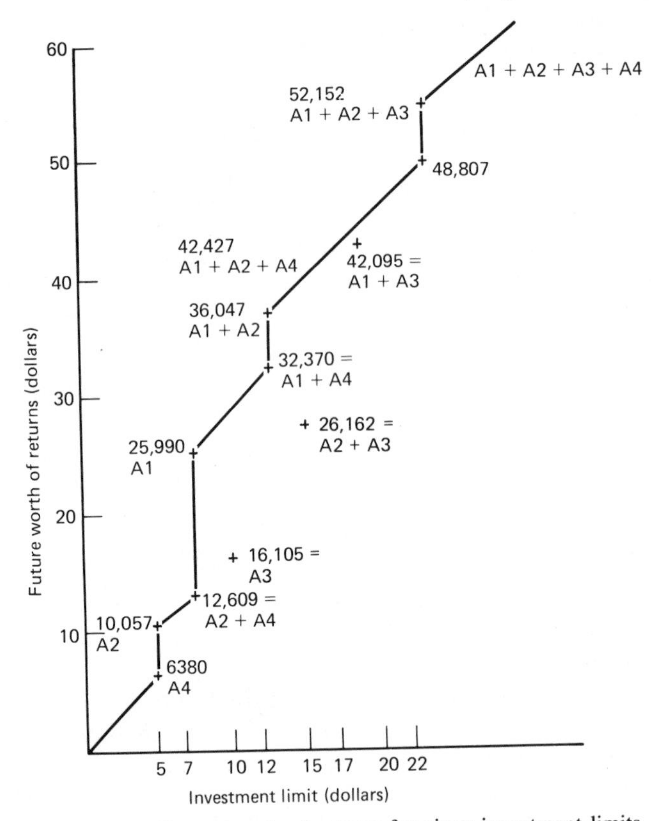

Figure 6.9. Future worth of returns for given investment limits.

planning case in which the system alternatives were designed so that only one could be selected (such as 10 A1 machines or 3 A4 machines but not both, since overcapacity would result). In this case consider the three alternatives of Table 6.4, which all do the same job, so that only one of the three can be selected. In this case the alternatives are ranked by the total investment required. Alternative A0 is just like our base case of 10 A1 machines. No investment is required. The future value (F) of the total returns obtained over the 5-year period is $15,000. Note that no rate of return can be calculated, since no additional investment is required to yield the future worth of $15,000. Alternative A1 has the same characteristics as Alternative A1 of Table 6.3. As an independent investment having a ROR of 30%, this $7000 alternative yielded a total future worth return of $25,990. However, now as a dependent investment, this $7000 investment would yield an incremental return over A0 of only

Table 6.4 *Evaluating Rate of Return for Dependent Alternatives*

(1) Investment Alternative	(2) Investment Cost	(4) Total Return from Investment (Future Worth)	(5) Incremental Return (Future Worth)	(6) Incremental Investment	(7) (SPCA) Single Payment Compound Amount Factor	(8) Marginal ROR
A0	0	$15,000	—	—	—	—
A1	$ 7,000	$25,990	$10,990	$ 7,000	1.57	9.4%
A2	$ 5,000	$10,057	$ 4,943	$ 5,000	—	No payback of investment
A3	$10,000	$16,105	$ 1,105	$10,000	—	No payback of investment

$10,990, yielding a *marginal* rate of return of only 9.4%. Alternative A2 as an independent investment yielding a ROR of 15% formerly provided a total future worth payout of $10,057. As an independent investment it does not even provide the $15,000 that can be obtained from A0 with no additional investment, so this alternative is immediately rejected.

Alternative A3 does provide a return greater than the $15,000 obtainable from A0. However, the incremental return is even less than the investment of $10,000 required, and hence A3 is also rejected. Hence, although A1 may appear to have a ROR of 30%, its true (marginal) ROR is really 9.4% and thus it should also be rejected if any other independent investment having a ROR greater than 9.4% is available. Funding such an independent investment plus continuing to operate A0 would yield a higher total cash return for the $10,000 investment than funding A3 alone.

This example illustrates why our zero investment alternative was designed, since it serves as our base case. All other alternatives requiring investment must be compared to it, taking into account other independent options for the same investment.

In summary, one can think of a business as having two components, each capable of generating a financial return:

- The main business component in which investments have already been made,[5] which can continue to generate a financial return for some future time period even with no additional investment. This is defined as our base case.
- A treasury of other funds available for investment. Such funds will also generate some return if the money is placed in a bank or invested outside the main business component under analysis.

Thus the total return is the sum of the two returns, and the primary function of the business managers is to maximize the total return, taking into account all investment possibilities available.

6.3.4 *Profitability Index Method*

The profitability index method is another way of overcoming the main deficiency of the present value method: it can be used to compare several alternatives, each of which has a rate of return that exceeds the minimum acceptable rate of return. In this method a profitability index (PI) or

[5]Such previous investments are *sunk costs* and do not enter into the analysis of future investments.

benefit-to-cost ratio is calculated as the ratio of the present value of the future net cash flows to the initial cash outlays, or

$$PI = \frac{Q}{I}$$

where Q = present value of the positive cash flow
I = present value of investment

Thus in the previous Example 1, the profitability index for bond 1 (PI_1) is

$$PI_1 = \frac{914.08}{800.00} = 1.143$$

and the profitability index for bond 2 (PI_2) is

$$PI_2 = \frac{1073.34}{950} = 1.130$$

Thus the profitability index can be used as a way of ranking alternatives. All other things being equal, the decision maker would select the alternative having the highest PI from this ranked list.

Hence bond 1 would be the preferred alternative. However, recall that following the incremental value of net returns method, bond 2 was selected as the preferred bond (having the higher return). This example illustrates that the various decision-making models may not yield the same final result, particularly if the returns from the two alternatives are reasonably close to one another.

Although the profitability index can be used as a way of ranking alternatives, it is not as accurate an indicator as the internal rate of return model. However, the profitability index model is easier to implement, since no iterations are required.

As an aside, I have seen one author who defines the profitability index as

$$PI = \frac{Q - I}{I} = \frac{Q}{I} - 1$$

rather than as previously defined (Q/I). Using either of these definitions would provide the same decision-making result.

6.3.5 Payback Period Model

This model uses the cash flow to determine the number of years required to pay back the original investment. Thus investments are ranked by the time required for payback. Some companies specify a threshold level of acceptable investments (e.g., no investment will be approved unless payback is achieved in 3 years or less). The main advantage of this method is that it simplifies the calculation effort, since no interest tables or present value calculations are needed.[6]

There are two main disadvantages to this method. First, the *timing* of the payback is not taken into account. For example, two investments, A1 and A2, may have the same payback period of 3 years. But A1 may produce paybacks that are fairly uniform over this time; whereas A2 may produce most of its payback in the third year. Present value analysis would calculate that A1 is the better investment. The payback model would state they were equal. An even larger source of error is that no account is taken of the cash flow after the payback is completed. For example, one investment may complete payback in 3 years but continue to produce income or savings for additional years. Another investment may complete payback in 2 years but then have no cash flow afterward. The first investment is probably the better one.

Applying the payback period model to our bond examples, we find the following:

- Bond 1 pays back $150 of interest (after taxes) in 10 years and then the aftertax principal of $950. Thus it requires 10 years to recover the original investment.
- Bond 2 pays back $500 of interest in 20 years, then $987.50 of the principal. Thus it takes 20 years to recover the original investment.

This example shows that the payback period model is not a good way to compare bond investments because the recovery of investment generally must wait until the end of the bond period. The model does work better for business investments where equipment depreciation[7] can be recovered each year, thus speeding up the payback.

6.4 MODELS FOR EVALUATING COST CENTER INVESTMENTS

There are two basic models used by the Department of Defense in economic analyses that should be mentioned since they apply to all cost centers:

[6]Some organizations calculate the payback period using a discount rate. Case 2 and Case 4 show numerical examples of this.

[7]To be discussed later in this chapter.

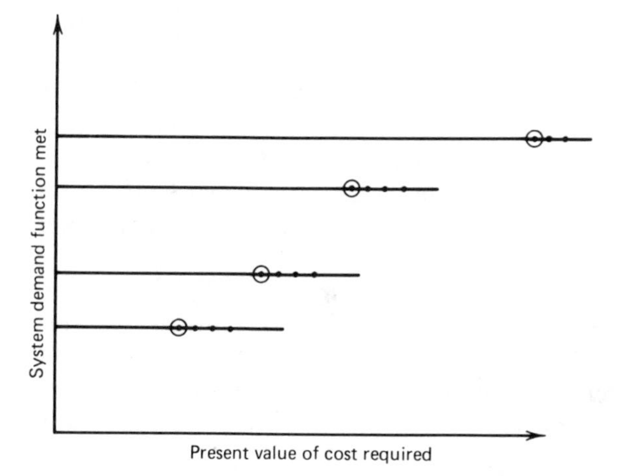

Figure 6.10. Results of present value of cost model (Pivoting on a given system demand function).

6.4.1 Present Value of Costs Model

Many systems used for defense or other government services do not provide revenue (positive cash flow) but do require various costs over time to develop, procure, operate, and maintain the total system (only a negative cash flow is involved). A preferred way of analyzing such systems is to

- Design all system alternatives to meet the same objective or system demand function. This function may be treated parametrically for different levels of effectiveness, as illustrated in Figure 6.10.
- Calculate the cost stream for all years included in the time span of the analysis.
- Calculate the present value of the cost stream, using an agreed-on discount rate[8] (currently, 10% is used by DOD for most cases).
- Select the system alternative with the least present value of cost.

An example of the use of this model would be for a government organization such as a government arsenal producing arms or a government hydroelectric dam producing electric power. In such a case we are concerned with the least costly alternative for producing a given number

[8]Although the term *interest rate* represents the cost of money to an organization, the term *discount rate* represents the organization's time value of money, and is a more appropriate term for capital budgeting problems.

of units per year, where production size may be a variable, as shown by the system demand function of Figure 6.10.

Of course, another approach is to design all systems for the same level of cost (present value of cost) and choose the system that provides the highest benefits for the given level of cost, as shown in Figure 6.11. This is also called *design to cost* and is generally a more difficult design problem, requiring a trial-and-error solution to meet exactly the total cost requirement.

A more detailed example of this present value of costs method is obtained in Case 4, the computer-managed instruction via satellite case. The main limitation of this method is that it can only compare the total costs of alternative ways of accomplishing the same objective. It cannot be used to compare systems having different objectives.

6.4.2 Department of Defense Savings to Investment Ratios (SIR)

A number of proposals made by government organizations involve systems that promise to provide cost reductions or savings for doing essentially the same job as a system currently operating (like our production planning case). An example of this would include a new, faster computer to replace an existing computer. In this case because the new computer is much faster, it can do the same jobs now requiring three shifts of opera-

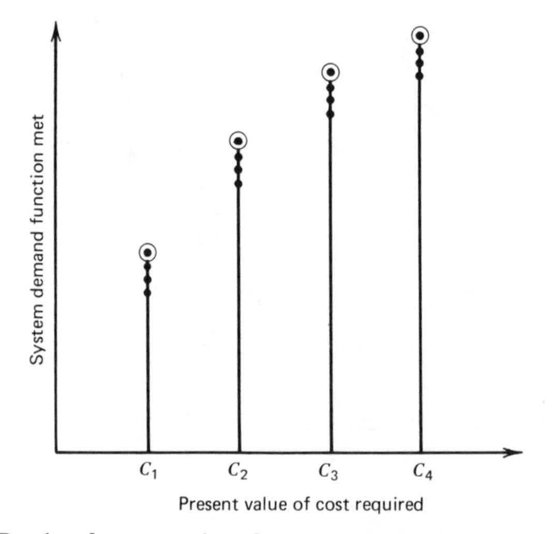

Figure 6.11. Results of present value of cost model (Pivoting on a given cost level).

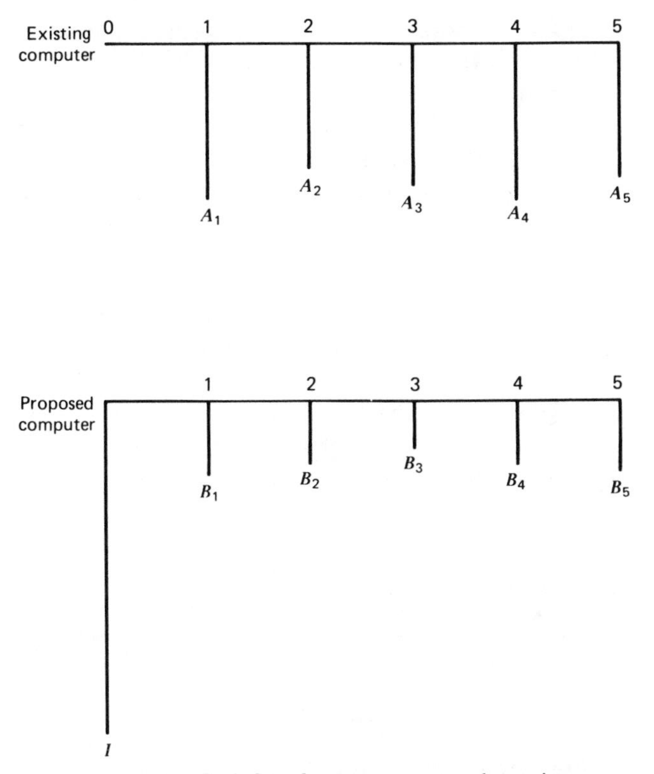

Figure 6.12. Cash flow for two computer alternatives.

tion in only one shift, thus greatly reducing labor costs. However, the new computer requires an initial investment, whereas the old computer investment is a sunk cost and is not to be considered in the analysis. The cash flow for each of the two alternatives is shown in Figure 6.12.

In this example we would like to compare the net *savings* obtained each year by using the proposed computer ($A_1 - B_1, A_2 - B_2, A_3 - B_3, A_4 - B_4, A_5 - B_5$) with the computer investment (I) required to obtain these savings. If we used the incremental present value of net returns model, previously described, and the discount rate specified by DOD, we could compare the present value of the net cost savings with the present value of the investment and see whether the savings exceeded the investment. However, constructing the savings-to-investment ratio (SIR), where

$$\text{SIR} = \frac{\text{present value of savings}}{\text{present value of investment}}$$

DOD can compare and rank alternative cost-saving proposals not only for the same objective, but even for different objectives, since cost savings is the major criterion.

Note that this model is basically the same as the profitability index method, except that the numerator of the ratio is the present value of the entire savings rather than the net cash flows as in a profit-making organization.

6.5 APPLICATION OF ECONOMIC EVALUATION MODELS TO PRODUCTION PLANNING CASE

To apply these decision-making models to the production planning case (a profit center), we need to calculate the total net return after taxes for each of the 5 years and to determine the residual value of the equipment at that time.

6.5.1 Calculating Income

We assume that the annual total income to the business is the annual sales revenue[9] (SR), where

$$\text{SR} = (S)(P)$$

where S = annual sales
P = average price during year

We further assume that the noninflated price stays fixed over time and that the annual sales are in accordance with the sales prediction of Figure 4.1.[10]

Thus the sales revenue for the first year, for example, is

$$\text{SR}_1 = (S_1)(P) = (160{,}000)(\$10) = \$1{,}600{,}000$$

The sales revenue for the other years is found by the same approach.

6.5.2 Equipment Depreciation and Residual Value

These two factors are treated together because of their relationship. Depreciation is the reduction in the value of equipment, as carried on the

[9] In a real business the income would be obtained some period of time after sales, taking customer billing and collection into account. Such time delays could be included in the cash flow diagram.

[10] This assumption will be challenged in Chapter 7.

company's accounting books, because of wear and technological or other obsolescence and is an allowable business expense on the company's profit and loss statement. Thus it consists of a yearly return of capital on all investments having an economic life longer than one year. It is important to understand that depreciation is not an out-of-pocket cost. However, depreciation is an important cost element for organizations paying taxes, since it increases the net return in two ways:

- The annual value of depreciation immediately reduces gross profits by this amount and hence it reduces taxes accordingly.
- The entire sum represents an actual return of capital, immediately available for reinvestment by the business.

There are four primary parameters whose values must be chosen by the business in calculating the depreciation of a machine:

- Original purchase price.
- Economic life.
- Residual value.
- Depreciation rate.

Figure 6.13 is the past and projected value of each of the A1 machines as carried on the company's books (its "book value"). In our case the current production equipment (Alternative A1) was purchased 5 years ago at $50,000 for each unit. At that time an economic life of 10 years was forecast, with the added assumption that the equipment would have zero value at the end of its life. It was further assumed that the value of this equipment would decrease linearly over time. This functional relationship, as shown in Figure 6.13, tells us the book value of each unit of equipment decreases $5000 per year (its annual depreciation) and after 5 years is currently valued at $25,000. Although the company must conform with IRS standards regarding the depreciation function used, there is some flexibility allowed and the company can consider the following factors in choosing a depreciation function, as illustrated in Figure 6.13.

- *Original cost.* Fixed as the purchase price.
- *Economic life.* This was assumed to be 10 years in our example, but a company can consider not only the equipment's physical life, but also technological obsolescence.
- *Salvage value.* In Figure 6.13 the salvage value at the end of the 10 years was assumed to be zero. Sometimes a positive salvage value is

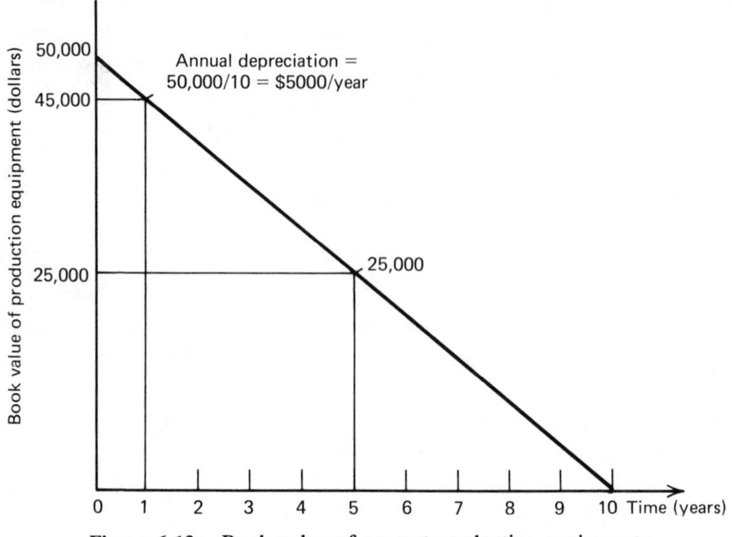

Figure 6.13. Book value of current production equipment.

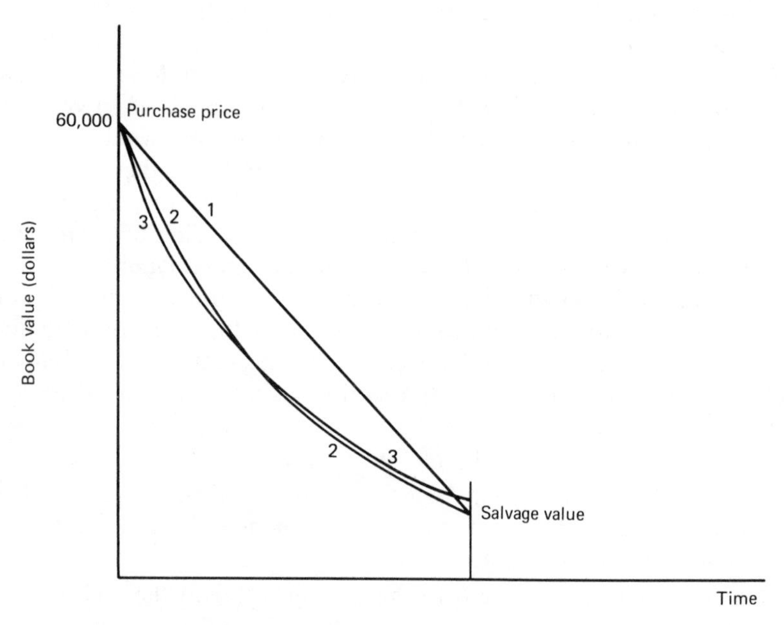

Figure 6.14. Different depreciation functions.

assumed, as shown in Figure 6.14. Sometimes it actually costs money to dismantle the equipment and take it away at the end of its economic life, and this is treated as an added expense.

- *Depreciation rate.* In Figure 6.14, curve 1, a linear function was assumed between the original cost and the salvage value. This assumption certainly simplifies the depreciation calculation, since depreciation now remains constant for each year (at one-tenth the difference between the two values). In general, the annual depreciation charge (ADC) is

$$\text{ADC} = \frac{P - S}{n}$$

where P = purchase price of the equipment
S = salvage value at the end of the equipment's economic life
n = economic life in years

However, for equipment with an economic life greater than 3 years, the IRS generally allows a more rapid rate of depreciation during the early years (as shown in Figure 6.14, curve 2). Many types of equipment (such as an automobile) have an actual market value that varies over time in this way. There are two methods commonly employed to calculate the more rapid depreciation to be taken early in equipment life, as shown in curves 2 and 3:

1. The sum-of-digits depreciation method takes as its annual depreciation a varying fraction of $(P - S)$. This fraction depends on the sum of the digits representing the total economic life of the equipment [equal to $n(n + 1)/2$] and the years of economic life remaining [equal to $n - N + 1$], where n = the economic life in years and N = the end of year age of which depreciation is being calculated. Thus the variable annual depreciation charge (ADC) for the Nth year is

$$\text{ADC} = \frac{n - N + 1}{n(n + 1)/2}(P - S) = \frac{2(N - N + 1)}{n(N + 1)}(P - S)$$

   For example, in our case, the 10-year machine A2 has a digit sum of 10 + 9 + 8 + 7 + 6 + 5 + 4 + 3 + 2 + 1 or by the formula 10(11)/2 = 55. The digit associated with the first year is 10 − 1 + 1 = 10 and the last year is 10 − 10 + 1 = 1. Thus the first year's depreciation is (10/55)($60,000) = $10,909, and the tenth year's depreciation is (1/55) ($60,000) = $1091, assuming zero salvage value ($S = 0$).

2. The double declining balance depreciation method (curve 3) applies to the book value an annual depreciation proportion to the book value,

which is twice the annual depreciation proportion calculated by the straight-line method. Its formula is

$$\text{ADC} = \frac{2}{n}\,(\text{book value})$$

where the book value $= P(1 - 2/n)^N =$ purchase price minus the accumulated depreciation. The $2/n$ factor sets the fraction by which the book value is reduced each year. Thus the first year's depreciation for A2 is (2/10)($60,000) = $12,000. The second year's depreciation may be found as two-tenths of the book value at the end of the first year: (2/10)($60,000 − $12,000) = $9600. Using this method, salvage value is disregarded since the book value of the equipment is generally small at the end of its economic life.

There are three main reasons for using the more rapid depreciation methods:

- During the first half of the equipment life they provide higher tax deductions and hence a faster return of capital, thereby increasing the cash available for other investments.
- They tend to provide a more constant sum of expenses, since maintenance expenses tend to increase with time.
- They provide a more realistic book value, as compared with actual asset worth.

These advantages must be weighed against the apparent disadvantages as compared with using the straight-line method:

- In the early years the higher depreciation reduces the company's profits, which may adversely affect the market value of the company's stock. This happened to the airline companies after they purchased jets and then "jumbo jets" and began their depreciation. However, this shows the fallacy of only looking at comparative profits as an indicator of stock value, as compared with cash flow.
- On cost-reimbursable government contracts the extra depreciation cost may result in the company's prices being too high and hence noncompetitive.

All these factors must be considered in choosing the preferred depreciation method. It is particularly important to consider the possibility that the machine will become obsolescent and of no value to anyone by the end of the fifth year.

Having decided on the appropriate and IRS allowable depreciation rate to be assumed for each machine over the next 5 years, the depreciation for each year is calculated and inserted as a cost in the system resource analysis matrix (Table 5.1) for each system. In addition, the book value of each system at the end of the 5-year period is inserted into the resource analysis matrix as assumed positive residual value. Since this is assumed to be a return of capital, no taxes are involved. The planner should also state in his report that in the case of the new machine investment alternatives, the equipment may have a useful life of an additional 5 years, and this has not been included in the quantitative analysis. It is important to emphasize this, particularly if a rapid depreciation schedule has been used, resulting in a low residual value at the end of 5 years.

6.6 COMPLETING THE ANALYSIS

The next step is to construct the net cash flow diagram for each alternative. This is done by accumulating all costs and all revenues and calculating the aftertax profit for each year, as illustrated in Figure 6.15. The net return after taxes during each of the 5 operating years will then be equal to

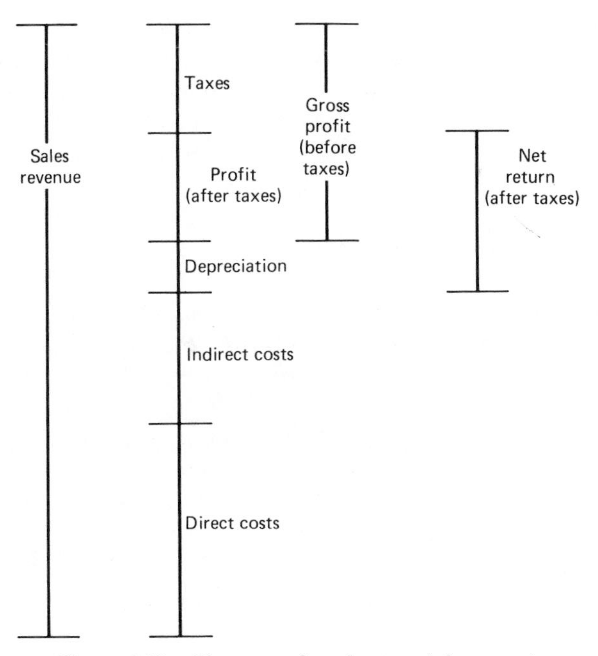

Figure 6.15. Elements of total return (after taxes).

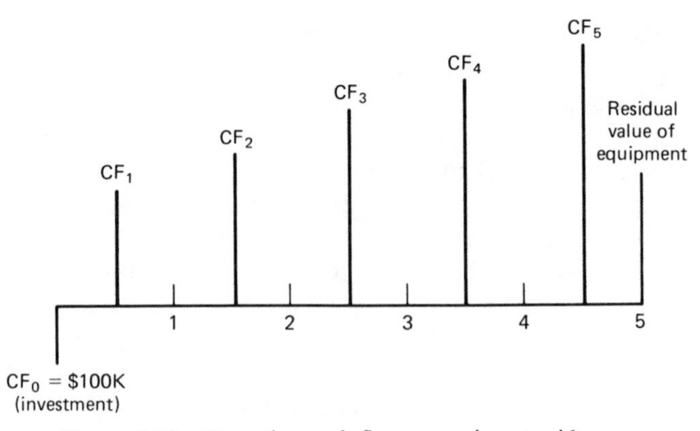

Figure 6.16. Operating cash flow occurring at mid-year.

the sum of the aftertax profit and the depreciation, as shown in Figure 6.15.

The present value calculation can be made more accurate by assuming that for each year all sales revenues are obtained, and all costs occur, uniformly throughout the year.[11] Thus all operating net cash flow is assumed to occur at midyear, as shown in Figure 6.16.[12]

For these assumptions the proper value of the midyear discount factors (SPPW) listed in column 3 of Table 6.5 can be calculated as the arithmetic average of the two end-of-year values on either side of midyear, listed in column 2, for a discount factor of 10%. Similar uniform series present worth (USPW) discount factors for a uniform series of payments occurring at midyear are given in column 4. As shown, these factors consist of the accumulation of the factors of column 3.

The next step is to decide which of the decision-aiding models we described will be used in the evaluation of the alternatives and perform the calculations described. Assume the planner would use the internal rate of return model. One word of caution should be reemphasized in performing the calculations. We are always evaluating the *incremental* payouts against the *incremental* investment, where the current A1 system (requiring zero investment) is the base of comparison. A number of other systems will be designed to meet the system demand. The impact of not adding additional equipment during the latter part of the 5-year period will

[11]For greater accuracy the analyst may wish to calculate the amount of revenues *received* during the year, which may be less than sales revenue because of delays in customer payments, or even bad debts.

[12]Investments are assumed to occur at the nearest half year, and residual values at the end of the 5 years.

Table 6.5 *Program/Project Year Discount Factors (Discount Rate = 10%)*

(1) Project Year	(2) Present Value of $1 (SPPW Factors) End of Year	(3) Present Value of $1 (SPPW Factors) Midyear	(4) Present Value of $1 (Cumulative Uniform Series) (USPW Factors)
0	1.000		
½		0.954	0.954
1	0.909		
1½		0.867	1.821
2	0.826		
2		0.788	2.609
3	0.751		
3½		0.717	3.326
4	0.683		
4½		0.652	3.977
5	0.621		

also be analyzed (not meeting the entire demand) for all systems such as the base case that may be incapable of doing so. The incremental cash flow for each year is then converted into a present value equivalent using a prescribed discount rate.

Applying the appropriate discount factors to the cash flow of each alternative being analyzed, the internal rate of return of each alternative may be calculated (as compared with the base case). These are then ranked in order of superiority. Before making a final selection, however, the planner must reexamine his analysis from the viewpoint of the uncertainties in the analysis. This is described in Chapter 7.

6.7 SUMMARY OF KEY PRINCIPLES

This chapter continued the systems evaluation process by describing the various ways of comparing costs with benefits, depending on the type of organization involved, a profit center (profit-making organization) or a cost center (non-profit-making organization). With a profit center we can compare all incremental revenues to the incremental costs by calculating the total cash flow (incremental revenues minus incremental costs) as a function of time for each alternative and compare these cash flows. With a cost center we need compare only the incremental costs as a function of time among alternatives since all system alternatives are designed to provide the same benefits. Thus the systems evaluation problem has been converted into an investment analysis or cost analysis problem.

Since these various costs and revenues occur at different times, some form of present value analysis must be performed to combine all costs and relate them to all net returns. Several examples were presented to illustrate how this is done.

A number of models were presented for evaluating investment alternatives, including

- Conventional present value of net return model (for profit centers).
- Incremental present value of net return model (for profit centers).
- Internal rate of return model (for profit centers).
- Profitability index method (for profit centers).
- Savings/investment ratio (SIR) method (for both profit and cost centers).
- Conventional present value of cost model (for cost centers).
- Payback period model (for profit centers).

An illustration showed the amount of analytical effort required for each method. All these decision-making methods use the system cash flow (or cost stream) as the basic input for decision. All but the payback period model use present value analysis. These models differ mainly in the accuracy of result and amount of calculation required.

Three methods for calculating depreciation, book value, and residual value of equipment were presented:

- Straight-line method depreciation.
- Sum of digits depreciation method.
- Double declining balance depreciation method.

The advantages of each type were discussed. The main reasons for considering depreciation are the tax implications when analyzing a profit-making organization.

7

Dealing with Uncertainties and Other Factors

7.1 SUMMARY OF ANALYSIS PERFORMED THUS FAR

At this point it would be useful to summarize where we are in the analysis and what work remains to be done. In Chapters 4 and 5 a systematic method of identifying a number of alternatives was described. These were then designed to meet the same objective (producing the same number of units as the estimated sales demand). The basic evaluation then consisted of determining all the costs required by each of the system alternative designs being considered to meet this demand. These alternatives were then compared using the most appropriate decision-making model of the ones described (such as, "choosing the alternative that offers the maximum rate of return, taking into account all appropriate constraints"). After inserting the appropriate performance and cost data provided in Tables 4.1 and 4.2 into the appropriate equations that were constructed to obtain the final numerical answers for each system alternative, the analyst should consider the common type of pitfalls that lie in wait for analyses such as this and that can destroy their credibility. The analyst must ask himself whether there is anything in the analysis that his client or anyone in the decision-making hierarchy could take exception to; if there is, the analyst must cope with this.

7.2 PITFALLS IN A SYSTEMS ANALYSIS

A systems analysis basically consists of these elements:

- A set of alternatives.

- A set of equations or algorithms comprising the decision-making method used.
- A set of numerical data describing performance, cost, and environment characteristics.

Thus any disagreement or controversy about any one of these elements may lead to the rejection of the analysis by the decision maker. Hence the systems analyst should consider the following pitfalls.

7.2.1 Insufficient Number of Alternatives Considered

The preferred solution is nothing more than the best of those alternatives that have been considered, and the decision maker should insist that a sufficient number of alternatives be identified, designed, and evaluated. Alternatives are obtained in several ways. As indicated in Chapter 3, personnel who are part of the current system usually have a store of good ideas about what improvements might be made. These ideas should be gathered, analyzed, and evaluated. They generally form a good starting point for a more detailed functional analysis and design effort. Alternative sources of ideas should also be considered.[1]

7.2.2 Inadequate Models

The reviewers of the analysis may not agree with some of the models used, or they may feel that these models do not properly represent the activities. Examples of this commonly include the set of scenarios used to describe the environment and the system in operation, or the operational flow models derived from these scenarios and used to design the system alternatives. These models may omit one or more factors that some reviewer feels are important. There may be some disagreement concerning one or more equations used to describe the operations. There may even be a disagreement about the objectives to be accomplished.

7.2.3 Inadequate Data

The data being used may be in question. Here the old expression GIGO (Garbage In, Garbage Out) applies. If there is not common agreement on the data used (or the logic by which the data have been combined), there is not apt to be agreement on the results.

[1]See Chapters 14 and 15 for additional details on this.

7.2.4 Inadequate Description of Analytical Approach Used

Finally, the analysis may be a good one, but it may not have been clearly described. The assumptions made and the logic and data used may not have been made explicit; hence reviewers may have difficulty in following the analysis.

7.3 AVOIDING PITFALLS

One of the primary tasks of a systems analyst is to uncover and reconcile controversies, whether from proposers of competing alternatives or from those who look at the problem differently. Thus it is important that the analyst conduct what might be called an open analysis, in which he makes available frequent progress reviews, showing the current information available, data being gathered, and where the analysis is heading. In this way feedback regarding omissions or differences of opinion can be obtained from reviewers at an early stage, and the analysis refined appropriately.

Early in the project the form that the final results will take should be shown to key monitors of the project. In this way they can see what type of information they will be receiving, thus minimizing surprises at the end of the analysis.

It is important that trade-offs be made between the accuracy of the results to be obtained and the time and analytical resources available to the planning effort. This is where the art of systems analysis comes in. The accuracy and credibility of the results obtained will be determined by

- The number of scenarios examined.
- The number of alternatives examined.
- The number of factors included in the models developed.
- The amount of data available and to be gathered.
- The amount of computation of data to be made (whether by analytical calculations or computer simulations).

Expanding any of these generally requires additional analytical resources; yet such resources are normally fixed (in time, manpower, etc.). Hence priorities on the work to be done must be set. Many times this is done solely by the analyst. This is a mistake because invariably the client will be disappointed by the choices made. Hence the analyst must press for client involvement in making such choices. The best way to do this is to "modularize" the analysis, by indicating the estimate of additional resources required for additional factors included. This will result in setting priorities of what should be included in the "first cut" analysis for a given amount of resources and what can be done during a follow-on phase using additional resources.

Having done this and obtained client approval on the scenarios, models, and alternatives to be considered, now consider Step 7 of the evaluation process: "Identify any differentiating characteristics among the system alternatives that are felt may influence the comparison and yet have not been included in the quantitative decision-making model. Determine the economic impact attributable to such characteristics." For example, the final performance (and perhaps even cost) characteristics of Alternative 4 (the automated equipment now completing development) are certainly not known to the same degree of accuracy as the other alternatives, particularly A1, the current system. This differing degree of uncertainty in these characteristics should be taken into account. This introduces the general topic of risks and uncertainties and how these can be included in the analysis as well as in the total planning effort. Now consider

- The types of risks and uncertainties that occur in problems of this type.
- The effects risks and uncertainties can have on system operations.
- How to take risks and uncertainties into account in evaluating these effects and redesigning the system accordingly.

7.4 TYPES OF UNCERTAINTIES

It is important to realize that the entire analytical process was based on gathering data and structuring them in such a way as to give a reasonable prediction of how an activity will perform in the future and at what cost. The only real data available are those that have been collected in the past. However, these data also can be extrapolated to represent some future operation by including the past experiences (called judgment) of people familiar with related activities. Although some degree of error is invariably introduced in this "extrapolation" process, the analysis should take into account errors that may be introduced and attempt to determine the impact of such errors on the course of action being recommended.

Consider a number of examples that illustrate the types of uncertainties and how to cope with them, starting with the simplest and extending to the more difficult.

7.4.1 Statistical Uncertainty

The system we know the most about is the current system A1, since this has been operating for the past 5 years. Performance and cost data describing this system were presented in Tables 4.1 and 4.2. (Production rate, yield, and time to repair are examples of performance data used in the analysis.) To determine how accurately the values of these characteristics are known, the analyst needs to go back to the source of the original data.

These data were no doubt obtained by collecting samples of all workers' performance during some past period considered representative of future performance. Generally a performance characteristic, such as production yield, varies between operators and from day to day. Statisticians call this type of characteristic a random variable. Although it is impossible to predict exactly the future value of this characteristic for a particular operator or a shift, statisticians can draw certain inferences from this limited set of data. For example, it can be shown mathematically that the maximum likelihood estimate of the true yield proportion is the mean of the data samples collected. Thus the mathematics verifies our intuitive feelings. Based on the data samples collected, a statistician can also formulate rules that can be used to state the accuracy of the data estimate. For example, Clopper and Pearson have published the chart shown in Figure 7.1, which provides confidence limits for sample data that are part of a binomial probability distribution for a confidence coefficient of 0.95.[2] Thus suppose that our production samples yielded 8 acceptable units from 10 trials ($x/n = 8/10 = 0.8$). Using Figure 7.1, the analyst can estimate that the true yield lies between 0.43 and 0.98, with 95% correctness. An estimator using this figure would be correct 95% of the time over the long run. On the other hand, if the production samples yielded 80 acceptable units from 100 trials ($x/n = 80/100 = 0.8$), the analyst could state that the true yield can be estimated to be between 0.70 and 0.88. We can never *guarantee* that any of these statements are true, only that 95% of them are true over the long run. Similar formulas quantifying the confidence limits are also available for other types of frequency distributions or for cases when the distribution is unknown.[3]

This is an example of the first type of uncertainty, called statistical uncertainty (also called risk). We can deal with risk as follows. Having

[2]See B. H. Rudwick, *Systems Analysis for Effective Planning*, John Wiley and Sons, New York, 1969, pp. 446–451, as well as the entire Appendix 1 for a summary of these concepts involving data uncertainty.

[3]See B. H. Rudwick, *Systems Analysis for Effective Planning*, John Wiley and Sons, New York, 1969, pp. 296–297 for an example of this.

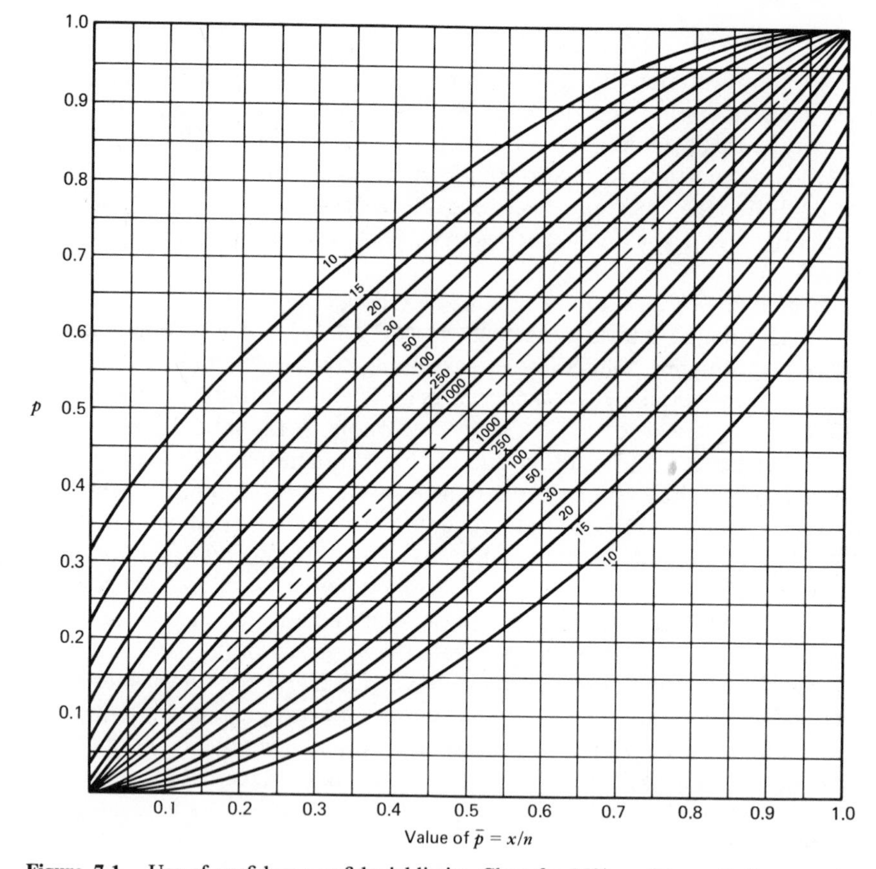

Figure 7.1. Use of confidence or fiducial limits. Chart for 95% confidence limits on p, the probability of success on a single binomial trial, assuming confidence coefficient = 0.95. To obtain confidence limits for p enter the horizontal axis at the observed value of p. Read the vertical axis at the two points where the two curves for n cross the vertical line erected from p. [By permission of the Biometrika Trustees this chart has been reproduced from C. J. Clopper and E. S. Pearson, "The use of confidence or fiducial limits illustrated in the case of the binomial," *Biometrika*, Vol. 26 (1934), p. 410.]

estimated the range of uncertainty in some performance characteristic, we can determine if the final result changes very much as the value of the characteristic is changed over its range of values. For example, if a given alternative were found to be the superior one based on the use of the most likely value of the characteristic (say, using a yield of 0.8 in our example), the analyst could perform a dominance or *a fortiori* analysis by inserting the worst value for certain data and showing that this alternative is still preferred. A numerical example of this type of analysis, called sensitivity analysis, is given in Chapter 13.

In this type of analysis the value of the computer becomes apparent, since the same computer program can be used, rerunning the evaluation using the worst case value of each uncertain performance characteristic considered important. Now the analyst must see how much the final result varies with each new value of a characteristic. If it is found that the range of uncertainty in a performance characteristic affects the system decision (i.e., this uncertainty changes the selection from one system to another), this uncertainty would have to be resolved by further data collection. Thus the final procurement decision may have to be delayed until additional data can be obtained.

7.4.2 Development Uncertainty

Since Alternative A4 is still in the development phase, its characteristics are known with much less certainty then the current system. To find each of these performance characteristics and the amount of uncertainty associated with each, the analyst would have to go through the following reasoning: First, what performance is currently being achieved by the development prototypes that have been constructed? Here the "hard data" that have been collected can be analyzed in the same way as was described in Section 7.4.1.

Second, how might the production model to be delivered perform in the factory's operational environment? Here real uncertainty, as opposed to statistical uncertainty, is present, because this production model has not been built yet. Furthermore, as developers know, machines tend to work better in a development shop, where more skilled personnel are available and adjustments may be made as required, than in an operational environment. Thus the opinions of experts familiar with the various technologies involved constitute the primary source of information that can be used to extrapolate from the performance characteristics provided by the development models to the performance that may be achievable in the operational unit to be provided. But there are two other factors that must also be considered:

- When will the operational model be available and installed in our plant?
- How much will the final selling price be?

For programs under development all three of these factors must be considered together. This evaluation can be made in one of several ways, depending on which of the three characteristics the analyst wishes to hold fixed. He could assume the mean value of performance obtained will be that listed in Table 4.1 if the expert opinion finds this a reasonable goal for the production model. He could then ask his experts to indicate the range of uncertainty in their estimates of the delivery time and final selling price of this equipment, each expressed as a three-point subjective estimate: the "most likely" value, a "pessimistic value," and an "optimistic value."[4]

A second way is to pivot on constant delivery time. This method more closely approximates defense system procurement. Since time and cost overruns are more readily detected (by higher-level managers or Congress) than failure to meet predicted performance and since there is generally heavy pressure to get equipment into operation, the government often accepts equipment even if it does not meet the initial performance specifications. This occurs particularly if the developer is having unforeseen problems and is already losing money on the contract. In this form of evaluation the experts would assume a predicted time and sales price and then estimate the range of uncertainty for each performance characteristic to be delivered following the assumed time and cost constraint.

In both these cases the data collected can be analyzed in the same way as was described in Section 7.4.1.

7.4.3 Cost Uncertainty

The same techniques used in estimating uncertainties in performance characteristics also apply when estimating future system costs. Particularly if he gets the cost estimate from someone else, such as a cost analyst, the systems analyst should make certain he understands how the cost model was generated, including the source of the original data used in the model. The cost model developed in Chapter 5 translates the performance characteristics of production rate, yield, downtime, and so on directly into costs through wage rates, cost of materials, and so on. Thus the cost formulas themselves are "deterministic" (assume no uncertainty), whereas the input data (performance characteristics) consist of random variables. Hence by assuming a worst-case situation on performance, the cost estimates automatically reflect the higher costs required under the worst condition.

[4]The pessimistic value is defined as that value of the characteristic in which there is only a small chance (say, 5%) that the characteristic will be less than this value. The optimistic value will only be exceeded 5% of the time. Both values constitute the confidence limits for a confidence coefficient of 0.90.

If the historical cost records accumulated by the cost accounting department had been used for the costs of direct labor, materials, and so on, the mean value of these costs would have been obtained. Only by conducting a statistical analysis of the actual historical data available could the range of uncertainty be estimated.

Sometimes a cost estimate is obtained for a product that is somewhat like past production equipment. Here "expert opinion" is used to translate historical cost data into the estimated cost of the somewhat different product by focusing on the cost impact of the differences between the components of the new product. In this case each expert should be asked to provide a range of uncertainty of his contribution to the new elements of the cost estimate.

When a design has not been completed in sufficient detail to generate an engineering estimate of cost, statistical techniques are used to relate cost to one or more performance characteristics of the system. These are called cost-estimating relations (CER). Two CER's are (1) aircraft cost as a function of speed and (2) electronics cost as a function of weight.[5] Obviously, these CER's generally change as the state of technology changes. Again, the analyst should ask for the degree of uncertainty accompanying the estimate.

7.4.4 Market and Competitive Uncertainties

The system demand function (in this case the amount of sales expected over the planning horizon at the assumed selling price) is perhaps the most important characteristic affecting this analysis. If sales were not predicted to increase radically, as shown in Figure 4.1, there would probably be no need to consider any changes in the production system, since the savings in premium wages would probably not be large enough to cover the investment costs required for any new machines.

The future sales estimate of Figure 4.1 was probably obtained from the marketing department by extrapolating from past sales data (including the new contract just received). But what will the competition (both competing plastic manufacturers and those making containers from other materials) do in the future? And what will happen to the sales price? These are all questions involving real uncertainty.

To analyze capital budgeting problems such as this one, some estimate of the future demand must be made. This uncertainty in sales can be expressed as a range of values in the same way that development uncer-

[5] See G. H. Fisher, *Cost Considerations in Systems Analysis*, American Elsevier, New York, 1971, Chap. 6.

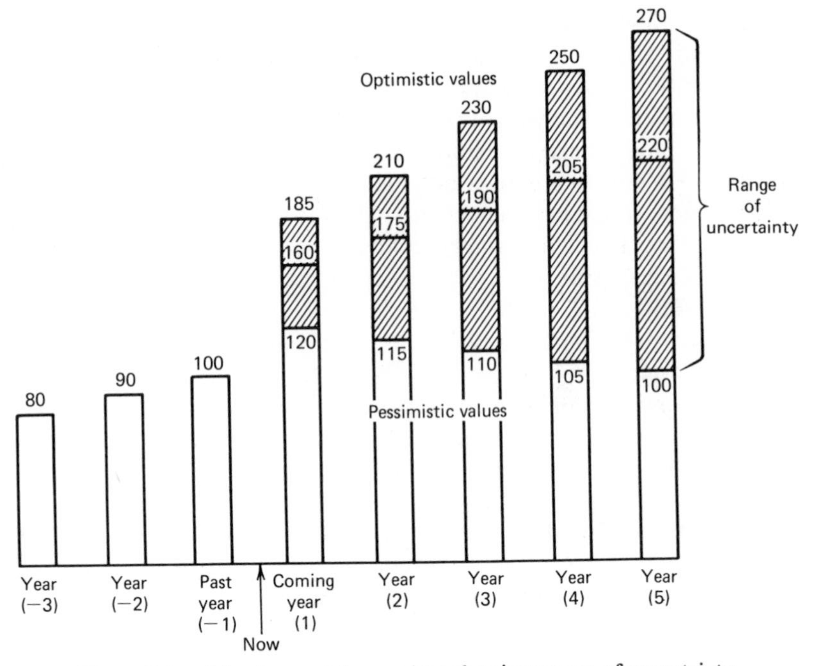

Figure 7.2. Sales forecast for product showing ranges of uncertainty.

tainty was expressed (see Figure 7.2). Now sensitivity analyses, as described previously, can be performed using this range of values.

7.5 COMBINING MULTIPLE UNCERTAINTIES

Performing a series of sensitivity analyses of various characteristics is better than ignoring the uncertainties in these factors. It does show which factors the final decision is sensitive to and hence enables the planners to focus on them and to find ways of dealing with these particular uncertainties. However, there are several disadvantages to this treatment of uncertainties. First, it does require a number of calculations. Thus the analyst should consider the use of a computer, since the computer program could remain the same, with only the input values changing. The second problem involves the interaction of a number of factors whose values are uncertain. The analyst could consider the "worst, worst" case, in which he uses the pessimistic values of all factors whose values are uncertain. However, the only time this type of analysis makes sense is when, in spite of such worst-case assumptions, a particular alternative is still superior to all others. But this is rarely the case. Furthermore, if one considers the small probability of the worst case occurring for just one characteristic, the probability of the worst, worst case occurring is exceedingly small. So this type of analysis is not very realistic.

David Hertz, a pioneer in the field of operations research, addressed the problem of dealing with multiple uncertainties in a capital budgeting problem in a very systematic manner.[6] Instead of using the average values or the worst-case values of the characteristics for the internal rate of return model (as described in Chapter 5), he proposed that the planners in the organization estimate not only the range of values for each characteristic but the likelihood of each value within that range occurring. This, of course, constitutes the complete probability distribution for each characteristic. As one of the more difficult examples, reconsider sales demand (Figure 7.2). One probabilistic model for this (Figure 7.3) treats sales as a discrete probability function over time, where future sales depend on past sales. In this model there are five possible "paths" that sales may take over the 5-year period. Each has the probability of occurrence shown. Thus there is a .2 likelihood that future sales will be 175,000, 200,000, 210,000, 235,000 and 245,000 units for the 5 years indicated. Another model (Figure 7.4) treats sales as a continuous probabilistic distribution for each of the 5 years. This model is a cumulative probability distribution and gives the probability that sales will be greater than the value shown. Thus in year 1 there is a probability of .8 that sales will be equal to or greater than 140,000 units and a probability of .5 that it will be equal to or greater than 160,000 units. Thus it can be deduced that there is a probability of .3 that sales will be between 140,000 and 160,000 units. In using this model it can be assumed that the probability distribution is uniform between the two limits. This model assumes that each year's sales are independent of the previous year's sales. However, another sales model could include some dependency on the past year's sales.

Another way of doing this would be to dissect the problem further and construct separate probability functions for total industry sales and for our company's proportional share of this total market. Expert opinions would be used to construct these functions. Again the worst-case situation could be applied to the problem to see if the preferred solution previously obtained still holds under the reduced market conditions. Each of these probability functions is then inserted into the internal rate of return model; the output is obtained as a probability function. Thus the final output for the rate of return for each of the alternatives would appear

[6]See "Risk Analysis in Capital Investment," *Harvard Business Review*, Vol. 42, No. 1, January–February 1964, pp. 95–106.

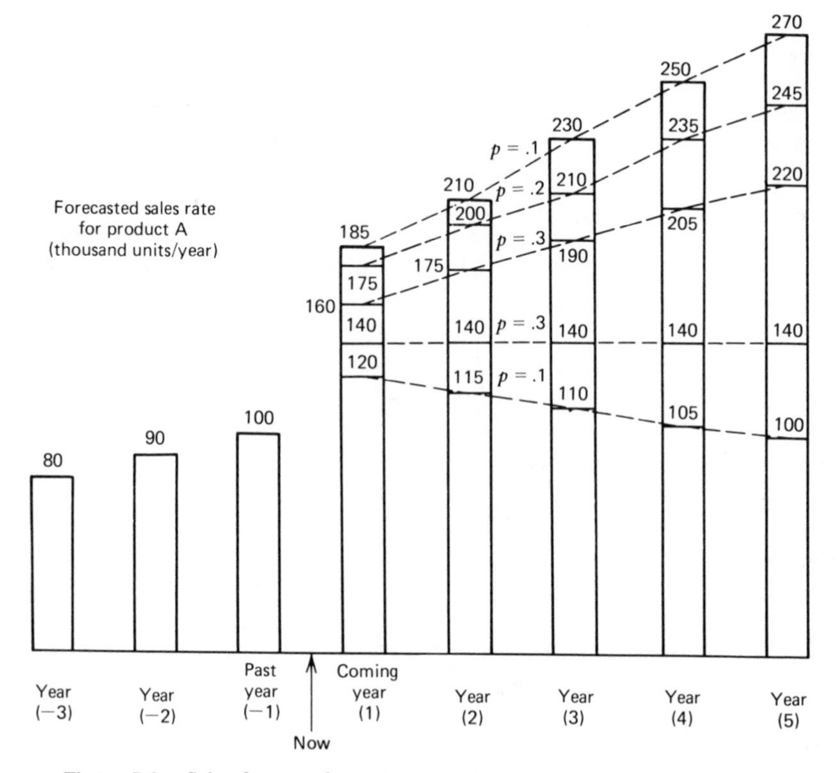

Figure 7.3. Sales forecast for product (a discrete probabilistic distribution).

as shown in Figure 7.5. In this way the decision maker would be given not only the expected value of ROR of each alternative but the additional information of the uncertainty associated with each alternative (i.e., the probability that the ROR will be greater than a given value). Thus the decision maker can use this information to intuitively balance his desire for additional ROR with his tolerance for risk, and select that alternative with which he feels most comfortable.

Obviously, it would be difficult to combine the large number of probability distributions involved analytically. Hence Hertz uses a computer-run Monte Carlo simulation to combine these distributions and obtain the probabilistic output in the form shown.

The major advantage of this approach is that the final output takes into account all the information currently available so that the possibility of good luck as well as bad luck is combined in the way that the experts have indicated. In fact, Hertz also indicates that the expected value of the rate

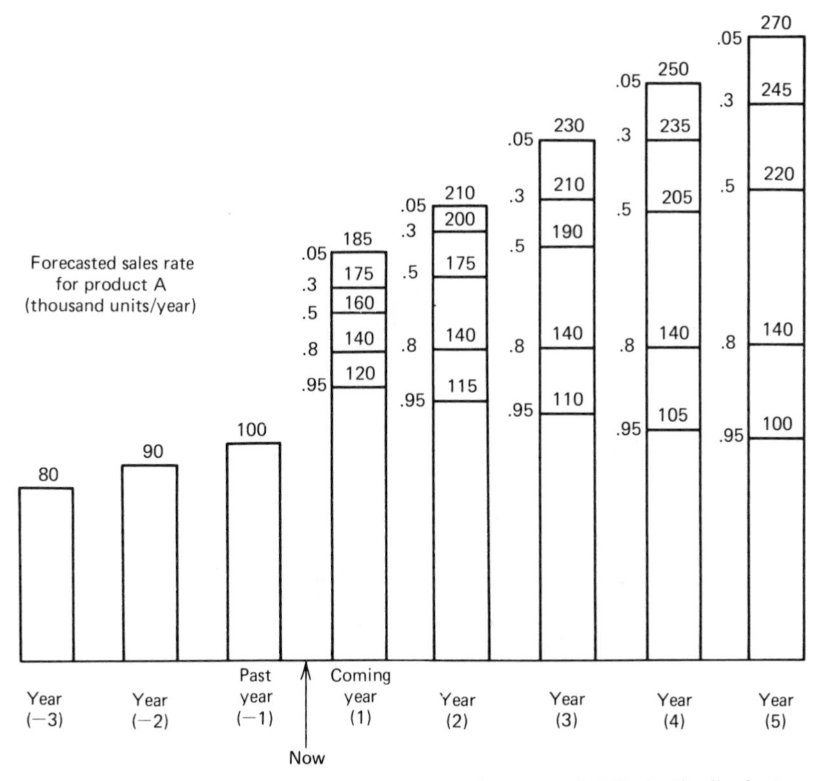

Figure 7.4. Sales forecast for product (a continuous probabilistic distribution).

of return cannot be obtained by merely inserting the expected value of each characteristic into the equations, as is generally done. The only correct way to obtain the expected value of ROR for each alternative is to combine all the probability distributions and find the expected value from the probabilistic results, as he does.

7.6 CONTINGENCY ANALYSES

We have described how the various uncertainties could be combined to provide a more accurate way of comparing alternatives. Now we shall review the basic assumptions behind this method, so that misinterpretations of these results can be avoided. Figure 7.5 should be interpreted as presenting the results that would be obtained if System A or System B were operated a large number of times. Obviously, such opera-

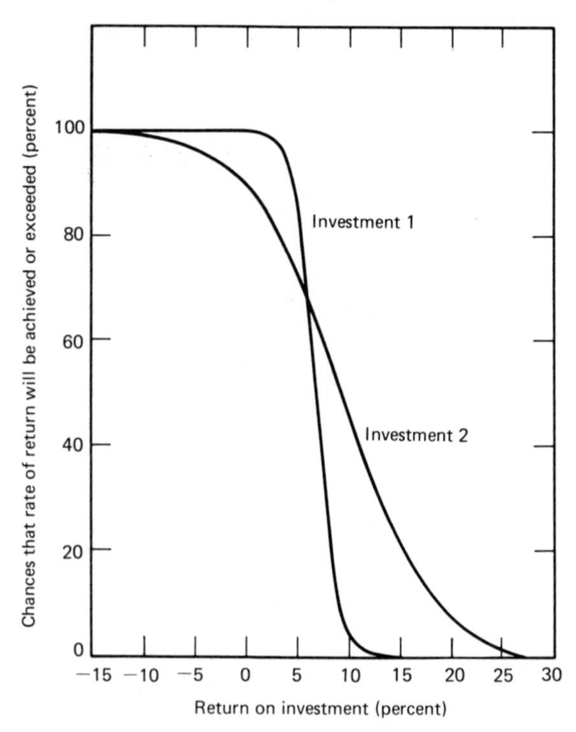

Figure 7.5. Calculated rate of return for two alternatives.

tions would only occur once. Hence these results can only be interpreted as showing the *chances* that certain results will occur. This is not to downgrade these results, because it does enable the decision maker to compare two alternatives in a much better way than is generally done. However, he still must be prepared to deal with "bad luck" or to exploit "good luck" if either should occur.

There is a second shortcoming of this approach. Certain characteristics, such as product sales, are treated as random occurrences but really have competitive uncertainties associated with them. Here the competitor does not act randomly, but gathers data about the market situation and intentionally acts to frustrate his competitors.

For both of these reasons the planner should go back to the worst-case situations and see how he could modify his design to cope with the times when this situation occurs. This is called *contingency analysis*. Unlike the probabilistic approach, whose objective is systems evaluation, the objective of contingency analysis is to modify the various system designs so that they are relatively insensitive to the uncertainties involved and choose the proposed alternative on the basis of superior performance or minimum cost even under variable conditions.[7] Here are some examples of contingency analyses.

7.6.1 Coping with Fluctuating Performance

For many of the reasons previously discussed, the effective production rate may vary. In any one day the number of acceptable units may be less than the number required for delivery to the customers or as input to a next stage of production. One way of meeting such short-range deviations from the mean is to establish a certain size inventory of acceptable units that can be used at any time that the demand for units exceeds the short-term supply. Thus this inventory buffers a fluctuating production from a fluctuating demand. The proper size of the inventory can be determined analytically, basically by considering the amount of uncertainty in the supply and in the demand, as well as the cost of inventory and the cost of not meeting the demand (or the amount of risk of not meeting the demand the manager will assume).[8]

Of course, the use of overtime production is another possible backup, particularly if one or more machines malfunction for any extended period of time. Malfunctions are particularly troublesome to Alternative A4, since a malfunction may result in a complete loss of production, unlike the "graceful degradation" in effectiveness afforded by 10 machines in the current system. Hence if Alternative A4 is chosen, it may be desirable to retain one of the 10 A1 machines as a backup, particularly if the trade-in offered on A1 is not very high.

7.6.2 Coping with Development Uncertainty

Sometimes the user is forced to make an immediate decision involving development uncertainties. For example, you may have to negotiate among several suppliers for equipment that must still be developed and then select the supplier on the basis of what you believe will be delivered in terms of performance, delivery time, and cost. Although the type of analysis previously described is useful, it is helpful to find ways of influencing the supplier chosen to act in accordance with your set of objectives. One way of doing so is the incentive contract, in which the final selling price paid is a stated function of the performance characteristics achieved by the developer. In this way the developer is given an

[7] See Rudwick, op. cit., pp. 193–210, for a further description of contingency analysis.
[8] Ibid., pp. 268–301, for an example of how to determine the size of inventories.

incentive to increase each level of performance over a given nominal value, or a lower price can be paid if the performance is less than the nominal value. Here the buyer's challenge is to design the incentive functions so that the buyer achieves good value for the incentive. The performance incentives should be developed by taking into account the relationships among characteristics. Otherwise the developer may apply his resources to the easier, perhaps less valuable characteristics. In our example the key incentive characteristic should be effective production rate, which combines the various terms, including average production rate, yield, downtime required, and so on. This would permit the A4 developer to optimize his equipment for this higher-level characteristic rather than have incentives for each characteristic. Incentives on delivery time should be based on the reduced operating cost obtainable by faster delivery (or the additional cost associated with a delay in delivery).

Unlike the commercial markets, the federal government regularly pays for development programs for defense systems acquisition, since a developer generally has no other market that can use the output of the development phase. In these cases the government has learned to help protect itself from development uncertainties in two ways. First, to minimize uncertainties in the delivery time, two or more parallel development programs are often initiated. These continue until additional laboratory or other test data are obtained that can narrow down the uncertainties of the critical characteristics. In addition, the government delays a decision on the remaining (much more expensive) phases of full-scale development—testing, evaluation, and production. In this way the decision makers can see which of the parallel developments is preferred and have a much better estimate of what the remaining phases of the acquisition program will cost before making the final decision on the more costly production phase.

7.6.3 Coping with Competitive Uncertainties

When the problem involves the uncertainties of a commercial market and the operation of competitors, we have a situation where events are not completely random but also involve a competitor's deliberate attempt to try to exploit any perceived weaknesses in his competitors' system. In the production planning case, we assumed the price and costs would all remain the same, and that only production volume was uncertain. In reality, since market share is an important measure in most businesses, if a company's market share goes down, a likely short-term response might be to reduce the selling price, increase the marketing efforts, or both. Such actions could reduce the sales price and raise costs.

Contingency analysis involves an analysis of different actions and reactions that competing sides may take in responding to one another's actions.[9] What would be the effect of reducing sales price as a marketing strategy? Competitive games may be played simulating the actions of perhaps 5 or more years of actions in a relatively short amount of time, so that one can see what might happen in real life. Competitive gaming could also be introduced into the Hertz simulation (at an increase in time to perform the simulation if human intervention is required), thus adding further realism to the game and to the end results obtained.

7.7 SUMMARY OF KEY PRINCIPLES

Prior to the end of the analysis the analyst should recheck the completeness of the systems evaluation in the following way:

- Review the differences between the different system alternatives in meeting the system objective. Such differences consist of

  Differences in performance characteristics, affecting the yield or time required to produce units. Such differences affect the total cost of obtaining the required amount of acceptable units of production.

  Differences in the investment cost of each machine, reflected in the cash flow.

  Differences in the other costs of each system (costs of operation, maintenance, and other support). These are determined by the differences in performance of each type of machine and the number of machines of each type available or purchased, since the latter determines the amount of work outside the first shift. These are also reflected in the cash flow.

  Differences in the risks or uncertainties involved either in the job to be done, in the performance system alternatives, or in the availability of the system, as in Alternative A4.
- Determine if any of these differences have not already been taken into account in the analysis thus far.

In our case the major differences between systems were in the risks and uncertainties in the performance characteristics to be obtained, and hence in the resulting total cost of each system alternative in meeting the objectives.

[9]Rudwick, op. cit., 193–210, "Systems Analysis for Effective Planning," which discusses contingency analysis in the context of planning for defense systems.

Table 7.1 *Steps in Systems Planning*

1. Recognize the problem
2. Define the problem (problem definition)
   a. What information do we have about the problem?
      (1) What is the generic type of problem?
   b. What are the objectives involved?
      (1) What is the system objective?
      (2) What is the system demand function and job to be done?
      (3) What is the hierarchy of objectives?
   c. What is the current system and how would it be used?
      (1) Operational concept
      (2) Activities or functions to be performed
   d. How is the present system inadequate?
   e. What alternatives have been identified thus far?
      (1) System entities that could be used
      (2) Operational concepts that could be used
   f. What operational constraints have been identified?
   g. Provide a complete definition of the problem
   h. Summarize the analytical approach to be used
3. Create system alternatives (systems synthesis)
   a. Synthesize the current primary system to meet the system demand function
      (1) Construct operational flow diagram showing functions to be performed
      (2) Determine which system entities could implement each functional activity
      (3) Determine system performance characteristics of each element
      (4) Construct the operational flow model
      (5) Calculate effective production rate of each system activity
      (6) Fix operational work schedule of system
      (7) Calculate number of system elements required in parallel to meet total system demand function
   b. Synthesize support systems using the same design steps. Demand function for each support system comes mainly from the performance characteristics of the primary system
   c. Synthesize other system alternatives as follows:
      (1) Identify each alternative system to be analyzed
      (2) For each, determine the operational concept to be employed
      (3) Construct the operational flow diagram showing each functional activity
      (4) Determine all system entities available for implementing each activity and their times of availability
      (5) Complete steps 3a and 3b above for each alternative design
4. Evaluate the system alternatives (systems evaluation)
   a. Make certain each alternative has been designed to accomplish the same defined objectives (job to be done)

Table 7.1 *(Continued)*

- b. Determine all incremental resources required over the alternative system for the entire defined system life
- c. Calculate the total cost stream to meet the objectives
- d. Determine the value of any other differences among alternatives not considered in the analysis thus far
- e. Relate costs to benefits:
  - (1) Profit centers (maximize net return over time span)
    Conventional present value of net return model
    Incremental present value of net return model
    Profitability index model
    Payback period model
  - (2) Cost centers (minimize costs to do job over time span)
    Present value of costs model
    Savings/investment ratio (SIR) method
- f. Cope with uncertainties
  - (1) Performance
  - (2) Technological
  - (3) Competitive
  - (4) Cost

5. Create additional alternative systems, if needed.
6. Select the preferred system, based on evaluation and any other information available but not used in evaluation.

Each of these identified risks or uncertainties can be evaluated in the following way:

- Identify each of the characteristics that has some uncertainty associated with it.
- Determine what data were used to estimate the numerical value of the characteristic:

  Statistics, leading to statistical uncertainty (risk).

  Uncertainties in the mean value.

  "Fluctuations" in the random variable.

  Opinions, leading to real uncertainty.
- Determine the range of uncertainty in the numerical value.
- Determine what steps could be taken to redesign the system alternatives, including taking appropriate action in the operating procedure to compensate for "bad luck" if it should occur.
- Reevaluate the preferred system taking into account the uncertainties in performance or cost such as in any of the following ways:

Using sensitivity analysis, determine the impact that the maximum range of the value of each characteristic will have on the decision criterion (total cost to meet the objective, or return on investment). Will this maximum value be enough to affect the choice of alternatives, and by how much?

Consider the uncertainties in all characteristics simultaneously. This is done by estimating each characteristic having uncertainty in the form of a probability distribution. Redo the same systems evaluation calculations using a Monte Carlo simulation technique and a computer to combine all the separate probability distributions and determine their combined effect on the final result (e.g., the aftertax return on investment expressed in probabilistic form).

If the alternatives examined are still found wanting, create additional system alternatives that have a different mix of performance and cost characteristics, using the same system design method previously described.

The final selection of the preferred system should be based not only on the quantitative analysis performed, but also on any other information that could not be quantified. Such information should be included in the analysis as a qualitative description.

7.8 CONCLUDING COMMENTARY ON CASE

This case concludes the major discussion of the models and tools that can become part of a formalized systems planning work process. Table 7.1 offers the reader a complete summary of the main steps in the process for use as a checklist. This same process is used to structure the case situations examined in the remainder of the book.

Although these steps are listed in a logical and ideal order, the work process should be considered to be iterative. The planner may proceed a few steps and then recycle to previous steps as new data become available.

As indicated previously, it was not my objective to provide a complete numerical solution to this case, but rather to describe and illustrate the approach to such a solution. However, it is suggested that such an exercise would be quite beneficial to those readers who really wish to understand the process described. Since there is a large amount of repetitive calculation required, the use of a computer program for this purpose is also recommended, as indicated at the beginning of this case.

Part III

DESIGNING INFORMATION SYSTEMS FOR MANAGEMENT CONTROL

CASE 2: THE APEX CORPORATION PRODUCTION REPORTING AND CONTROL SYSTEM

Whereas the first case dealt with the design and evaluation of a production type of operation, the second case extends the planning process to the design of information systems that inform management about the performance of a production system, thereby facilitating corrective action when required. This case focuses on the following objectives:

- Extend the systems design process for material flow systems into the area of an information system designed for control purposes.
- Show how an information system can be designed to meet the needs of various information users.
- Set forth the key principles of management control systems.

8

The Apex Corporation Production Reporting and Control System

8.1 PROBLEM AS GIVEN

You are a management consultant. Charles Mack, the general manager of the brake linings division of the Apex Corporation, one of your clients, asks you to attend a meeting with Ron Messenger, the division controller. At this meeting, Mr. Mack states,

As you know, we are currently installing a UNIVAC 9300 with disk files to replace the punched card machine accounting system we have been using for many years. Our EDP department has hired some computer programmers to develop the employee payroll and order entry systems, which are fairly conventional data-processing applications. While they are doing that, I would like you to analyze and design a labor reporting system for this business.

We manufacture our brake linings through a sequence of operations, including material mix operations, material slab production, sawing, bending, oven curing, grinding, drilling, and inspection. Each one of these separate operations is called a department, operating over three shifts. Each department and each shift is headed by a foreman. We currently gather information regarding labor costs in the following way:

1. Each job lot of material-in-process inventory is stored on a cart along with a job lot card showing the complete production record of the lot as it passes through the sequence of operations I described. This record includes the job lot number, the original number of pieces of material started in the job lot (sometimes over 1000 units), and the number of pieces remaining as it leaves the department.
2. When a worker is ready to work on a new job lot delivered to his station, he goes to the job ticket time clock and inserts into the time clock one of his job tickets representing this lot. The clock prints the starting time on the card. He

then writes the job lot number onto the job ticket, which has already been prepunched with his name, clock number, shift number, and department number.

3. The worker then performs his operation on each unit of the lot. During this time there may be some nonproductive time, for the following reasons:
   - A different machine setup may be required for a new job lot (for different sawing dimensions, for example).
   - The machine may break down, requiring repair.
   - The machine may have to be readjusted during the job lot (for example, readjusting grinding wheels for some critical dimension).

   The worker records the start and finish times of this nonproductive time on the job ticket, as well as the reason for this downtime. If the machine is down for any abnormally long time, the foreman indicates this on the job ticket and signs for it. Units spoiled in the process are put to one side.
4. When the operator completes the lot, he goes back to the time clock and punches the time on the original job ticket. He also records on the ticket
   - The number of acceptable pieces completed.
   - The number of rejects.

   He also records the following information on a separate job lot card, which stays with the lot in the cart:
   - Starting time.
   - Stop time.
   - Number of acceptable pieces completed.
   - Number of rejects.
   - Operator number.

   This provides a complete record of the entire job lot.
5. At the end of the shift, each operator gives his set of job tickets, representing all the work he did on that shift, to his foreman. The foreman reviews the data for completeness and turns the cards over to one of two timekeepers. The timekeeper compares the times on the job tickets with each worker's attendance card, which is used as the basis of his pay. When he finds a discrepancy (invariably, the sum of the job tickets is less than the attendance time), he changes the times on the job ticket to agree with that on the time cards. In this way, cost accounting can account for all times during the entire shift.
6. The job tickets are then processed to tabulate data showing what production occurred and how much time was taken. The data are sent to cost accounting, to the department foreman, and to higher management.

Mr. Mack indicated several concerns with the current system.

Right now it takes more than a week to get the printouts from the data-processing department. This does not bother the accounting department, but the manufacturing department claims it needs the information the next day. It is particularly concerned with closely monitoring departments where new production standards have been set or departments where production problems are occurring. Because of this need, some manufacturing people have been assigned to gather the same data from these departments manually (their so-called production report). The manufacturing department obviously cannot monitor all departments. This would take too much effort. Manufacturing fears that something may go wrong in a department that it is not monitoring and that it will not know about it for over a week. For my part, I don't like to pay for two separate labor reporting systems.

My second problem is that some job tickets get lost. The foremen and timekeepers have to hunt for the tickets or try to reconstruct what has happened. In fact, the foremen are always griping that they just have too much clerical work to do.

Can you develop a system to measure labor time and output and give us the next-day turnaround that we need? I would like each foreman to have a report on his shift's production on his desk at the beginning of the shift on the next day. I'm also interested in measuring labor efficiency. If we have workers who don't perform well, we need to know who they are. We can either transfer them to another department or terminate them in accordance with our union contract. Can you devise a way to measure worker efficiency and have the system report this?

One word of caution. We have had some poor experience in the past with computer "monsters." I can recall when the quality control people assembled data on the quality of production of the various departments. We had reams and reams of printouts coming out of our ears. The various department and product line managers never even looked at the data, which were too much for them to comprehend. The data weren't useful to them. So when you design your system, make sure you don't repeat that mistake. We don't want another "monster."

With this background information from the general manager, describe how you would develop an information system to meet his needs. Try to make use of the information in Chapter 2 and Case 1.

8.2 DEVELOPING THE ANALYTICAL APPROACH

8.2.1 Understanding the Problem and the Current System

One of the best ways of gaining an understanding of the problems posed by Mr. Mack is to begin with an analysis of the current system. Study its strong points and its deficiencies. Apex Corp. has two information systems to provide production reports: (1) the EDP department's job ticket system, which generates data about the production of all workers, and (2) manufacturing's production report system, which generates data on the production of only selected departments. Since the former system is automated and comes closer to what the future system will look like, let us analyze it first.

Four basic data input forms are used in the total current information system:

- The *job lot card* shows the entire production history of one job lot as it moves through each of the production activities (departments), in terms of who worked on the lot and on what days and the attrition of the lot as unacceptable units are removed.
- The *job ticket* shows only the work done on one job lot by one worker during one shift. During each shift each worker collects the data described previously by Mr. Mack and inserts them on each job ticket.
- The *time card* shows the attendance record of each worker for each day of the week.
- The *production report* shows the entire work done by all workers on each shift being monitored by manufacturing. It contains essentially all the data contained on all the tickets during one shift, but assembled on one sheet of paper.

The current job ticket information system is modeled (Figure 8.1) using the generic operational flow diagram of Figure 2.9.

Near the end of the shift each worker's job tickets are collected and reviewed by the shift foreman. He must certify whether each worker's downtime is reasonable or not. He checks to see if all job tickets are in. Finally, he scans the data to get an approximate idea of how well his workers have performed, which may not be a simple task.

Each foreman gives the timekeeper his entire set of job tickets for further review and comparison. The timekeeper compares the chronological times on each worker's job tickets against his time card to make certain that all job tickets are in. In addition, he may change various times indicated on the job tickets to ensure that the sum of these times exactly equals the total time on the time card.

The job tickets, arranged by worker, by department, and by shift, are then given to a keypunch operator who enters the information on an 80-column card. Then a tabulator prints out a report listing the production for each worker and for each department, shift by shift. This report is distributed to the various levels of manufacturing management and to the accounting department.

The basic deficiency of the job ticket system is that, in the opinion of the users of this information (the manufacturing department), the information takes too long to reach them. Because of this a manufacturing representative also collects the same data (for only selected departments), using the production report form. The production results for each shift observed (but *not* for each worker) are reported the very next day.

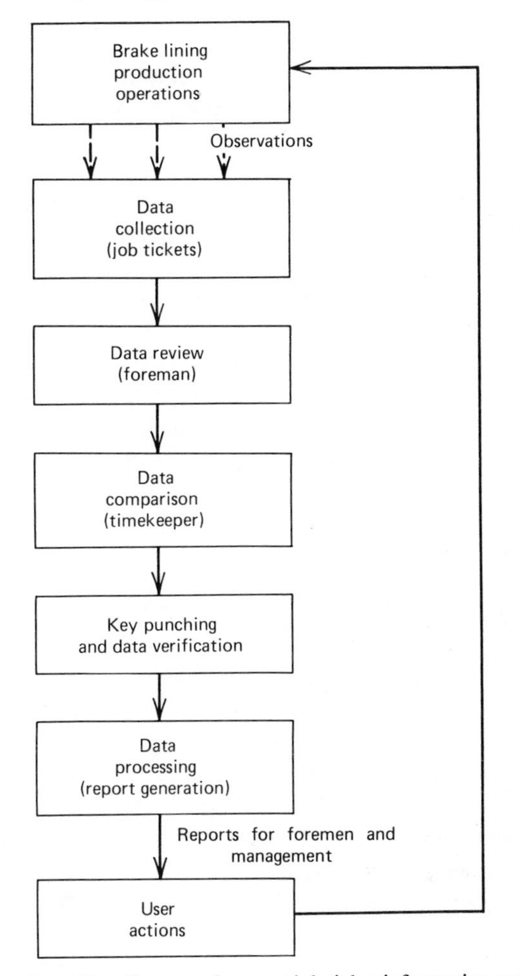

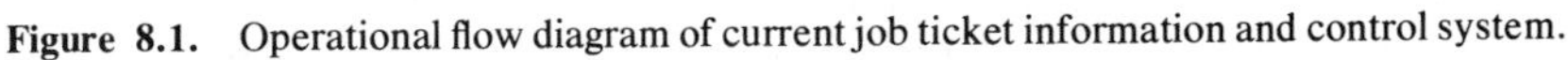
Figure 8.1. Operational flow diagram of current job ticket information and control system.

8.2.2 Identifying the Type of Problem

Comparing this problem with the steps in system planning (Table 1.1) shows that the problem has been recognized. Although further definition of the problem is required, there is a need to design an information system to provide more timely information to Apex management, using the newly installed UNIVAC 9300. Hence this can be classified as an information systems design problem.

8.2.3 *Approach to an Information Systems Design Problem*

The key steps leading to the design and evaluation of production systems, as presented in Case 1, will now be interpreted to produce a preferred design of an information system:

STEP 1. Determine the primary objective of the activity; construct the "Hierarchy of Objectives"

A "means to an end" model can be constructed (Figure 8.2) to show the hierarchy of objectives associated with this problem. Management has demanded that more timely information be distributed so as to facilitate better control of production. Specifically, management has to have information to help it evaluate the three key characteristics of a production activity: (1) the quality of product output, (2) the production completion times, and (3) the costs of production. It wishes to compare this against its original goals and to take corrective action if a discrepancy exists between goals and actual output. Such action should lead to higher sales revenues or lower operating cost to the organization, leading to high aftertax cash return and higher return, even after all costs of the new information system are considered.

STEP 2. Choose a specific job to be done or level of effectiveness (the systems demand function) as the basis of the information systems design.

In Case 1 this was easy to do. The system design function was to produce enough units to meet both the quality level (the specifications of an acceptable product) and the quantity level (the postulated sales demand

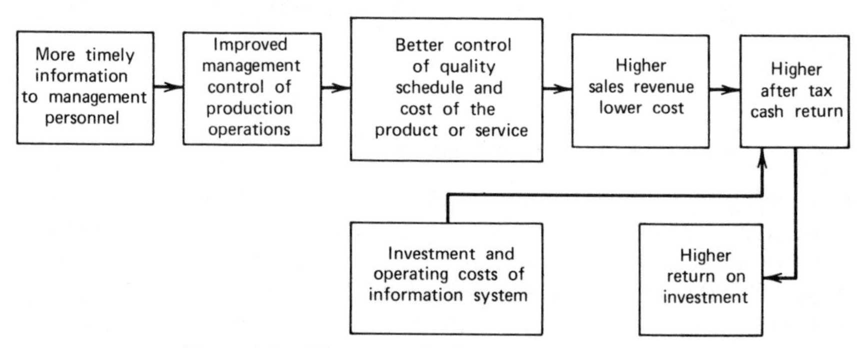

Figure 8.2. Hierarchy of objectives in organization.

for each year) at lowest total cost and to maximize the net return to the firm.

The new information system can also be thought of as a quasiproduction system, one that gathers, processes (transforms), and communicates (delivers) information outputs (Figure 8.1). Now the information is provided (not sold) to information users who control the primary production system. They need to know how the production system is operating and what corrective action is required to improve the production costs or delivery schedule.

However, since the information system itself costs money, we need to determine which set of information output improves the net return to the company. Thus the analysis and design of an information system must proceed along the following paths:

- Determination of the proper information specifications (i.e., what information output should be provided to each user, in what form, and how often).
- Analysis and design of the data-handling system providing this information.
- Determination of the cost of the data-handling system.

Since none of these has been specified as yet but all three are related, the analyst must develop the information system in an iterative fashion, in the following way:

- Postulate several levels of information that he feels would be of benefit to the users. Specifically, he details the characteristics and accuracy of different sets of information that could be provided.
- Design alternative information systems for each level of information postulated.
- Select the preferred alternative using the most appropriate systems evaluation method, as described in Case 1.

8.3 POSTULATING THE ALTERNATIVE INFORMATION SYSTEM DEMAND LEVELS

The term *user needs analysis* is sometimes used to describe the iterative process of postulating different levels of information demand that might be provided to the users of the information—in this case to management. User needs analysis can be performed in various ways. One might start with the information output of the current (job ticket) system. Identify the current users of this information and interview them to obtain the following information from each:

- What deficiencies have you identified in the information currently provided you (e.g., late information)?
- What information would you like to have provided, in what form, and how often?
- How would you use this information (to improve the control of your operation)?

This would be helpful in validating the stated need by giving at least a qualitative description of the benefits that may be derived from this information.

8.3.1 Analysis of User Needs

The first step in a user needs analysis is to identify all those who might use this type of information if it were available on time. The second step is to identify how this information would be used.

Analysis of the company's organization chart (partially shown in Figure 8.3) helps identify those individuals who would find this type of production information of benefit to them:

- *Shift foremen.* Each shift foreman is responsible for evaluating each of his workers and identifying those that are not performing satisfactorily.
- *Department foreman.* Each is responsible for the total operations of the three shifts constituting his department. The department foreman must (1) evaluate the total output of each shift, (2) identify any shift that is not performing satisfactorily.
- *Higher-level management.* Each product line manager (as for auto brake products, aircraft brake products) is responsible for the entire set of operations of all departments contributing to a particular product line. Thus he is responsible for evaluating each of his departments and identifying which ones are not performing satisfactorily. In addition, the manager of manufacturing and the general manager may also wish to review some of this information to confirm that corrective action has been taken by lower-level managers or foremen.

In all these cases it is also necessary to determine the reason for any

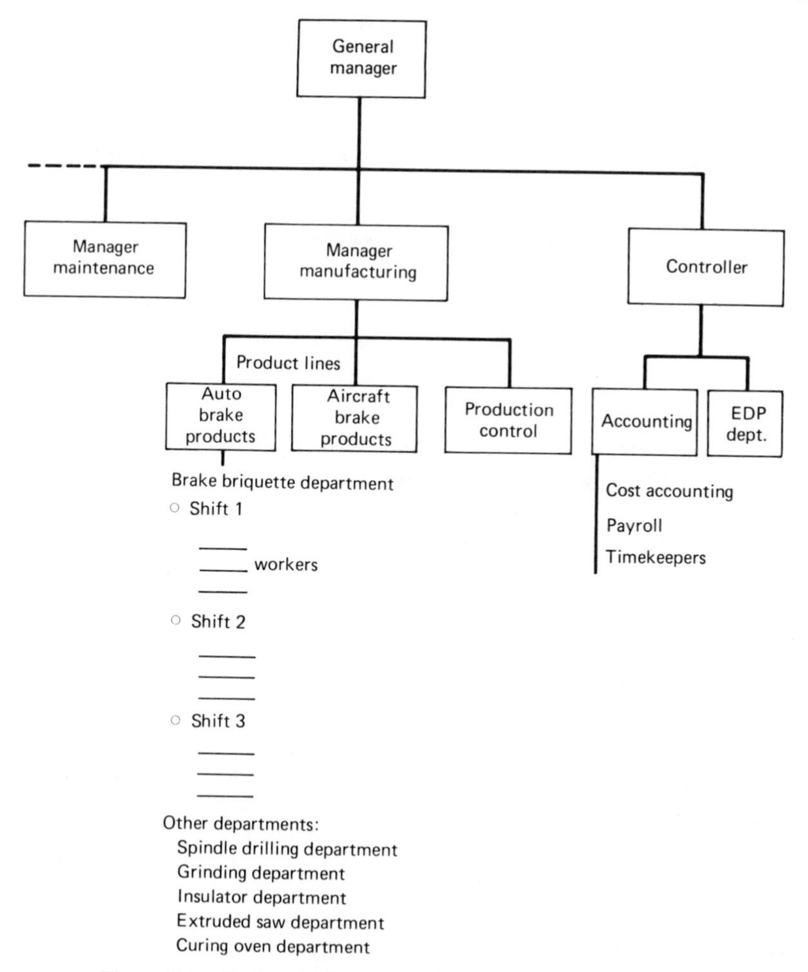

Figure 8.3. Pertinent elements of Apex table of organization.

unsatisfactory performance and the corrective action to be taken. It is also necessary to determine what information each of these managers needs, and when he needs it.

- *Cost accounting*. This department must report the average unit cost of each operation over time. This is compared with some budgeted cost. Since the average unit labor cost is the product of the average labor time used and the wage rate, both must be measured. Both can be obtained from this information system.

- *Production control.* This department is primarily responsible for scheduling the production through each department in accordance with a given overall production schedule. It also monitors the actual production against plans. If a discrepancy occurs, it reschedules the work flow. Production outputs would be provided by our information system.
- *Timekeepers.* Timekeepers compare the times recorded on the job tickets of each worker with the time shown on the time card, the primary indicator for worker payment. Any discrepancies between the two must be reconciled.

8.3.2 Feedback Control System Analysis

Each of these users can be thought of as controlling some part of the total manufacturing organization. To understand better the type of information required for controlling an operation, consider the characteristics of a management control system using feedback control. Consider the following questions:

- What is a management control system?
- What data and information does it require to operate successfully?

Figure 8.4 is a model of a feedback control system for a production system and illustrates the operation of such a system. The material flow is shown as solid lines beginning with the input and yielding an output. The rest of the total system constitutes the feedback control system with its information flow shown in dotted lines. The first element of control is the

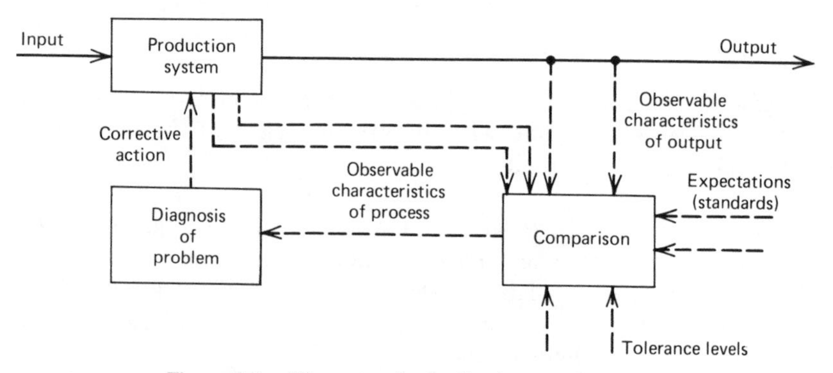

Figure 8.4. Elements of a feedback control system.

identification of the observable characteristics of the *output* and the *process*. Some of these become the basis of the control of

- The quality of the output.
- The time required to complete the output (or rate of production).
- The cost of the output.

Other characteristics of the output are used to determine if the system is operating satisfactorily. For example, the width of a brake shoe is an output measure of quality and such a measurement is the ultimate test that the brake shoe is the right width. A periodic check of the dimension of the wheel that grinds the width in one of the final steps of production might be considered one test of the process. Although we might infer that if the wheel were set at the proper dimension, the brake shoes would generally have the correct width, sometimes other problems can develop. Hence feedback control systems rely on the measurement of output characteristics as the most important test of process performance. Measurements of the process, however, are very important in diagnosing the cause of a product problem once that problem has been detected.

To monitor the system performance in terms of the characteristics chosen, we need some standard value for each of these characteristics. Often this standard is obtained from historical data. For newly installed production systems it may be an "expectation" based on time and motion measurements or analysis. Although the value of a characteristic of an acceptable unit does not have to match the standard exactly, *acceptability* is defined using tolerance limits on either side of the standard.

If characteristics of one unit of output fall outside the limits, that unit is deemed to be unacceptable. If enough of these units are found to be unacceptable, the process itself may come under question and in extreme cases be halted. A proper diagnosis of the reason(s) for the condition must be made. Proper diagnosis leads to corrective action designed to modify the system in some fashion that will cause more future units to be acceptable.

Let us now apply this theory to the control needs of the identified users, leading to the development of an appropriate management control system for Apex.

8.4 ANALYSIS OF SHIFT FOREMAN'S NEEDS FOR INFORMATION

The key management control questions faced by each shift foreman in evaluating the performance of his shift are as follows:

- Did the entire shift produce the amount of acceptable units expected during the work period under observation?
- Did each worker produce the amount of acceptable units expected from him during the work period under observation?
- In what particular area(s) is the entire shift (or particular workers) having difficulty?

For a number of reasons the total performance analysis will be begun by measuring the performance of each individual worker. The second question was the one originally posed by the management: how to measure the productivity or "efficiency" of each worker. The performance of the entire shift is equal to the sum of the performances of each worker. Thus an understanding of how to evaluate the lowest element often leads to an understanding of higher components.

8.4.1 Analysis of Individual Worker Performance

Table 8.1 lists the various data elements collected on the job tickets each day for each worker, available as production standards for each job. Now consider various ways of converting these data into useful information. This can be done if the data are processed in such a way as to answer the key management control questions previously posed. For example, consider the second question, dealing with the quantity of worker production. This question can be answered by totaling the set of acceptable units of each worker and comparing the total with the number of units that should have been produced for the time the worker was paid.

There are several possible ways of relating these data to a worker's productivity. The method chosen calculates worker efficiency (WE) as

Table 8.1 *Data Elements Required to Evaluate Production*

(1) Worker Data Collected Each Shift Production Characteristics	(2) Time Characteristics	(3) Standards
Total number of units Total number of unacceptable units (scrap)	Uptime Downtime Acceptable Unacceptable Reason for downtime	Product type Department (process) Acceptable production rate Acceptable downtime Acceptable yield

the ratio of the number of acceptable units actually produced by a worker during a shift to the number specified by the production standard. The formula is as follows:

$$WE = \frac{APS}{SPW}$$

where APS = actual number of acceptable units produced by a particular worker in the shift
SPW = standard number of units expected to be produced by the worker in the shift.

For this worker $\quad SPW = SEAPR\ (WRS)$

where SEAPR = standard effective acceptable production rate expected, measured in acceptable units, which should be produced by one worker per hour (all lost time included)
WRS = total hours of the worker's time paid for in the shift (generally 8 hours)[1]

There is one major problem in calculating SPW in this way. The downtime may vary radically from shift to shift, depending on the amount of readjustments to be made or the number or type of machine malfunctions that occur. Because of this possible variability in SPW, it was decided to calculate SPW excluding downtime, but controlling worker downtime in other ways. Using this approach:

$$SPW = SNAPR(TTS - ADT)$$

where SNAPR = standard normal acceptable production rate of acceptable units, measured in acceptable units per hour (no time lost)
TTS = total sum of worker's time charged to that job in which the worker is receiving pay
ADT = acceptable downtime during which the worker is not available for production, as approved by his foreman

The production time "charged against" the worker in this calculation does not include downtime for machine maintenance, setup, or adjustment, since the machine is not available for production during this time. Each downtime is recorded, however, on the job ticket along with its cause. Both entries must be approved by the foreman, who knows the

[1]This formula assumes that a worker works on only one type of product during the shift, which is generally true. If he works on more than one type of product, the formula is modified slightly to include the time taken for each product type.

approximate time that each type of machine task should take. Should the worker run into a problem that takes much longer than the time normally required, he is expected to notify the foreman. If he fails to notify the foreman, the worker runs the risk that all of his downtime may not be allowed. This reduces his worker efficiency. Thus the downtime of a worker can be controlled.

The computer can be programmed to perform this simple calculation of each worker's efficiency. There is no need to require the foreman to calculate it. Since the input data are collected following each shift, a tabulation of worker efficiency could be made on a daily basis, if desired, and made available to the foreman at the beginning of the next day's shift.

8.4.2 Management by Exception Reporting

The clerical dimensions of the foreman's job can be further reduced by automatically evaluating the results of the worker efficiency calculations. Instead of reporting the performance of all workers, the computer can be instructed to print only the names of workers whose performance is exceptional (i.e., outside specified tolerance limits). For example, we may list all workers whose worker efficiency during the shift was less than 80% or greater than 110%.

8.4.3 Measuring Other Worker Performance Characteristics

Although worker efficiency is a good overall indicator of exceptional (particularly substandard) performance, the data contained on the job ticket can be used to provide the shift foreman with the following information as additional diagnostic information if his performance is calculated as substandard:

- Ratio of normal production rate (total input including rejects) to the standard nominal production rate. This alerts the foreman if a substandard worker is too slow.
- The production yield (i.e., the ratio of the number of acceptable units to the total produced) versus the standard yield. Alternatively, the computer might calculate a "yield index," the ratio of the actual to the standard yield for this product for all workers in this department. This index would identify workers that were too careless.
- Similarly, the ratio of percent indirect labor (i.e., total time for machine adjustments to total shift time) to the standard ratio selected

for this department. Some analysts define indirect time to include downtime.

Although these data could be used to calculate each worker's single-shift performance, an additional calculation of value would focus on the worker's longer-term performance. It would utilize data from the worker's job tickets from the past week or month.

8.4.4 Frequency of Reporting

Since the input data are collected following completion of each shift, the output report could be available to users 16 hours later, before the start of this shift on the following day. However, the basic question is, "Do the users really need to receive a report each day?" In operations that have been changed recently, a daily report is required while worker "learning" is taking place. During this time, as a statistical history of production is obtained, management needs to be sure that production is under control and, in fact, improving over time.

With routine operations, worker output may vary. If a worker has an unsatisfactory day, he realizes this and increases his production over the next few days to make up for it. In fact, to reinforce this tendency, the union contract specified that worker production be measured on a weekly basis. If a worker's average performance over a week's time is found to be unsatisfactory (i.e., below a given threshold),[2] the worker is transferred to another department. If his average weekly performance is also unsatisfactory at the second department, this is reason for termination. When confronted with this use of the information, Apex management indicated that for routine operations, worker efficiency could be calculated on a weekly basis.

The next question asked was, "Would providing this same information to the foreman on a daily basis be of help to manufacturing, as was originally requested?" Obviously, having knowledge that a worker did very poorly on one day would be of some value. At a minimum, a preliminary warning could be given to a worker. Thus it would be possible to generate a daily worker production report providing the same information for only those workers who were very deficient in their production (say, a worker efficiency below 60%). However, such information is not free. There are additional out-of-pocket costs for additional daily com-

[2]If the feedback control system is to work, it is important that each individual know (as quantitatively as possible) what the expectation is.

puter operations, including daily printouts to all foremen. The foreman would also have to spend additional time "prodding" deficient workers.

It was the judgment of the manufacturing department that the weekly management-by-exception reports were sufficient for their purposes.[3] Note, however, that all job tickets would have to be keypunched daily. It still may be beneficial to print out any of the results of daily production, so that they are available as a "library record" (or stored on magnetic tape) if they are needed later. But the management control system for detecting worker deficiencies was to be based on weekly production averages.

8.4.5 Analysis of Entire Shift Production

The objective of this evaluation is to answer the question previously posed: Did the *entire shift* produce the amount of acceptable units expected during the work period under observation?

In this case, *true* productivity or efficiency of a given shift was measured, the first way described for measuring worker efficiency. That is, the data from all the workers' job tickets are used to calculate shift efficiency (SE) as the ratio of the number of acceptable units produced by all workers in the entire shift to the number they should have produced in the available time according to the production standard.

$$\text{SE} = \frac{\text{APW}}{\text{SPS}}$$

where APW = actual number of units produced in the shift by all shift workers

SPS = standard number of units expected to be produced by all workers in the shift

For all workers $\text{SPS} = (\text{SEAPR})(\text{WST})$

where SEAPR = standard effective acceptable production rate expected, measured in acceptable units per hour (all lost time included) for one worker

WST = total man-hours of all workers' time expended in the shift

Note that all downtime (whether acceptable or not) is included in calculating SPS. There are two reasons for calculating shift efficiency in

[3]In fact, Apex used the weekly report to detect unacceptable worker performance. At first it warned the worker; a transfer occurred only after consistent underperformance. Such a policy saves money, considering the cost of worker learning in a new department or the cost of termination (i.e., hiring and training a new employee).

this way, rather than ignoring downtime, as is done in calculating worker efficiency. First, this method is a more accurate measure, since all downtime is included. With more data included in the calculation (job tickets for 1 week rather than only one shift), an abnormal downtime will have only a small effect on the total shift efficiency. In addition, holding the shift foreman responsible for obtaining a full 8 hours of work from the entire shift reduces the possibility of any collusion between the foreman and his worker. It also constrains the foreman from being too lenient in approving abnormally high downtimes (such as allowing unauthorized "break time" to be included as an extra part of legitimate downtime). Each foreman knows that he is being measured by all downtime in his shift. He is apt to exercise a greater degree of control over this performance characteristic. Again, a management-by-exception reporting system can be used by comparing the shift efficiency against two acceptability thresholds (say 80 and 110%) and placing an appropriate "flag" next to any shift whose efficiency falls outside the acceptance range.

8.4.6 Measuring Other Shift Performance Characteristics

If the shift efficiency falls outside of the tolerance range (particularly if it falls below what is acceptable), the shift foreman needs diagnostic information to understand why. There are two possible explanations for insufficient shift performance. First, the efficiency of one or more workers may be sufficiently low to cause the shift efficiency to be below threshold. This information would show up on the list of substandard workers on the worker efficiency report. Second, the worker efficiencies may be high but worker downtime may be large enough to cause the shift efficiency to be below threshold. To confirm, it is useful to calculate for the entire shift the same set of other performance characteristics relating to downtime and yield as described previously. In addition, a list of workers with deficient performance should be generated by a similar management-by-exception reporting mechanism for each performance characteristic versus its tolerance threshold.

8.4.7 Looking for Information System Deficiencies

The planner should analyze the proposed information system for flaws, particularly ways that an individual may "beat the system." One possibility of error can be seen in the management-by-exception reports. Plotting such data could produce a worker's cumulative production record (Figure 8.5). This particular worker has an average efficiency of 90%,

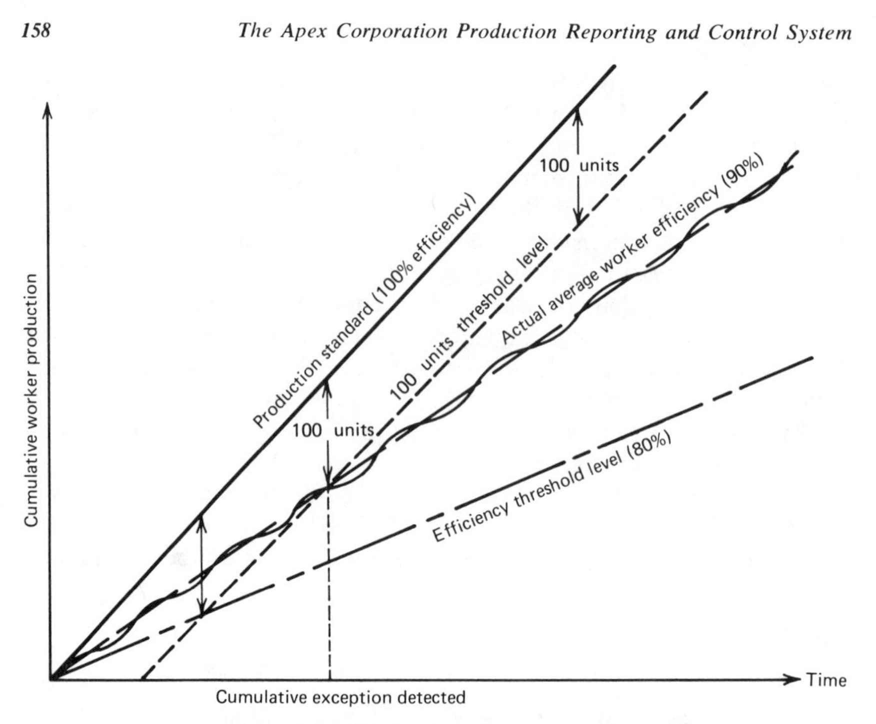

Figure 8.5. Worker cumulative production.

consistently above the worker efficiency threshold level of 80%. Hence his performance is, by definition, acceptable. However, as time goes on, this worker's *cumulative* production falls below the standard, but the foreman is not alerted to this fact by the management-by-exception reporting system. This situation can be detected by inserting a second threshold level, one that measures the deviation in cumulative production from the standard. In our example, suppose this threshold were set at 100 units, as measured from some starting point in time. Thus even if the worker's production rate for the week is acceptable, if his cumulative production is greater than 100 units from the standard,[4] his name would appear on the shift foreman's management-by-exception report, along with the following data describing his production data since the starting time:

- Total number of production units deficient.
- Average worker efficiency for the period.

[4]The standard cumulative production is the standard production rate times the number of hours over a given period to date (less acceptable downtime).

The other worker information, such as average yield ratio and average downtime ratio, for the entire period would also be useful for diagnostic purposes.

The entire measurement period could be 1 year, 1 month, or any period desired. At the end of the period the cumulative accounting could again begin at zero, and past cumulative deficiencies could be ''forgiven.'' This calculation can readily be made by the computer from the data at hand. In this case the worker would be asked to ''catch up'' by improving his performance in accordance with the deficiencies noted.

8.5 ANALYSIS OF DEPARTMENT FOREMAN'S NEEDS FOR INFORMATION

The department foreman's need for information is the same as that of his shift foremen. His key management control questions are the following:

- Did the entire department produce the amount of acceptable units expected during the work period under observation?
- If not, which shifts failed to meet their quota?
- What is the cause of the difficulty?

In the proposed system each department foreman would receive copies of all three shift efficiency reports as well as the cumulative production reports for his department. The asterisk would flag those characteristics that are outside the standards. In addition, the department foreman would be given a total aggregated report showing the same characteristics for all three shifts taken as one unit. In this way he would see how the department as a whole is doing (in department efficiency and cumulative production). Then if any of his shifts had unacceptable characteristics he could follow its future performance to see that it had ''caught up'' with the standard, consulting with the shift foreman as required. The reports given to the department foreman should also be available to the higher-level management.

8.6 ANALYSIS OF COST ACCOUNTING NEEDS

The production reporting system must provide cost accounting with a record of labor costs. In this way total labor cost by product can be compared against labor standards and corrective action can be taken. Total labor cost in dollars by product and by each department may be calculated from the job tickets, since the wage rate for each worker is stored in the computer data base. Any "missing time" contained in a worker's time card not shown in his job tickets would also have to be accounted for. This missing time is calculated as the difference between the total time registered on the worker's time card (for which he is paid) and the sum of the times on his job tickets. This time would be treated as another indirect labor cost.

Thus the primary cost accounting report would consist of both the direct and indirect labor costs, by product and by department (1) for the past period (say the past week) and (2) since a given starting point (e.g., the start of a particular contract). The budgeted labor cost for the period is also shown, and some comparison is made with the actual costs (ratio or variance).

To diagnose a higher than budgeted labor cost, the manager must refer first to the labor hours spent and its variance from budget. The remaining variance must be due to a variance in wage rate.[5] This can also be calculated (the quotient of the total labor cost of the lot divided by the total man-hours required for the lot).

8.7 ANALYSIS OF PRODUCTION CONTROL NEEDS

The number of acceptable units produced is also needed by production control to schedule production for each day. Any variation in the day's output from one department will affect the work-in-process inventory feeding the next department. In fact, the major objective of this inventory is to accommodate any random differences in the effective production rates between the two departments. Hence the computer can be instructed to keep a daily record of inventory between departments, based on an initial level plus the day's acceptable production output of the preceding department minus the day's production output of the succeeding department. Thresholds above and below an acceptable inventory level can be set. If a threshold is crossed, the computer can notify production control so that the necessary change in production scheduling can be made to keep the inventory within acceptable bounds.

8.8 ANALYSIS OF TIMEKEEPERS' NEEDS

The primary purpose of the timekeepers in the current system is to compare the production times reported by each worker on his job tickets

[5]This is the case of using a higher-wage worker than the production was budgeted for.

with the time recorded on his time card, the basis of his pay. This is an ideal job for the computer and can be done in the following way. Total the times from each worker's job tickets and compare this with the work time listed on the time card. If the difference is greater than a given amount (say 20 minutes, an acceptable tolerance level), print the worker's name and both times on a daily hours monitor report. The foreman will use this as the basis of discussion with the worker on the next day.

Differences between the two times smaller than the threshold level are to be treated as unaccounted time, an additional indirect labor cost. In this way cost accounting has a complete record of total labor time spent without having the timekeepers manually change the job tickets, as they must do in the current system. The one problem with this proposed approach is that the time card data are currently fed into the computer at the end of the week. Hence the deviations could readily be detected at that time. This should not be a problem. Any deviations could be handled before the paychecks were issued. With this option the monitor report could be issued weekly instead of daily.

If it is desired to make this cross-check on a daily basis, as is done in the current system, the timekeeper could check the time cards against a daily hours monitor report for any large deviation. This report would save the timekeeper from having to do all the addition he currently has to do.

8.9 ANALYSIS OF MAINTENANCE NEEDS

Finally, the basic data relating to machine failures or machine adjustments over time may also be accumulated by machine for a statistical analysis of each machine. Using these data, similar machines may be compared against one another to see if any condition is developing that would require overhaul or replacement.

8.10 SYSTEM DESIGN PROCESS

The preceding analysis up to now has focused on the user analysis and has specified different levels of a postulated system demand function, whose parameters are as follows:

- The quantities of information delivered.
- The types and forms of reports delivered.
- The frequency of these reports.
- The time lags in delivering these reports.

Now we must design alternative systems to provide these parameters.

The starting point is an analysis of the current system, already performed. The design procedure we follow in synthesizing other alternatives is as follows:

- Reexamine each of the functional activities in the operation flow diagram (Figure 8.1), and determine those still requiring implementation. This provides the new operational concept.
- Determine what implementation alternatives exist for each functional activity, providing the information level specified.
- Select the preferred alternative for each activity on the basis of the lowest cost required to provide the information level specified (taking into account any other differentiating characteristics not considered among alternatives).

The difficulty with this approach is the large number of system alternatives that need to be evaluated. For example, if each of five functional activities can be implemented in three different ways, there are 5^3, or 125, different combinations that theoretically are possible alternatives requiring evaluation.

There is an easier way of evaluating the various alternatives. This involves evaluating the various alternatives for each functional activity separately, choosing the preferred one for each.[6] This set of preferred alternatives then constitutes the preferred system. This method greatly simplifies the evaluation process, since in our example the evaluation process has been reduced to five separate evaluations of three alternatives each. Hence this is the evaluation method that will be used in this case.

8.10.1 Analysis of Data Collection Function

All the data currently being collected by the current job ticket system are needed for the improved system too. The designation of each operator's machine should also be added. If we find that all three operators use the same machine on three shifts and experience difficulty in their production (high failure rate, downtime, or low yield), one inference is that the machine is causing the difficulty. Such information would be of use to the maintenance department.

The two different data input documents used were (1) the punched cards for the job ticket system and (2) a production report for the manual

[6]Obviously, any relationships between different activities must also be considered.

manufacturing system. These were the two alternatives considered for implementing the data-gathering activity.[7] Since lost tickets were a problem with the job ticket system, it was decided to use one production report to record all the work done by one worker (or a production team) during one shift. Since the generic information requested of all departments is essentially the same, most items on the production report could be common to all departments. Only the name of the output would change to meet the particular department. Thus one common form design could be used for all departments, with any information unique to a particular department added.

8.10.2 Analysis of Foreman Data Review Function

Near the end of each shift, the shift foreman would review all production reports for completeness. Had all data been recorded? Was there continuity on all time entries? Had all downtime been approved by the foreman? If the entries were satisfactory, the production reports were collected and made available for data processing.

8.10.3 Analysis of Editor Data Review Function

In the current system the final data review is done by the timekeepers. In the proposed system this review would be done by an editor, who would make certain that all entries had been made and add certain edit codes. These entries currently are made by the key punch operators themselves. However, this means that each key punch operator has to be experienced with the processing system. Thus when these experienced operators are absent, the total work is delayed and the reports are not available on time. By simplifying the keypunching job by having the editing function done by an editor, temporary key punch operators could be hired to take the place of absentees, thereby meeting the data-processing time schedule.

8.10.4 Analysis of Keypunching Function

In the proposed system the edited production reports would then be keypunched onto cards and verified by key punch operators. Enough key punch operators would be available so that any report could be made available within one day's turnaround of the latest data provided.

[7] Another alternative briefly considered was the use of remote entry terminals. Here each worker could enter the desired information directly into one of the computer disk files, eliminating the key punch operation. Apex management did not feel this alternative would have been compatible with its environment.

8.10.5 Computer Data Processing

Computer programs following the logic described are needed. Input data to be stored in the computer would include production standards for direct labor, indirect labor, and scrap for each different product and operation; management-by-exception thresholds for each desired characteristic also need to be specified. These would be developed by the manufacturing department. The computer operators would then run the system so as to meet the report due dates.

8.11 COST-BENEFIT ANALYSIS

A comparative analysis was made between the proposed production reporting system and the current system on the basis of

- The incremental costs required to develop, implement, operate, and maintain the new system.
- The benefits achievable (operating cost savings and other benefits).

8.11.1 Analysis of Incremental Costs

An estimate was made of all costs associated with developing, implementing, and operating the proposed system. Specifically, a base case estimate was made of the cost of producing management reports on a weekly basis. To enable the decision maker to see the additional cost of providing reports more frequently these costs were also calculated separately. In this case the input data are already collected and stored in the computer. The only additional resources required are for computer operation and report distribution.

Each estimate had some uncertainty associated with it. The uncertainties were quantified and their cost implications made explicit. These estimates for each of the cost elements considered are shown in Table 8.2.[8]

8.11.2 Analysis of System Benefits

There are three types of benefits that the new information system could provide

[8]The derivation of these costs is contained in Appendix 8.A to this chapter.

Table 8.2 *Analysis of Net Savings*

	Low	Expected	High
Development costs			
Systems development	$ 2,100	$ 2,800	$ 3,500
Programming	5,500	7,150	8,800
Test and evaluation costs			
Computer use—test	1,200	1,400	1,600
Computer use—evaluation	1,600	2,000	2,400
	$10,400	$13,350	$16,300
Operating costs			
Computer equipment time	$ 240	$ 320	$ 400
Editor	220	330	440
Additional keypunching	20	20	20
Forms	10	10	10
Hardware rental	16	20	24
	$ 506/mo	$ 700/mo	$ 894/mo
Gross savings			
Timekeepers wages	$ 1,760/mo		
Net operating savings			
Gross savings	$ 1,760	$ 1,760	$ 1,760
Operating costs	894	700	506
	$ 866/mo	$ 1,060/mo	$ 1,254/mo
TOTAL net savings	$10,392/yr	$12,700/yr	$15,080/yr

- Cost savings in the information system.
- Improved timeliness in receiving reports.
- Improvements in the manufacturing process that can be translated into cost savings.

The following shows how each of these savings could be quantified.

- *Information system benefits*

  Elimination of timekeepers. The major saving resulting from this system is the cost of two timekeepers who would be reassigned to other positions. At a rate of $5 per hour, this saving amounts to $1760 per month, assuming 22 workdays per month at 8 hours per day.

  Elimination of job tickets. No saving. Computer cards must be punched regardless of what system is used.

  Keypunch operation simplified. By having the editor insert the edit codes, keypunching is easier with the new form. Hence substitute keypunch operators can be obtained if a regular operator becomes sick, thereby ensuring that the reports are available on time. This provides an improvement in performance, not cost savings. The computer will convert any formula-type alphabetic description to a numeric one automatically. This eliminates the need for referring to a conversion table, thereby saving time.

  Fewer information errors. Errors will decrease because of the increased responsibility of the foremen and editor in checking the production reports for accuracy. This was also treated as an improvement in performance, not cost.

- *Manufacturing system savings*

  Increased worker productivity. Efficiencies are to be calculated for each worker and each department. This allows manufacturing personnel to concentrate on evaluating departments that are performing below standard. In the past, time had been spent working in one department when attention was needed in another area. Since a 1 to 2% increase in productivity has been noted in the departments currently evaluated, it is estimated that such increases in productivity could be gained in all departments.

  Future improvements. The new source document would collect information relevant to labor variance reporting, scheduling, and inventory control. These areas need to be automated in the future. If the scheduling procedure were automated, this not only would reduce labor costs but could provide the kind of accurate inventory control necessary to reduce in-process and finished-goods inventory.

8.11.3 Calculation of Cost Savings

In terms of the decision-making problem facing Apex management, the following factors were generally agreed on by the management. Intuitively, all felt that if the proposed system worked as predicted, a great improvement in the management capabilities would follow. What was desired was an economic analysis to estimate the total costs of developing, implementing, and operating the new system. Such information could be used to justify these expenditures on the new information system.

One of the immediate benefits was the elimination of the two timekeepers. Thus, to minimize the analytical resources required, it was decided to do the cost-benefit analysis in an iterative fashion. That is, to see if the savings obtained by eliminating the timekeepers are enough to pay for the entire information system costs. If so, the additional benefits listed could be treated as "bonuses" and would not have to be quantified to justify a decision to implement the system. Thus the analysis continued by calculating each cost element and then calculating the cost savings.

8.11.4 Comparing Costs to Benefits

The total costs for development, test, evaluation, and operations were calculated using the three-point estimate to consider uncertainties (Table 8.2). A cost-benefit analysis was then made considering only the cost savings in operations resulting from the elimination of the two timekeepers (bottom of table).

Although the system evaluation could have been made in a number of different ways, the payback period approach was used, since it was compatible with the company's approach to investment analysis. The payback period was calculated in two ways. Example 1 used the standard method and ignored any discount factor. Example 2 repeated the calculations, with a discount rate of 30% to satisfy a request of Apex management. In both cases the payback period results were well within the 3-year requirement that Apex used as a standard for approving investment proposals. Both cases included the cost uncertainties involved.

The detailed calculations follow.

Example 1. No Time Discount Rate

Table 8.3 is a matrix showing the nine possible outcomes, taking into account the three possible values in development, test, evaluation, and operations costs (columns) and the three possible values in net operating savings per year (rows). The payback period required under each of these nine conditions was then calculated as shown. From this it can be seen

Table 8.3 *Calculating Payback Period (Years) for Example 1: No Time Discount Rate Assumed*

Breakeven analysis using no discounting factor: (based on total resource costs)

		Possible development, test, evaluation and operation costs		
		$10,400	$13,350	$16,300
Possible net operating savings per year	$10,392	1	1.28	1.57[a]
	$12,700	0.82	1.05[c]	1.28
	$15,080	0.69[b]	0.89	1.08

[a]Worst case.
[b]Best case.
[c]Expected case.

that the payback period will vary from 0.69 (the most optimistic condition) to 1.57 years (the most pessimistic condition), all well within the 3-year requirement that Apex uses as a standard in approving investment proposals.

Example 2. A 30% Time Discount Rate

To determine the number of years of savings required to pay back the investment (for development, test, evaluation and operation) costs, using 30% as the discount rate, the formula relating present value of an investment (P) to future net annual savings (A) was used:

$$P = R\left[\frac{1 - \dfrac{1}{(1+i)^n}}{i}\right]$$

where P = present value
R = yearly savings
i = interest (30%)
n = number of years

Taking the worst case, where net savings are the smallest ($10,392 per year) and investment costs the largest ($16,300), we performed the calculations in an iterative fashion. First, we let n equal 2 years and then 3 years. Then we interpolated to obtain the estimated value of payback period.

For $n = 2$:

$$P = (10{,}392)\,\frac{1 - \dfrac{1}{(1 + 0.30)^2}}{0.30} = \$14{,}250$$

For $n = 3$:

$$P = (10{,}392)\,\frac{1 - \dfrac{1}{(1 + 0.30)^3}}{0.30} = \$18{,}900$$

Interpolating between the preceding figures for the desired value of $16,300, we obtained 2 years and 5 months, which was still within the 3-year payback requirement. It is important to note that although the cost estimates in this analysis were on the high side, the savings still easily offset them.

8.12 SUMMARY OF PROBLEM-SOLVING APPROACH: Production Reporting and Control System

This section summarizes the key principles of planning as they were applied to this case. Since many of the principles summarized in Case 1 also apply here, the focus is primarily on those principles unique to this case.

8.12.1 Problem Definition

In this case the task was to design an information system. This design proceeded along two parallel but interrelated paths:

1. The analysis and specification of the proper system demand function. In this design of an information system, the system demand function was interpreted to be the information to be provided to each information user, its form, and its timing (what information will be provided).
2. The design of the data-handling system to provide the level of information postulated (how this information will be produced).
3. Determination of the level of information that is cost-effective.

We proceeded in an iterative fashion toward the preferred system, taking into account the following:

1. The users' perceptions of the information they need.

2. How the users intend to use this information during the planning horizon.
3. The costs of providing the information.
4. The benefits that can be achieved by having the additional information.

Analysis of the current information system led to the identification of

1. The various data-handling functional activities required, the data sources, and the data-handling procedures currently in place.
2. The equations for converting the data into information useful in controlling the production system.
3. The identification of information users, the deficiencies they perceive with the current information, what improved information they desire, and how they would use such information. Such an analysis would provide answers to the following questions:
   a. Who are the users of this information?
   b. What information do they desire?
   c. How do they use the current information?
   d. How would they use the improved information?
   e. How often do they need this information?
   f. In what form should the output information appear?

From this analysis the planner can postulate a set of possible alternative information outputs for each user. These outputs range from the information provided by the current system to the client's stated desires (postulated "requirements") or even to an "ideal" system that generates the best possible information. An ideal system can serve as a standard, against which the analyst can compare the current system and any intermediate designs he generates.

The next step is to design a feedback control system (Figure 8.2) around the following data:

1. Data indicating that the operation is proceeding in accordance with expectations. This requires
   a. Measurement of some key characteristics (performance or cost). In general, the output characteristics of rate of acceptable production and cost are best. Sometimes, some other lower-level characteristics having to do with the process rather than the output are also used for control purposes.

   b. The expectation or standard against which each output characteristic is measured.
   c. Tolerance levels around the standard that indicate when the system is operating sufficiently outside the standard to justify some corrective action.
2. Additional data concerning the process that might indicate the cause of the difficulty (diagnosis of the problem or process) and the proper corrective action to be taken.

The following desired output can be presented to the users:

1. *The production output achieved:* A report for each production period measured (like the 8-hour shift). A tabulation or plot of this output data corresponds to a series of "snapshots" taken during some period of observation. This report enables the user to recognize any exceptional activity during this period by comparing the output with some standard. An abnormal event may be a precursor of some deficiency that will continue.
2. *A tabulation or plot of the cumulative output:* Production for the entire period, starting from some initial reference point (like the start of the contract or the fiscal year). This information can be compared against the planned cumulative function and from this can be determined:
   a. The long-term production actually accomplished as compared with that expected.
   b. The total deviation in output to date, as compared with the total output expected.
3. From both of these types of information the following can be extrapolated:
   a. The expected deviation in the final cumulative output as compared with that expected, assuming that the recent average value of production rate is maintained.
   b. Conversely, the average production rate required to meet the final goal can be calculated.
4. A "management-by-exception" reporting system can be designed by establishing expectations or standards plus an acceptable threshold level for each characteristic being monitored. This would permit managers to concentrate their attention on only those parts of the organization whose outputs deviate sufficiently from the standard so as to be considered a possible threat to meeting the expected goal. Such a system could report on the following deviations:

a. Production rate (time).
b. Production quality (yield).
c. Production cost (derived from time, wage rates, and other cost elements).

Such exception reports could be based either on "instantaneous" outputs for the minimum period of observation or on the cumulative outputs. Once an exception report is generated, more complete information, including appropriate diagnostic information elements, should also be generated as an aid in determining the cause of the trouble and what corrective action should be taken.

8.12.2 System Design

The next step is to synthesize alternative ways of gathering and processing the data to provide the information to each user, as defined by each of the information system demand functions postulated. Each of these designs should take into account possible economies of scale that may be achieved by any similarity of information among users. Each design alternative can be based on

1. An operational concept of the information system, translated into an operational flow diagram that shows how the input data are transformed into the desired information output. The operational flow diagram of the current information system is a good starting point for generating the alternative operational flow diagrams. See if it is possible to improve the effectiveness or efficiency of the current system by "pruning" or eliminating any functions that do not seem to be really needed in providing the new information.
2. New system operational concepts.
3. New system entities that are available within the time constraints of the problem.
4. System constraints, including those elements with which the new system must interface.
5. The use of as many elements as possible that are already part of the current system. Their use in the new system reduces the total out-of-pocket investment cost required and may reduce total system cost.

Thus the systems synthesis process is essentially a combinatorial problem, where the systems designer is synthesizing all (or a large enough number of) combinations of available entities so as to meet the information system demand function within the constraints.

8.12.3 Systems Evaluation

The total cost of each system alternative can now be estimated. The preferred system design for obtaining each level of the system demand function should be identified, generally on the basis of lowest cost. From this can be obtained the incremental cost for advancing from one level of information provided to a higher level, such as providing each type of report or varying the frequency of providing the report (daily, weekly, etc.).

The benefit to each user of the information supplied him should be estimated under the following set of conditions:

1. *Benefits*. Estimating the total benefits accrued to the entire organization (the information system, the material flow system, or other parts of the total system that the information system is designed to control). These benefits can be increased financial returns or cost reductions in the total system.
2. *Losses*. Another way of obtaining this information is to estimate the losses that would be incurred by not having the timely information proposed. This could be estimated by analyzing the past records and determining which expenses incurred in the past could have been avoided had the proposed information been available. Each expense can be extrapolated to the future operations.

The preferred information system design is selected by relating the gross benefits obtainable through the information package to the incremental cost of the system required to provide the information (plus any other cost in achieving the new form of operations). Any of the appropriate decision-making models described in Chapter 6 can be used for this purpose. The evaluation should also include a consideration of the major uncertainties associated with each system. These should be identified and quantified as uncertainties in time or resources (and thus in costs) required to develop, procure, operate, and maintain the system.

APPENDIX 8.A. DERIVATION OF SYSTEM COSTS

The following system costs were considered.

Systems Development

The EDP department would have the responsibility of finalizing the timing of the production report collection and output procedures. In addition, they would direct the efforts in regard to programming, debugging, and evaluating any necessary hardware that might be required initially or in the future. In discussions with the EDP department it was estimated that 1½ to 2½ man-months would be required to complete the preceding tasks. At approximately $1400 per month, this would result in costs ranging from $2100 to $3500.

- *Programming.* Programming efforts would include the development of new programs, and the reprogramming of current programs resulting from the use of a new production report. EDP estimates these tasks would require from 5 to 8 man-months at $1100 per month. These costs would then be from $5500 to $8800.
- *File set-up and maintenance.* Computer files would have to be established to handle this new system. Two possible costs were considered: establishment of the manufacturing standards by manufacturing and programming these standards into files. Manufacturing stated that standards exist for all active part numbers. Since new standards are established for each new part number as they arise, this cost need not be attributed to the proposed system. The EDP department included the file programming in its program development estimate. Since EDP can incorporate additional standards by making another entry to the established file by simply entering an additional computer card, the costs were treated as too insignificant to be considered.
- *Testing and evaluation*
- *Parallel system paperwork.* The proposed system would be gradually phased into manufacturing over a period of time. A likely procedure would be to run the new production reports along with the current system for some time to allow the workers to become accustomed to the new forms. The original production reports and job tickets would be eliminated, along with the manufacturing manpower now used for the current production report system, once management was assured of the accuracy and timeliness being sought.

  Manufacturing felt that no significant decline in production would be experienced. Hence it was decided to treat this as a nondifferentiating cost.
- *Computer usage.* Two periods of computer usage would be necessary to develop the proposed labor reporting system. First, a debugging period would be needed to eliminate errors in the initial program and to test its accuracy. It is estimated that 30 to 40 computer hours would be required to accomplish this. At $40 per hour this cost could range from

$1200 to $1600. Second, a parallel run period to locate errors in program logic that might otherwise be unforeseen would be necessary. An estimate of 40 to 60 hours has been made for this procedure and at $40 per hour, the cost would be from $1600 to $2400.

Operations

All operating costs to be discussed are "out of pocket" (i.e., costs incurred to obtain new resources). These costs are denoted by an asterisk (*).

- *Computer Equipment Time.** The reports to be generated by EDP include a daily monitor report, a department weekly efficiency report, and a weekly efficiency report by worker. Since the daily monitor report would replace a report currently being processed, the only incremental costs that would arise would be for the processing of the two weekly reports. These would require from 1½ to 2½ hours at $40 per hour, for a cost ranging from $240 to $400 per month.
- *Editor.** After considering the number of production reports to be handled each day, we established that an editor would spend from 2 to 4 hours per day processing and transferring the report to EDP.

  From the current Apex organization chart it was concluded that in the worst case some 425 production reports for the disk brake area might be processed, or one for each man on each shift. Fifteen of the 25 reports required that the original and the carbon copy be separated and transmitted to EDP and manufacturing. These 15 reports are 60% of the total, or 225 reports. Allowing 4 seconds per report to separate the copies, it would take approximately 17 minutes to complete this portion of the task. The remaining 170 reports involve team reports and require the editor to include an edit code on each entry of each form. If we allow 30 seconds per form for this process, the editor can complete the job within 1½ hours. This totals only 1 hour, 47 minutes, but to allow for errors, trips to EDP, and the addition of more production reports as the system is extended into other departments, we expected that the total time would range between 2 and 4 hours per day. Using a range of $5 per hour, the cost would be from $220 to $440 per month.
- *Additional Keypunching.** The EDP department did not foresee the need for additional key punchers, at least initially, because of the elimination of the daily reports and a decrease in the total number of

cards to be punched. Rescheduling might occur, however, for which a 5% night premium rate would apply. This would amount to a $20 per month cost.

- *Form Preparation.** Since we were simply replacing the current production reports with new ones, the total number remained the same. Consequently, the amount of time spent by Apex personnel in the preparation of these forms remained the same. However, an additional cost of $10 per month would be incurred, since 13 reports must be changed from half sheet to full sheet size. These reports represent approximately 52% of the total, or 4840 reports to be processed each month. At two sheets per report and at $10.35 per 5000 sheets, the total expense would be $20. Subtracting $10 as the cost required currently for the small forms, the additional expense becomes $10 per month.
- *Additional Hardware.** The EDP department estimated the additional hardware cost to be as follows:

Addition to core storage of computer	$400/month
Additional disk drive	$400/month

  It was not anticipated that both additions would be necessary. However, it was possible that one or the other might be acquired even if the new system was not adopted. The new system required 8 hours per week, or 4% of the current machine time. Thus if the additional hardware was required, the additional operations cost would be $16 to $24 per month.
- *Maintenance.* Any maintenance costs incurred are included in the computer rental fee.

Part IV

DESIGNING PLANNING SYSTEMS

CASE 3: TELECOMMUNICATIONS SYSTEMS PLANNING

Case 3 extends the design of information systems to a management planning system for the redesign of a complex telecommunications system. The major problem in this case was to gather data that would systematically relate the communications services provided by this system to its manpower cost, to provide the necessary data base for planning purposes. Such a planning data base consists of (1) the set of planning factors and (2) a planning logic needed for properly designing the system, making the necessary trade-offs between communications services available and allowable system costs.

Case 3 has several objectives that relate to the systems planning process followed in the solution of the previous two cases. In Case 1 most of the performance and cost data were provided to the analyst. His primary job was to develop evaluation models that could assemble the data so that the best evaluation of the alternatives could be made. But how are such cost and performance data obtained? This case illustrates how to go into a complex organization and perform a functional analysis so that data needed to design and evaluate various alternatives can be obtained. One of the important elements of the management control system of Case 2 was the set of quality, time, and cost standards used for determining whether the work was being done in an acceptable manner. The following case describes how such standards can be derived.

The specific example presented is the analysis of the Naval Telecommunications System. However, since this system performs similar functions to Western Union's telegraph operations, the same approach and planning principles also apply in analyzing such commercial operations. In fact, the principles described also apply to any type of manpower analysis.

One final word of guidance to the reader is in order. This case, as well as those that follow, illustrates the amount of detail that must be presented to the client to make the analysis accurate and credible. It also shows the type of "procedures manual" that must be generated to show the client in simple terms how to apply the results obtained. Since not all readers are interested in this degree of detail, some of the more detailed material has been separated into appendixes for those desiring to see the form that the final solution takes.

9

Analyzing the Transmission Systems

9.1 PROBLEM AS GIVEN

You are a consultant with a planning organization that in the past has participated in a number of planning projects for the Naval Telecommunications Command (NTC). You have been asked to attend a meeting with the commander, Naval Telecommunications Command (CNTC), RADM Walter Johnson. When the meeting begins, Adm. Johnson says to you,

As you may know, this command is currently engaged in several short-range and longer-range efforts to realign the Naval Telecommunications System. Over the next few years satellites are being phased into the system and satellite terminals will be installed on practically all Navy ships and at our major naval communications stations. They will replace much of the HF (high-frequency) transmitters and receivers. Now we know what resources are needed for this new equipment. The problem is how much of the older HF equipment to phase out. This is particularly a problem as the chief of naval operations has recently reviewed the 5-year defense plan for this command. He has proposed some large cuts in manpower based on some previous estimates in manpower savings that were predicted for the satellite system when it was proposed. I think those previous estimates of manpower savings were too optimistic. But unless I can show in a systematic fashion just how many resources are required for the backup HF system, I'm not going to be in a good bargaining position with the budget people.

During this past year we gave a system realignment study contract to Delta, Inc., a planning firm. They were to analyze the problem of just how much HF equipment we will need at the various stations when the satellites are phased in. Their reports recommend which stations should be completely closed down and which should continue operations with a reduced level of equipment and manpower.

I'm very unhappy with the draft of Delta's analysis. I do not find it very helpful in answering our basic question: "What should the mix be between satellites and HF in the realigned NTC system, in light of the pressure to reduce manpower?"

I also have a shorter-range, but related, problem. As I visit each of our 26 worldwide communications stations, I am constantly getting requests from different station commanders for additional resources, particularly manpower. They feel that they are handling a larger work load with less people than other sites. They don't think they are being treated fairly.

Furthermore, the Government Accounting Office (GAO) is currently studying the communications services of all three services. GAO reports that the Army and Air Force have developed a planning system that relates manpower required to work load for different work sites. For example, GAO studied the message centers of all three services and found that, in a message center handling a given number of messages per year, we in the Navy use about twice as many people as the Army or Air Force. We had a deuce of a time showing that in our message centers, we also perform functions, such as reproduction and distribution of messages, that the other services do outside their message center. This accounts for our need for the additional people.

What we need is a planning system that will relate resources needed to the various communications services or work done at the various sites. I would like you to develop such a system. Since manpower takes about 75% of the total equivalent dollar costs to operate and maintain our sites, start with manpower requirements. You can analyze the other resources we need, such as supplies, fuel, utilities, and so on, during the next phase of this project.

There must have been some manpower rationale developed when the stations were initially set up. But no one here today can explain what it was or prove that it still holds. It obviously is not being used today. Just compare the manning at similar types of sites today. It doesn't make sense. Furthermore, as we keep changing the requirements at each station, we are going to need some way of modifying the resource requirements of each site in a systematic fashion, if only to justify our budget to headquarters.

Can you help me in realigning our telecommunications system on some rational basis? I know that there are some at this command who will say that each of our sites is unique, different in layout and number and mix of equipment used and that, therefore, the manpower allocation job must be done strictly by experience and judgment. However, the DOD budget people don't believe this. They cite the work that the other services have done in developing manpower standards for their communications systems. Maybe having someone who is not familiar with our operations but who can ask the right questions can make some sense out of our current planning problem.

After some further discussion with the admiral, you agree to do a preliminary analysis of a part of the problem as defined in order to determine if, within the rather limited resources available, you can develop a solution.

The specific problem you propose to attack in the short term is as follows: The new satellite system is scheduled for test, evaluation, and then operation of the fleet broadcast channels in the Mediterranean. When the fleet is satisfied with the results, the naval command will shut down the HF broadcast transmitters. How much savings can be made when these transmitters are shut down?

We agree to study the problem for 1 month, using the voluminous data collected by Delta, Inc. Working on this specific subset of the problem will give us an opportunity to understand the overall problem better. Analyzing Delta's data will also give us a head start in understanding the problem, since it not only will show the approach they used but will show much of the data available. Finally, the exercise will enable us to estimate the total analytic resources required to solve the problem. We plan to meet with the admiral in 1 month and present both the cost savings involved in shutting down the broadcast transmitters and a proposal for a more comprehensive study.

9.2 PRELIMINARY ANALYSIS MADE

9.2.1 Descriptive Material Available

The four Delta, Inc., reports available gave an extensive amount of data relevant to gaining an understanding of the current NTC system. Volume 1 described each of the communications sites in the Mediterranean area, and included the following characteristics:

- Name of Mediterranean sites and their locations.
- Total dollar budget, by funding categories, such as utilities, materials, and so on.
- Total number of manpower billets authorized, including officer and enlisted men grades, and some indication of the type of jobs they do. However, the true manning on duty at each site was at a level of about 80 to 85% of the authorized billet levels.
- The equipment associated with COR (the communications operating requirement). This is the number and type of equipment required to provide particular types of communications circuits needed during peak operations, such as in a contingency operation or a major fleet exercise when maximum communication is required.
- Original equipment investment cost at each site. As "sunk costs," these were extraneous to the immediate problem.

Missing from the report were data on equipment usage.

9.2.2 Delta's Planning Approach

The planning approach taken by Delta could be summarized as follows:

- When the satellite system was fully operational in the Mediterranean, it could meet most of the COR requirements, even as they increased over time. For example, practically all of the Navy ships would have satellite broadcast receivers installed on them. However, Allied Navy vessels and U.S. maritime vessels will not. Hence some small amount of HF broadcast equipment is required to serve these ships.
- The difficult problem was coping with satellite failure. If the satellite failed, some backup HF capability must be available to carry the more essential messages until a satellite replacement becomes operative. It is obviously too expensive to carry 100% HF communication redundancy as a backup. Hence only a smaller load of the more essential, higher-priority, messages would be handled until a satellite replacement is launched and operating.

Delta was thus faced with the problem of determining what the reduced communications requirement (the system demand function) was apt to be. They claimed that they tried to obtain this information from the command or others but were unsuccessful. Since this was the most important input to their design process, they recommended that the smaller-capacity backup system should be able to handle just the "command and control" messages, leaving out all the lower-priority administrative messages. They claim that no one took exception to the assumption. At the same time, they did not get formal approval.

Using this key assumption, they configured several alternative systems to meet (or exceed) their assumed communications demand. They chose the one with minimum cost (maximum cost savings) over the planning horizon chosen.

9.2.3 Deficiencies in Delta's Planning

We were never able to find out specifically from the command why they did not like the Delta study. However, disregarding the possibility that the final answer was not the one desired, we analyzed Delta's study and noted two deficiencies:

- The communications demand that was assumed was never agreed on by the command. Hence the design results might also not be acceptable, since it followed from the system demand function assumed.

Delta might have done an analysis for several different levels of communications demand, thus giving the decision maker a set of options rather than only one.

- Delta supplied no supporting evidence for their estimate of the resources required to operate, maintain, and support each alternative system. If a site was completely closed, there was no estimation problem, since just about all the resources (manpower and others) would be saved. However, if the number of transmitters was reduced from 160 to 40, how many resources would then be required? Certainly not one-fourth of that used before. When Delta was asked how they arrived at the new resource requirements, they said it was based on "experience," a subjective estimating approach rather than a systematic one.

9.2.4 Analysis of the Mediterranean Problem

Having gained a better understanding of the characteristics of the current system and some of the analytical pitfalls to be avoided, the analyst could now proceed to the specific problem assigned.

Interview with User. Discussions with experienced communications personnel at the command indicated that if the transmitters currently devoted solely to the Mediterranean fleet broadcast were turned off, the following savings would be achieved:

- *Maintenance.* The current maintenance policy is to perform preventive maintenance (PM) on all equipment (operating as well as backup equipment) in accordance with designated PM procedures. These are listed on the maintenance requirement card (MRC) for each type and model of equipment. All standby transmitters are kept in a state of "hot iron" (filaments on but high voltage off) as the best means of minimizing the failure rate. Even with this policy, the failure rate of standby transmitter equipment is about the same as that for operating transmitters. Thus the consensus was that the resources required for PM and corrective maintenance (CM) would be the same as if the equipment was still operating. On the other hand, if the equipment was removed from the site, the proportional amount of manpower would be saved.
- *Operations.* The consensus was that very little manpower would be saved by eliminating the HF fleet broadcast. These transmitters always operate continuously and at the same frequency. Thus operator intervention is needed only when (1) PM or CM actions are required, (2) the

old transmitter must be turned off, and (3) a spare replacement must be made. In addition, a small amount of man-hours is required for quality control checks. Conversely, most other types of transmitter circuits must be turned on and off when needed. When they are operating, they must have their frequency changed at least twice daily to meet changing propagation conditions. This tuning function requires much more operating manpower per hour of usage than that needed for the continuous circuits such as the fleet broadcast.

- *Power.* Major energy savings could be achieved by discontinuing the HF fleet broadcast. The planner calculated the input power required from all the transmitters involved and converted this to generator power and finally generator fuel costs per year.

Thus the answer to the initial problem was that the major manpower savings in eliminating the HF fleet broadcast would be in maintenance. The only way that these could be achieved would be by removing the transmitters from the site.[1] The amount of such savings could be estimated using Figure 9.1. It relates the average total of man-hours per year required to maintain a given number of transmitters of a given type (direct labor only, no supervisors or support). This model assumes that the average maintenance man-hours required for PM and CM are directly proportional to the number of transmitters of a certain type at a site. This is a reasonable assumption, similar to that used in the production problem of Case 1.

Using the preceding information, describe the approach you would use to solve the short-term planning problem.

9.3 UNDERSTANDING THE CURRENT TELECOMMUNICATIONS SYSTEM

To broaden his understanding of the problem, the planner next looked at the question, "What is the current telecommunications system and how does it operate?" Figure 9.2 illustrates the communications flow through the main elements of the system. The "backbone" of the system is the global Defense Communications System (DCS) operated by the Defense Communications Agency (DCA) on behalf of all three services. The DCS

[1]Communicators at the command felt that if the transmitters were left completely off at the site, it might take several weeks to repair them if they were needed during a major contingency or exercise, or they might be cannibalized for parts, again delaying their availability until they could be repaired.

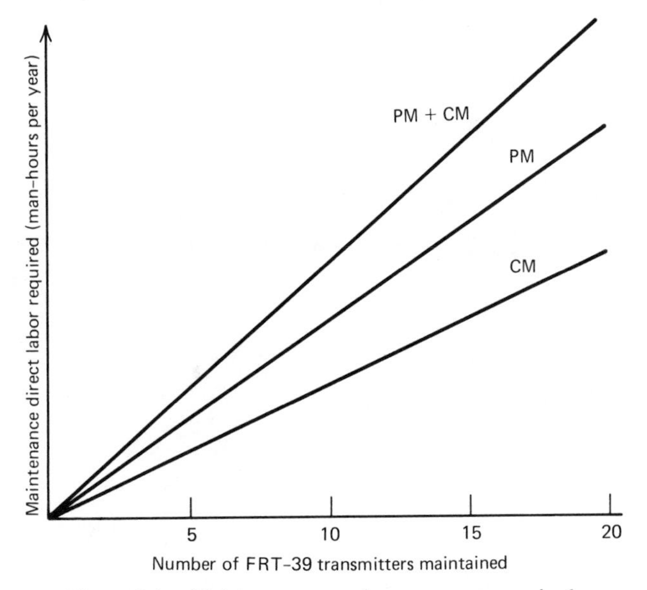

Figure 9.1. Maintenance man-hour per year required.

is a set of global "trunk circuits," consisting of a large number of wire land lines, microwave links, satellite paths, and undersea cables, each of which connects two geographical points. One or more links connected in series makes up a path between two points desiring communications. Some links are interconnected through automated switching centers. Thus if the preferred path (a set of links in series) between two points is busy, the switching center may choose an alternative path that is available at that time to arrive at the final destination. Two basic communications systems use these DCS paths:

- The Automatic Digital Network (AUTODIN), which handles all record message traffic (similar to telegrams). It was developed by Western Union and patterned on its telegraph system.
- The Automatic Voice Network (AUTOVON), which handles all long-distance voice communications between local exchanges.

Here is how a typical user sends a naval record message from his shore office to an addressee located either at another shore station or on a ship at sea using these long-distance communications paths. First, the desired message is typed on a sheet of paper in a standard Navy message format. The top of each message contains a "message header," which includes (1)

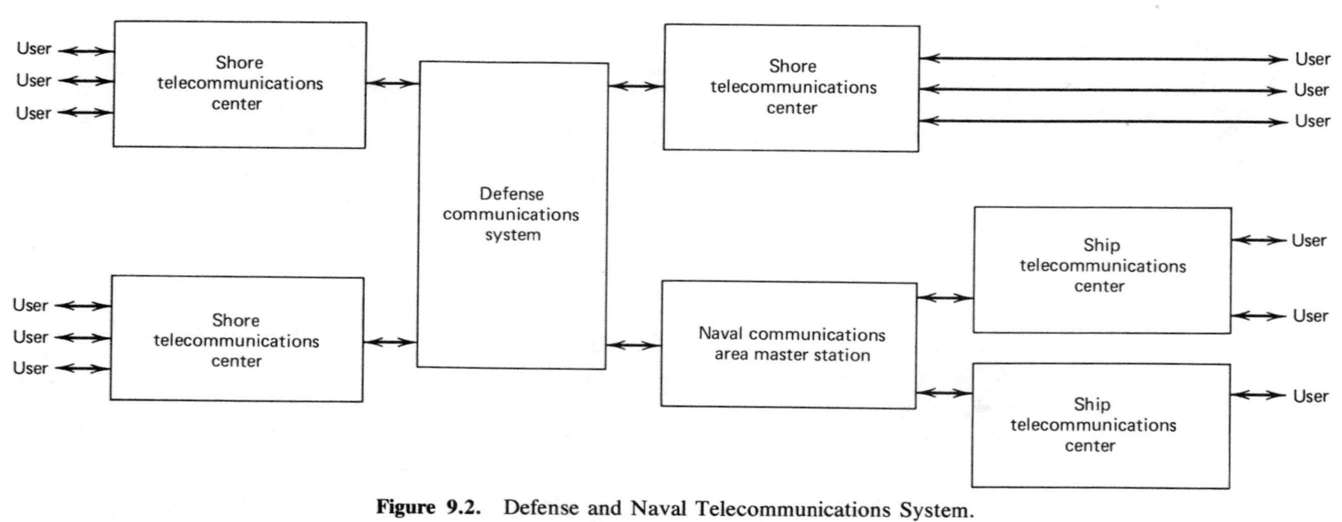

Figure 9.2. Defense and Naval Telecommunications System.

the receiver's complete address, (2) a "routing indicator," a set of symbols that tells the various AUTODIN switches the proper way to route the message over the AUTODIN system, (3) the date and time the message was composed, and (4) the security and precedence (priority classification) of the message. Following this is the message itself, and then a "trailer," or concluding nontext information.

The message is then hand-carried by special courier or is sent by the regular, periodic base delivery system to the local telecommunications center (TCC) serving the user at his shore station. For convenience, there are several TCCs at large shore stations. The typed message is presented at the TCC's message counter, logged in, and converted into suitable electrical form for entry into the system, generally by typing it on a teletype, which produces a paper tape that is then fed manually into a paper tape reader. Automated telecommunications centers may require that the original message be typed by the user in a given format on a special typewriter so that the message can be read by the TCC's optical character reader (OCR). This automatically converts it into the proper electrical form, thus eliminating the teletypewriting and tape reading functions. If the message addressee were located at a land site, it is routed automatically through the DCS AUTODIN system using the routing indicator information to the addressee's TCC. There it is converted into a paper copy multilith master using either a teletypewriter (at a normal TCC) or a high-speed printer (at an automated TCC). The required number of reproductions are made, multipage messages are collated, and all messages are distributed to all addressees on periodic trips around the station by the TCC's delivery system.

If the addressee were on a ship, the message routing indicator automatically sends the message to one of four naval communications area master stations (NAVCAMS) servicing that area of the world (Figure 9.3). The four NAVCAMS are located at:

- Norfolk, Virginia, serving the Atlantic.
- Naples, Italy, serving the Mediterranean and Indian oceans.
- Honolulu, serving the East Pacific.
- Guam, serving the West Pacific.

Thus any message addressed to a ship is directed through the AUTODIN system to the appropriate fleet NAVCAMS message processing center (also called a fleet center[2]). There it enters the Naval Communications

[2]See Chapter 10 for a more detailed description of the message processing operations at the fleet center.

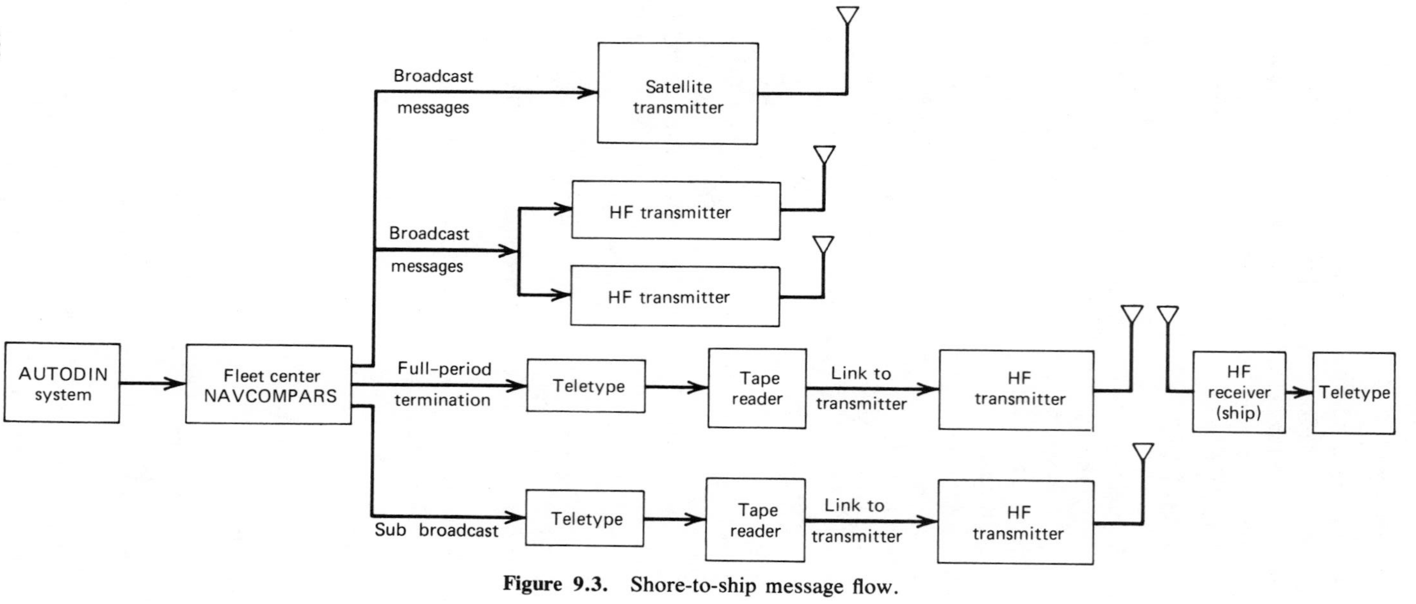

Figure 9.3. Shore-to-ship message flow.

Processing and Routing System (NAVCOMPARS). This is an electronic computer that logs in and stores all messages, keeps track of them, and distributes them to their assigned transmission channels in order of their precedence and by time of receipt. For example, a smaller ship does not receive a large number of messages per day. To obtain maximum utilization of the communications channels, messages to such ships are transmitted in a "party line" fashion. Thus all destroyers in a given area tune to an assigned frequency and know that any message addressed to that ship will be transmitted on one of the 16 channels that can be carried simultaneously (in parallel) by each transmitter.

In fact, to minimize the possibility that the message is not received because of a malfunction of the broadcast transmitter, the ship's receiver or teletype, or because of HF propagation difficulties, a large number of redundancies are employed. Thus the set of messages goes to:

- Perhaps five or more transmitters, each tuned to a different frequency.
- Two different channels on each transmitter.

In addition, each smaller ship has two complete broadcast channels (receivers and teletypes), each tuned to a different HF frequency and channel. Thus if the ship's radio operator finds that one of his channels is receiving a weak signal (evidenced by garbled messages received), he will switch that receiver to a different frequency (still receiving messages on the other acceptable channel). If his own equipment malfunctions, he switches to a spare.

Furthermore, all broadcast messages are automatically repeated in 1 hour on a different channel. Any message not received properly the first time can be received on the rebroadcast. Finally, all broadcast messages are sequentially numbered. It is the responsibility of each ship operator to monitor all numbers and make certain that he has received all messages in his channel. If he does miss any messages, he can ask a sister ship if any of the missing numbers were addressed to him. If so, he sends a service message to his NAVCAMS requesting an additional retransmission of this numbered message.

Large ships, such as aircraft carriers, have many more messages sent to them and hence they are assigned several "dedicated" channels that handle only their message traffic. These dedicated channels are called "full-period termination" channels. The NAVCOMPARS computer is programmed to recognize all such messages for this ship and routes them to a particular teletype station, where the message is reproduced in the form of a punched paper tape. The fleet center operator assigned to this teletype station takes the tape and inserts it into a paper tape reader assigned for messages just for that ship. It converts the tape input to electrical signals and transmits them through land lines or a microwave link to the transmitter site, where they are routed to the particular channels of the HF transmitters used for this ship's traffic. Again, redundant channels and transmitters are used to increase the communications reliability.

The electrical signal comprising the message modulates the assigned transmitter channels, modulating a radio signal to the ship. This signal is received by the ship's receiver, demodulated, and converted to a paper tape message by the teletype connected to the receiver. The ship's radio operator then reads the message for accuracy. If it is acceptable, he transmits a receipt for the message to the fleet center operator. Both then log in the message. If the message is not received in acceptable form, the ship's radio operator asks for a retransmission. The acceptable message is then reproduced and distributed to the addressees on the ship.

When the ship wishes to send a message to shore, the reverse path is followed: the ship's radio operator types the message on the teletype, which converts it to paper tape. The tape is fed into a tape reader, just as it is at a shore station TCC. The electrical signal then modulates a ship transmitter channel and the signal is received by the shore receiver, which demodulates the message, obtaining both a paper tape and a typed message. The NAVCAMS operator checks the typed message for accuracy. If it is acceptable, he feeds the tape into a tape reader, which sends it into the AUTODIN system. If it is unacceptable, he asks for a retransmission.

Further details of fleet center operations are given in Chapter 10.

Discussion of the table of organization of a NAVCAMS with a member of the command produced the following additional information:

- Each transmitter site and receiver site has a large number of transmitters and receivers, so that during a contingency operation or a fleet exercise, a sufficient number of communications channels will be available from all sites under control of the NAVCAMS control center. These resources required to meet the communications needs of the number of ships at sea are then activated. The NAVCAMS control center obtains these resources by contacting the technical control unit (tech control) at each of the appropriate communications sites to make the assignments. If the quality of a circuit becomes unacceptable, the ship notifies tech control to correct the problem, generally by changing the transmission frequency.
- The transmitter and receiver sites are generally located in rather remote locations from the main NCS headquarters to obtain good an-

tenna locations and, in the case of the receiver site, to be away from interference signals. Thus these two sites must have their own maintenance units and must provide other services and facilities, such as a galley for eating, administration, fire fighting, security, public works, and so on.

- Maintenance for the central communications facility is provided by the central electronic maintenance division (EMD). It also provides maintenance on some specialized equipment to remote sites where the amount of maintenance does not warrant specialized maintenance on site.
- The public works division provides centralized utility maintenance functions such as building repairs, antenna repairs, and so on.

9.4 UNDERSTANDING THE PROBLEM

After further discussions with Adm. Johnson and his staff, it became apparent that the command was faced with a hierarchy of related problems. Ultimately, the command needed a communications plan to submit to Navy headquarters stating how much HF equipment would be needed at each of the communications sites that would operate after the satellite system became fully operational. One could not produce such a plan without first performing a number of contingency analyses using different operational scenarios, each with varying amounts of communications. The element of change would have to be dealt with in the analysis. The different equipment, including the satellites, could malfunction and have to be repaired. The effect of communication time delays on the mission during such operations would have to be determined. In short, it would be time-consuming and expensive to develop a communications plan in this way. Also, the results obtained would be based on the scenarios chosen, which might not be accepted by all the reviewers. Furthermore, a client such as the command will generally not fund such an analysis.

The analyst was therefore forced to adopt another analytical approach. The communications system planners could generate a number of different, feasible system alternatives, each differing in the level of communications services provided. Such differences would include geographical coverage, the maximum message rate and types of communications service that can be handled for both peak operations and the entire year, division of responsibilities among communications stations, and the division of these loads between satellite and HF equipment. The analyst would then develop a systematic way of estimating the total resources required each year for each system design being analyzed and apply this method to each of the system alternatives proposed. This would provide the decision maker with the cost of a number of communication systems, each configured to provide a different level of communications service. Thus when a budget was proposed by Navy headquarters for the command, the admiral would be able to show what communications services could be provided for these funds. Furthermore, he could present the range of alternative systems analyzed, including their costs. Higher-level decision makers could finally determine the appropriate budget, based on the relationship of communications services provided and cost.

Based on the experience we gained in grappling with the simpler problem, we decided to extend our analysis to the broader planning problem using the approach just described. After briefing the admiral on our preliminary findings we proposed to pursue the problem in the following manner:

1. *Segment the problem.* We would begin the development of the resources planning system, a major part of the realignment planning system. The objective for the resources planning system would be to specify the resources required at each site for any designated set of communication services offered at the site. Here communications services are some function of (1) the number and types of equipment installed at the site (including spares), (2) how the equipment is used operationally, and (3) the amount of such usage. Thus, unlike the Delta approach, our system would provide the resource requirements for *any* planning option.
2. *Focus on key resource.* Since manpower is the most important resource used by the system (it consumes 70 to 80% of the entire cost of operations), this is where the real resource saving occurs. For this reason manpower is the first resource we analyze; the other resources, such as fuel, spare parts, supplies, and so on, are considered later.

9.4.1 Summary of Analytical Steps

From the preliminary analysis of the problem it became evident that the objective of the further analysis was to develop quantitative relationships among those major elements of the problem illustrated in Figure 9.4:

- Communications services are provided to users. The command defines these activities as:

  Transmission, which means the radiation and reception of radio signals, using the transmitters and receivers at these sites.

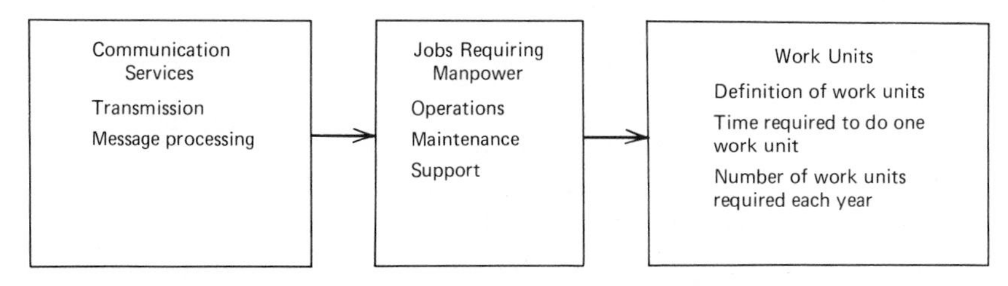

Figure 9.4. Steps in functional analysis.

Message processing, meaning all the handling of messages from the TCC through the fleet center, transmission, conversion of the transmitted message into paper copies of the message, and distribution.

- These services are provided by manpower working at various jobs. The initial classification of these jobs is shown in Figure 9.5, in accordance with a standard industrial model, as might be applied to the Apex Corporation of Case 2. All personnel in each work site could be divided into the seven work categories shown in Figure 9.5 based on the following work classifications:

All personnel can be classified as either "hands-on, direct labor" or supervisors/managers.

Most personnel are associated with any of three primary functions at a site:

Operations of equipment.

Maintenance of equipment.

Support, such as public works, galley personnel, and fire fighting.

There is also a fourth function, general management, whose responsi-

	Operations	Maintenance	Support	General Management
Direct labor				
Management				

Figure 9.5. Categories of personnel by function.

bility is the overall management of these three primary functions (operations, maintenance, and support). The general management function includes the officer in charge of the site, his assistant, secretary, and any person assigned to his office not doing a function that could independently be listed as support. Since this category is treated as an "overhead" item, we do not differentiate between direct labor and supervisor. Furthermore, an individual can spend his time in more than one category, such as a radio operator who also does certain support functions or a "working supervisor" who spends most of his time as the supervisor and also does "direct labor" as a radio operator during peak activities.

The next step in this model building process was to define more specifically the various jobs and work tasks each person does and to determine the time required per year for each job. Hence in this case representative sites were asked a number of questions concerning their work during the past calendar year and the personnel used for this:

- What direct labor jobs were done at the site within the functions of operations, maintenance, and support? (The analysis would start with direct labor jobs first, treating the supervisors as "overhead factors." To minimize the possibility that a site would omit any job, it was suggested that it also list all equipment on hand at the site and all personnel currently on duty [from the table of organization], using both of these as guides regarding jobs that are done.)
- How often were these jobs done?
- How many man-hours were needed to do each job each time it was done?
- If certain jobs were not done because of a manpower shortage, how many man-hours would have been required to do these jobs?

The next step in the model building process was to identify those factors that would tend to make their manpower requirements different among similar types of sites. These included:

- The different type of equipment and number of pieces on hand. (This was probably the most important factor.)
- The age of equipment. (Older equipment might require more maintenance.)
- Communications services provided (to distinguish the "busy" site from those that do not handle as much traffic). Communications service might include such factors as:

  Number of messages handled (the primary characteristics of service).

The type of communications service offered (such as full period termination or fleet broadcast).

Secondary characteristics, such as:

Equipment usage. (With message processing equipment there is a direct relationship between the number of messages processed and teletype usage, for example. However, a transmitter which is "on line" only radiates when a message passes through, yet it takes as many man-hours to tune and check its performance as one that is very busy. The same is true for a receiver.)

Equipment availability. (Equipment that is on "standby" until high traffic volume occurs may still require periodic preventive or even corrective maintenance.)

- The experience and training of the site personnel may differ, and this would be reflected in the man-hours required to do a given job.
- The physical environment of a site may influence man-hours in a number of different ways:

  Some sites are well ventilated or even air conditioned, others are not. This may affect equipment failure rate.

  Some sites have the equipment laid out in an efficient fashion so that the operator walks a minimum amount when going between equipment. At other sites the distances are greater.

  At some sites the same type of equipment may be located in more than one building. Since as a safety precaution there must be at least two people per shift when working with high-voltage equipment, decentralization of equipment tends to increase the number of people required. Travel time of a supervisor (between buildings) also increases.

Data Gathering. The next step was to gather these data at a representative site. Norfolk was chosen as the "analytical test bed." This information gathered at Norfolk was then used as the basis of a better understanding of the modeling problem and a "first cut" of the problems to be encountered in data gathering.

The results obtained were then written up to serve as both a detailed questionnaire for other participating sites and a first draft of the final report explaining to the sites how the data would be used.

Setting Priorities. Originally, we were hopeful that all the work functions performed at all communication sites could be analyzed. It soon became apparent that the job of collecting, validating, and correlating data among similar operations would far exceed the time and analytical manpower available. Hence it was decided to restrict the problem to those operations of a communications site that would most directly be impacted by the satellite system. These would be:

- The transmitter and receiver sites, whose complement of HF transmitters and receivers were expected to be greatly reduced.
- The fleet center, which handles a large number of messages manually when HF is used. There messages have to be manually checked for accuracy of transmission and many repeat messages sent because of the unreliable propagation path of HF during certain times of the day. When a high-quality transmission path provided by the satellite is introduced, much of this message processing will be done automatically by the computerized Naval Communications Processing and Routing System (NAVCOMPARS). Hence the manning at this type of site will depend on how much manual processing capability is desired as backup.
- The electronics maintenance division, which performed all the maintenance for the fleet center.

Other divisions that were not likely to be impacted by the satellite were:

- The telecommunications centers (TCC), since the same number of messages would probably still be sent.
- The support functions, such as administration and public works.
- Technical control units. Although a smaller number of communication entities would have to be controlled, this unit is fairly small now and hence not many resources would be apt to change.

In addition, it was decided to restrict the analysis to NCS Norfolk, Honolulu, Guam, and Naples, since these were to be the major stations of the system. They contained the largest number of equipment assets and personnel. Furthermore, any results obtained from these four stations would also apply to all other stations, since the equipment was essentially the same.

The set of detailed instructions showing how to gather the data and how the data were to be used in constructing the planning system was then sent to the other three sites. They were then visited by the planner and a member of the command's staff to explain the project, motivate them to participate, and answer any of their questions. Their data were then received and edited. From this the various planning factors were derived and the planning system logic was formulated. The total data were then published, along with recommendations for how the data were to be used.

9.5 ANALYSIS OF TRANSMITTER SITES

The first type of site analyzed was the transmitter site. This section describes the data collected and how they were used to derive planning factors for the operations, maintenance, and support functions. This description includes:

- What data were originally asked to be collected for each work function.
- How these data were intended to be used to generate the planning factors.
- What problems arose in obtaining the proper data.
- How these problems were overcome so that the data could be properly converted into planning factors.

The data originally requested from the Norfolk transmitter site and the way planning factors were generated are described in the following sections.

9.5.1 *Maintenance Requirements*

Preventive maintenance consists of a fixed amount of work to be performed on each transmitter on a fairly fixed schedule. Hence PM is a "deterministic" work function. Its man-hours can be estimated fairly accurately as a linear function (Figure 9.1). Corrective maintenance, on the other hand, consists of repairing different types of equipment malfunctions. Hence CM is a "random variable" and the actual man-hours required vary from year to year. Thus the linear function shown represents the mean or expected value of the estimated man-hours required.

9.5.1.1 Corrective Maintenance (CM)

The number, type, and description of all equipment maintained at the transmitter site over the past year were listed.[3] Next, all the work requests for CM action (i.e., repairs) over the past year were gathered and sorted by equipment type. From these data were obtained:

- The total number of failures for each type of equipment.
- The total man-hours used for CM for each type of equipment.

[3]If any of this equipment was not maintained for the entire 12-month period, the total number of equipment-months when it was available for use was to be indicated.

Dividing the total CM man-hours by the amount of equipment of each type would give the CM planning factor for each type of equipment (the average number of man-hours per year required for CM for each unit of equipment).

9.5.1.2 Preventive Maintenance (PM) Data

A similar approach to that used for the analysis of CM was employed. Again the available records were to be sorted by type of equipment and the following data supplied by the site:

- The total man-hours actually used for PM, by type of equipment.
- Any additional man-hours of PM required by the maintenance schedule but not actually expended because of manpower shortages.[4]
- The preventive maintenance standard (PMS), which is the man-hours per year that the Navy Electronics Systems Command had previously determined as the average time required to do the required PM. These times were to be taken from the maintenance requirements card (MRC), which delineated each of the required PM actions and the average time required. Each site had such a card for each type of equipment. We wished to gather these data for two purposes. First, to determine if the site knew what the standard was; second, to see how this standard time compared with the time the site said they required.

9.5.1.3 Data Problems

Even when the planner gives clearly written data-gathering instructions, which the recipients say they understand, the data may not be collected in the precise form anticipated. Hence the planner must double-check early in the data collection process so that any adjustments can be made before the data collection has been completed incorrectly. This was one of the advantages of using Norfolk as an "analytical test bed."

After the maintenance data requested from Norfolk were received and reviewed, it became apparent that the data collected were not what was anticipated. Several changes in the data collection procedure had to be made, and modifications were then incorporated in the instructions given to the other three stations. Similarly, when comparing the Norfolk data with those from Honolulu, the first overseas site visited, additional differences were noticed and also modified.

Here were the key data problems noted and how they were handled:

[4]It was explicitly assumed and validated that although some PM could be omitted, all CM eventually had to be done.

1. *PM work time records not available.* Unlike the record keeping for CM where a work request is completed for each CM action, the time actually taken to do each PM action was not recorded. To minimize the paperwork, the technician is given a PM work schedule and merely checks which scheduled PM work is done (or deferred until the next time if abnormally high CM activity prevents the work from being done). Since the maintenance cards indicated all scheduled PM that was supposed to be done, this PMS time could be used as the basis of the PM man-hours required. However, these time standards were only for working time and did not include time for other functions such as picking up and returning tools and travel time (called "make-ready, put-away").[5] The sites were asked to take work samples to confirm their estimates of the PM time required.
2. *Sites did not agree on PMS standard.* As a result of their measurements, several sites provided their own local PMS standards; these differed from the maintenance cards for the following reasons:
   a. Some sites did their PM on a batch of equipment and hence could do the total job in less time than the PMS standard allowed.
   b. Although some PM was specified on a periodic basis (such as monthly or quarterly), some was specified "as required" and hence was left to the option of each site.
   c. Although sites may have had the same equipment number, the model number might not be the same and hence the PMS standard might be different.
3. *Nonuniform definition of PM.* If a part replacement was required during a PM inspection, some sites counted this time as PM; others counted it as CM. The command was in the process of changing its accounting system to include this as CM (so that the PM requirements could be roughly the same). Hence all sites had to be instructed to use this same accounting system, so that data could be compared on a uniform basis.
4. *Additional PM done at some sites.* Some sites did more PM than was required on the maintenance card. For example, two sites overhauled a portion of their transmitters each year. The other two said they would like to but did not have the manpower. This problem was handled by defining three types of maintenance:
   a. Conventional PM work.

[5]This time is included in the total CM time recorded on the work request.

b. Other non-CM work.

c. CM work.

Conventional PM work was defined as the minimum PM actions specified on the maintenance card and did not include any additional non-CM work a site did because it felt it was necessary. The conventional PMS man-hours were defined to include all maintenance man-hours used, including the man-hours required for "make-ready and put-away." The operator and the technician each did part of the PM actions; consequently it was necessary to know the share of each because this division could differ at each site from that recommended on the card.

Other non-CM work consists of a number of maintenance jobs (such as overhauls) that are done at a number of sites but are over and above those listed on the maintenance card. To identify these jobs and still allow the systems planner at the command the option of including those work functions desired in the analysis, we structured all this nonstandard, non-CM work and the man-hours each job required as "additional jobs." But to obtain official credit for such work as part of the PMS system, the command would have to approve such recommendations and submit them to the Naval Material Command for its approval.

CM work consists of all CM actions, including replacement of parts during PM.

9.5.1.4 Analysis of Maintenance Data Collected

The objective of this analysis was to compare the data received from the four transmitter sites on some uniform basis. The first step was to compare the data received from each of the four sites against one another and then against the PMS standard. This analysis was done as follows:

- An inventory of the numbers and types of all equipment being maintained at each site was made. These data were also useful in showing the potential reductions in inventory (and thus in maintenance requirements) possible at each site.
- The total man-hours per year needed for PM maintenance, including extra jobs for one unit of each type of equipment,[6] and the PM standard used at each site were calculated. There was poor correlation between the PM required at different sites because of the man-hours spent on the extra non-CM jobs by the different sites. There was also

[6]An indication was made if the site did extra jobs (at additional man-hours) for this type of equipment.

poor correlation between the PM required and the PMS standard when extra non-CM jobs were done. For this reason a calculation of the conventional PM required was made.

- The total man-hours per year needed for conventional PM maintenance (not including the extra jobs) for one unit of each type of equipment[7] were studied. These data were compared against the PMS standard and good correlation was found, since the sites indicated that they estimated the PM required as a function of the PMS.
- The standard man-hours needed for each unit of equipment for conventional planned maintenance, as specified on the maintenance cards (as done by operators as well as maintenance technicians), were listed.
- The total man-hours per year needed for CM for one unit of each type of equipment were calculated. These data varied greatly among sites. Since the frequency of failure and the time to repair can both be considered random variables, we would expect such large variations among sites for any one piece of equipment.

To analyze whether one site consistently required more man-hours for CM (or the other maintenance functions) than the other sites, it was decided to analyze the total man-hours required for each of the maintenance jobs, at each site, for all equipment at that site. Basically, the analysis consisted of two types of data comparisons. First, the man-hours reported required by each site to do a common set of work elements were compared on a common basis. Second, official Navy standards (approved by Navy headquarters) were also identified, and these were compared with the requirements stated by each site.

Table 9.1 shows the results of this comparison.

First, consider the intersite comparison. The analysis consisted of calculating a number of ratios using the PMS standards for the particular set of equipment at each site as the uniform basis of comparison, thus eliminating differences in the numbers and mix of equipment among stations. In the analysis:

- Line 1 of Table 9.1 shows the sum of PMS standard man-hours for all equipment at each site.
- Line 2 shows the total man-hours required by each site to do all PM jobs, both the conventional PM and all extra non-CM jobs (both recurring and nonrecurring). Norfolk included a 20% factor for

[7] A list of such jobs and the man-hours required for each was also compiled and presented to the command for review.

Table 9.1 *Results of Maintenance Analysis*

(1)	(2)	(3)	(4)	(5)	(6)	(7)
	Honolulu	Guam	Norfolk	Italy[a]	SCA Approved	OP-124
1. PMS standard (man-hours/year)	20,536.0	17,058.4	8,341.4	1417.0	—	—
2. Total PM required (man-hours/year, including all extra jobs)	15,199.7	28,066.0	18,750.4	1694.6	—	—
3. Extra non-CM jobs (man-hours/year)	2,483.3	11,525.0	8,740.5	277.6	—	—
4. Conventional PM (man-hours/year)	12,716.4	16,541.0	10,009.9	1417.0	—	—
5. CM required (man-hours/year)	5,666.3	17,227.9	26,513.4	537.3	—	—
6. (PM required + CM required)/PMS	1.0	2.7	5.4	1.6	3.0	2.94
7. PM required/PMS	0.7	1.6	2.2	1.2	1.5	1.47
8. Conventional PM required/PMS	0.6	1.0	1.2	1.0	1.5	1.47
9. Extra jobs/PMS	0.1	0.7	1.0	0.2	—	—
10. CM required/(PMS × 1.47)	0.2	0.7	2.2	0.3	1.0	1.0
11. CM required/PMS	0.3	1.0	3.2	0.4	1.47	1.47

[a]Analysis based on incomplete data submitted.

"make-ready and put-away" and "work breaks" in its PMS requirements; the other sites estimated they do the conventional PM work in PMS time, including the breaks, make-ready, and put-away. All four sites indicated they took work samples as the basis for their estimates.

- Line 3 shows the man-hours used for the extra non-CM jobs done at each site.
- Line 4 shows the man-hours used to do the conventional PM jobs (the differences between lines 2 and 3).
- Line 5 shows the total man-hours required for CM.
- Line 6 shows the ratios of total maintenance requirements by each site (including all extra non-CM jobs) to the PMS standard (lines 2 plus 5 divided by line 1). This was the most important result.

These ratios of line 6 were then compared with Navy maintenance standards approved by Navy headquarters. Although these standards were constructed for communications equipment used by the fleet (rather than shore stations), they are the best data available. The standards were obtained as follows:

- The PMS standard listed on the maintenance card is the official requirement for the PM actions. But the PMS standard is for working time only; an additional 17% is allowed for personnel fatigue and delay (breaks and allowable reduction in production rate during sustained performance).
- The PMS standard is the average working time allowed and does not include make-ready and put-away time, which is allowed as an additional factor. The exact amount of this time is a function of the distance between where the tools and parts are kept and where the equipment is located and how many times the same tools are used in maintenance at that location. Navy headquarters permits an average factor of 30% for the fleet and has indicated it would also permit a 30% factor for shore stations until they could conduct a more thorough study.

Thus the total Navy PM requirements for work specified on the maintenance card is 1.47 times the PMS standard.

Although there is no Navy CM standard similar to the PMS standard, there is a headquarters policy used for fleet-manning purposes. This policy states that for every hour of CM action, 1 hour of PM action is needed for electronic equipment. Headquarters further interprets this policy for determining billet requirements by estimating the CM man-hours required for the fleet as being equal to the total PMS man-hours required. Again, they will permit this factor to be used as the Navy requirement for shore stations until they can conduct a more thorough study. Thus the total maintenance requirement is 2.94 times the PMS time. Additional man-hours for extra non-CM maintenance would appear on maintenance cards when officially approved by Naval Material Command.

Since the security cryptologic agencies use some of the same type of equipment and their maintenance standards had already been approved by DOD, these standards were also obtained for comparison. These were found to be three times the PMS man-hours, reasonably close to the headquarters standard of 2.94 times PMS standard.

With the preceding discussion in mind, we next calculated the ratio of each of the site's total maintenance requirements to PMS standard (line 7 of Table 9.1) and compared each ratio with the ratio for the derived Navy requirement (2.94). Honolulu and Italy require much less than the Navy requirement. Guam is 92% of the Navy requirement. Norfolk, by contrast, is 184% of the Navy requirement. All sites except Norfolk can do all their current maintenance jobs and stay under the Navy requirement. However, Norfolk indicated it could meet the Navy requirement in the future.

Because of large differences in ratios among the sites, several other diagnostic analyses were also made at the next level of detail. The first was to separate conventional PM work from the man-hours required to do the extra, non-CM jobs that were being done. A comparison among sites of the man-hours required for conventional PM is best shown by taking the ratio of these man-hours required to the man-hours associated with the PMS standard. These ratios are shown on line 8 and can be seen to be well within the headquarters or SCA approved ratios. The ratios associated with the extra jobs are shown in row 9 of Table 9.1. Although Honolulu does extra jobs (though not as many as Guam and Norfolk), its total PM is only 10% of the PMS standard. Italy requires 20% of the standard. Guam requires 70% of the standard, and Norfolk requires 100%.

A second analysis was concerned with finding the ratio of CM man-hours to the official Navy man-hours allowance for PM and then comparing this ratio with the Navy requirement (unity). This is shown in row 10 of Table 9.1. Notice that Norfolk is also very high in this respect. Row 11 of Table 9.1 provides a similar ratio of CM required to the PMS standard, rather than to the Navy PM required. Again Norfolk far exceeds both the standards and the other sites.

These results show that the four sites can be placed into three classes:

- Honolulu and Italy[8] perform about the same—that is, few man-hours for extra PM jobs—and their CM required only a small percentage of the Navy PM requirement.
- Guam spends 70% extra on non-CM jobs, and its CM requirement is 70% of the Navy PM requirement (well within the 100% requirement).
- Norfolk, by contrast, spends 100% extra man-hours on non-CM jobs, but its CM is 220% of the Navy requirement. This example seems to violate the rule of thumb that doing more PM reduces CM. Much higher CM is the main reason why Norfolk's maintenance manpower needs are 184% of the Navy's requirement and 540% of Honolulu's.

9.5.2 *Operational Manpower Requirements*

The operational manpower planning factors were derived using the same analytical and model building approach as was used in deriving the maintenance planning factors. That is, data were gathered answering the following questions:

- What operational work do you do?
- How often is this work done?
- How many man-hours are required to do each unit of work?

9.5.2.1 Description of Operational Work

The entire operational work load was found to consist of:

- Tuning and returning transmitters in use (not those on standby or unavailable).
- Quality control (QC) checks.
- Other operational activities, including tuning and readjusting a transmitter following power outage, on-the-job training, and excess travel of O&M personnel.

Each of these jobs was then analyzed in greater detail to apply the model-building techniques.

[8]Italy had maintenance data available on only three transmitter types and some minor equipment; the analysis was based on that equipment. However, this set of equipment accounted for 81% of the total maintenance requirement, as measured by the PMS standards, a good sample of the total load.

9.5.2.2 Analysis of Tuning and Retuning Job

The first step in analyzing this job is to understand the sequence of activities involved. The total transmission system consists of a number of transmitter and receiver pairs tuned to the same frequency. These are activated as needed to meet the communications traffic demand. This system demand depends on three major factors:

- The number of ships in a geographic area[9] requiring communications.
- The amount of communications required (primarily messages, but also voice).
- The type of service afforded (full-period termination versus fleet broadcast).

Based on this, the Naval Communications Area Master Station (NAVCAMS) serving that area assigns anywhere from a part of a channel to several channels to a ship, depending on its traffic. This allocation roughly relates to the size of the ship. For example, shore-to-ship traffic for all smaller ships, such as destroyers, is treated as a batch and sent in series over one channel (the fleet broadcast) using a number of different transmitters, each tuned to different frequencies.

Large ships, such as cruisers or aircraft carriers, have a sufficiently heavy traffic density and are sufficiently important to have several 500-characters-per-minute teletype circuits or channels dedicated to their use. These are called full-period terminations and are always available. If the ship's traffic increases, additional channels are assigned.

When an additional transmitter is required, the appropriate technical control unit assigns the frequency and informs the appropriate transmitter site of its decision over a teletype "orderwire" used for network control purposes. The transmitter site then tunes the transmitter to the desired frequency, selects an appropriate antenna, connects the message signal line into the transmitter, and informs tech control via the orderwire when they have completed the activation process. Periodically, an active transmitter may malfunction. When this occurs, a replacement transmitter is activated using the process previously described.

Although HF is an effective communications method, its propagation path changes with solar activity, its major deficiency. Thus a frequency that may be good during the daylight hours may not be good at night; hence its frequency may have to be changed.

All these conditions require tuning or retuning a transmitter. The model

[9]There are four major communications areas: Atlantic, East Pacific, West Pacific, and Mediterranean.

to be used in determining the manpower required for the tuning and retuning function incorporates the following features:

1. There are two different types of communication systems: those that operate continuously and on the same frequency (such as the fleet broadcast) and those that operate intermittently and whose frequency may be changed (the full-period termination).
2. A continuous system is operated until it malfunctions, at which time its replacement is activated and tuned.[10] Thus the number of tunings per year should be related to the amount of transmitter usage. Theoretically, it should be equal to the total number of usage hours divided by the mean time between failures for each type of transmitter.
3. Intermittent transmitters should require the same number of original tunings as continuous transmitters plus the number of times that the transmitter was brought into service for a ship. In addition, the transmitter would have to be retuned to a different frequency at least two times per day of usage to meet the different propagation conditions.
4. Thus the first planning factor for this operation would be the average number of tunings and retunings required per 1000 hours of usage for each system and transmitter type. Then by estimating next year's usage (using the current usage as a base), the number of tunings and retunings for each system-transmitter type could be calculated. This assumes that each future system-transmitter will require the same number of tunings and retunings per hour of usage as the previous year.
5. The man-hours required per year for each system-transmitter type will be equal to the product of the number of tunings and retunings per year and the average time required to tune a transmitter of this type.
6. Finally, the total man-hours required would be equal to the sum of the man-hours required for each system-transmitter type.

9.5.2.3 Data Requested from Sites

With this model in mind, each site was asked to provide the following data based on the past year's operations:

1. A complete list of each different type of communications "system" used (fleet broadcast, full-period termination, etc.) and the different types of transmitters used for each communication system. These two

[10]PM generally occurs when a transmitter is already inactive, decreasing the number of transmitter replacements required.

pieces of data yielded a descriptor called the "communications system–transmitter type."

2. The number of operational hours of usage and the number of tunings and retunings required for each communications system–transmitter type.

The basic source of these data was the station composite log, since every time that a transmitter is activated or deactivated, this must be recorded in the station log kept at the transmitter site.

Since the data collection process involved could be extensive, sites having a large number of entries were given a choice:

1. Record all pertinent data from the station log.
2. Record only a sample of the data and extrapolate the results. For this latter operation, the sites were given a set of 37 days (a 10% sample choosing the days randomly, using a table of random numbers). Sites were asked to describe the method they used in arriving at the data provided. Guam used a 6-month sample and doubled this.

In addition to the number of tunings, each site was also asked to submit the average time spent in tuning or retuning a given transmitter type, including the time spent on the orderwire, on logging the occurrence in the station log, and on selecting a new antenna when required. Norfolk's original submission of this time was based on their work samples. However, it was later found that since the orderwire messages record all the events from the initiation of the request by tech control for a tuning or retuning until the final confirmation of the completed job by the site, a more accurate estimate of the total time required could be obtained from sampling a number of orderwire messages.

9.5.2.4 Analysis of Data

The first step taken after receiving data from the sites was to review the process by which it was collected by each site to make certain that the data were comparable. Obviously, any arithmetic calculations that the site made were checked for accuracy. Sometimes the data had to be adjusted, as illustrated by the following two examples.

First, Guam indicated that the number of tunings and retunings they submitted was based on a 6-month sample (1 April through 30 September). Presumably, the sampled data were extrapolated to 12 months by doubling it. If this were the case, their only peak operations (a major fleet exercise during the first 3 weeks of July) would be counted twice. Therefore a factor to correct the sampling error was generated for Guam.

The second example involved the average total time required by each site to tune and retune each type of transmitter (Table 9.2). Guam generally took the least time, and Norfolk the most. Since Norfolk indicated that its data were based on sample work measurements rather than on orderwire logs, as the other sites did, an audit of Norfolk's log was also made to validate their data further. The log data results are listed in column 5. Since these times turned out to be appreciably smaller than the original Norfolk data (column 4), an arithmetic mean of both sets of data was used as the final Norfolk data (column 6). Finally, a weighted mean time for all four sites was calculated (column 8). The weighting was based on the number of tunings and retunings of that transmitter occurring at each site. When the entire set of data was reviewed by the command, it was decided that the lower Guam data would not be representative for all sites. Hence the final planning factor for the unit times was calculated as the weighted arithmetic mean of data from Honolulu, Norfolk, and Italy.

9.5.2.5 Simplifying the Model

Although the approach described may be the most accurate way of calculating the total man-hours required for the job, this approach not only required a large tabular data base but a large amount of calculations. Thus we also tried to develop a simpler way to relate the total number of tunings and retunings to total operating hours.

To develop this more simplified model, the number of tunings and retunings made for all systems at each site was plotted against the amount of transmitter usage.[11] Separate plots were made for continuous and intermittently operated systems (Figure 9.6).

Although the four data points plotted for the intermittently operated systems follow a linear function, it does not pass through the origin, as expected. More study is needed to determine why. But because of the good correlation obtained, this function apparently could be used (instead of the original set of planning factors for each system-transmitter type) as long as the mix of systems is not changed radically at a different site. Further analysis of this model is needed to obtain additional validation.

The model of continuous operations seems to hold for three sites but not for Honolulu,[12] which required fewer tunings and retunings than the function predicts. To determine why Honolulu was different from the

[11] Each 8760 hours of transmitter use per year is one full-time equivalent transmitter.

[12] Again this function does not pass through the origin for some unaccountable reason.

Table 9.2 *Tuning-Retuning Unit Times (Minutes)*

(1)	(2)	(3)	(4)	(5)	(6)	(7)	(8)
			Norfolk				
Transmitter type	Honolulu	Guam	As Submitted	Log	Mean	Italy	Mean all Four Sites
FRT-39	11.1	5.9	12.5	8.8	10.7	9.5	9.8
FRT-40	11.7	7.4	15.5	12.8	14.2	10.6	10.2
FRT-70	—	4.5	—	—	—	—	4.5
FRT-72	13.0	6.4	8.5	—	8.5	—	8.5
FRT-83	—	—	7.5	4.2	5.9	5.3	5.8
FRT-84	—	3.1	7.5	4.9	6.2	—	4.8
FRT-85	—	3.8	7.5	—	7.5	—	4.2

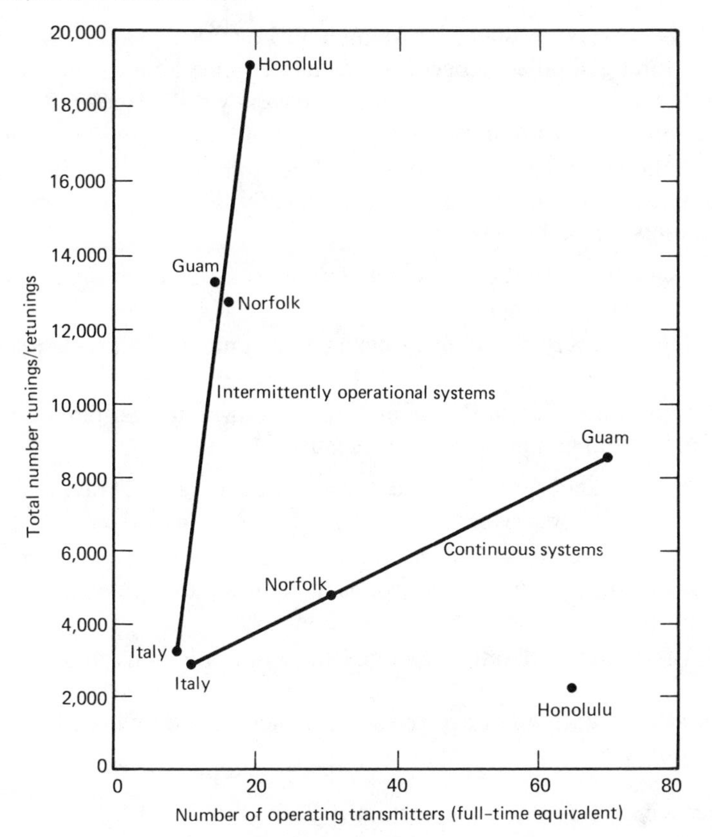

Figure 9.6. Number of tunings and retunings required for different types of transmitter systems.

other sites, the ratio of the number of tunings and retunings per 1000 hours of operation was calculated versus the number of operating hours for each system-transmitter.

Most of the systems unique to Honolulu have a much lower ratio of tunings and retunings to operating hours than do the other stations. It may be possible to treat these communications systems as a special category, thus permitting more simplified models than the tables to be used for all transmitter sites. Further work is needed for this validation.

A partial analysis of why these systems differ indicates that instead of dividing the entire set of systems into two classes (continuous and intermittently operated), three classes should be considered:

1. Continuously operated systems, such as multichannel broadcast, that

always operate on the same frequency. For this class the only reason for a tuning should be because of maintenance actions (CM or PM), and the only reason for a retuning is frequency drift. Hence the number of tunings and retunings per operating hour should be very low.

2. Continuously operated systems, such as some point-to-point circuits, that undergo frequency changes periodically. For this class tunings and retunings occur because of:
   a. Maintenance actions, expected at the same rate per operating hour as continuous systems.
   b. The number of frequency changes (retunings) occurring per operating hour.
3. Intermittently operated systems, such as full-period terminations, that undergo tunings and retunings because of:
   a. Maintenance actions, expected at the same rate per operating hour as continuous systems.
   b. The number of activations per hour of system operating time (i.e., the more often the system is activated, the more tunings are required).
   c. The number of retunings once the system is activated.

Thus all intermittently operated systems need to be reviewed and these factors introduced:

- Average uptime once the system is activated.
- Average uptime at a given frequency once the system is activated.

Three other factors were also considered:

1. Type of transmitter used. (This would influence how often maintenance actions [and hence additional tunings for the replacements] are required.)
2. Number of ships in the area. (Communications Area Master Station Norfolk indicates that as more ships enter the area, the number of transmitters operating is merely increased. The number of ships thus does not seem to influence the number of tunings and retunings per operating hour.)
3. Quality control checking. (The more QC checks that are made, the greater the chance that transmitter drift or other deviations will be detected, requiring transmitter adjustment [i.e., retuning as defined here]. QC checking policies differ among sites and may cause non-

uniformity among sites in the tunings and retunings needed for any mix of systems.)

Finally, once the number of tunings and retunings has been estimated at a site, the average time required per tuning and retuning needs to be determined. This will be calculated as a weighted average of the various times required for each transmitter type within each communications system class (as developed in the preceding discussion). In this case the weighting is directly proportional to the operating hours associated with that transmitter type.

For example, consider that within a class of communications systems at the site, we have estimated these number of hours of transmitter usage for all continuous systems:

FRT-39: 10,000 hours
FRT-40: 20,000 hours
FRT-83: 30,000 hours

Also assume that the command standards for tunings and retunings are

FRT-39: 10 minutes
FRT-40: 12 minutes
FRT-83: 6 minutes

The weighted average tuning and retuning time for this transmitter mix is

$$\frac{(10{,}000)(10 \text{ minutes}) + (20{,}000)(12 \text{ minutes}) + (30{,}000)(6 \text{ minutes})}{60{,}000} = 8.7 \text{ minutes}$$

9.5.3 Analysis of Other Operational Jobs

A number of other jobs associated with transmitter operations were also identified. These were

- Quality control of transmitter equipment and the communications circuits (either land lines or microwave) that carry the signal from the operations center to the transmitter.
- Retuning a transmitter following a power outage (which occurs several times a week in the Pacific island sites).

- On-the-job-training (OJT) for all new personnel assigned to both operations and maintenance.

In all cases the same approach as that used for previous analyses was again used:

- The system demand function consisted of a definition of what the job consisted of and how often per year the job was required.
- How many man-hours were required to do the job satisfactorily each time it was required.

9.5.4 Support Manpower Requirements

From the total list of billet titles classified as doing support jobs, three types of support work loads were identified:

1. From the table of organization, it is noted that many people were associated with nonoperations and maintenance jobs, but in the support function (cooks, bakers, public works, etc.). Since providing this type of support was their primary duty, this support work load was defined as support primary duty work load.
2. In addition, many nonsupervisory personnel, particularly those in operations and maintenance, do support-type work, such as cleaning, manning the volunteer fire department, and guard duty, in addition to their primary duty. Since this work is collateral to their primary duty, it was defined as support collateral duty work load.
3. Finally, the third support work element was that done by nondirect labor supervisors; this was defined as supervisory work load.[13]

To analyze all these different support jobs in the same degree of detail used in the previous analyses would have required a great deal of effort, more than was available. Consequently, it was decided to develop the planning factors for these support jobs in a simpler fashion:

1. Each job was to be clearly defined so that similar jobs could be compared across sites.
2. The man-hours required for each job were compared among sites. Any large differences that could not be explained in terms of the environ-

[13] Note that work of the officer in charge and his staff is classified as general management, a separate work category.

mental differences at the different sites would be questioned and some explanation required.

9.5.4.1 Support Primary Duty Factors

Table 9.3 is a partial list of all support primary duty billets filled at the four sites. This comparison of similar support primary duty jobs was accomplished as follows:

1. A list of all job titles and tasks associated with each job was assembled by the planner from data available from previous manpower evaluations.
2. The command decided on a standardized list of job titles preferred for future use. Column 1 gives these position titles.
3. The total list of preferred titles and work tasks associated with each was sent to all sites as a guide for standardization.
4. For each filled billet, each site listed the job title they used that most closely corresponded to the job title of column 1. These are shown in columns 2 through 5.

If the site uses the same title as that shown in column 1, "same" is indicated. Support billets that did not correspond to a billet from the master list were preceded by the letter used to identify the position submitted by that site.

After each site's billet title is the number of persons now in that billet (when that number is more than one). Also indicated is the percentage of time (when under 100%) that the person is involved in direct labor. Part of this direct labor time may be spent in collateral duty support jobs.[14]

None of these billets was analyzed in detail during this study. Instead this list was defined as the support primary duty planning factor.

Most of these billets were organizationally located in the support divisions of each site. Those that are in operations or maintenance at a given site are so designated.

Because the number of billets is relatively small, the total analysis effort to assign these support billets systematically can be reduced by using the following procedure:

- Validate that the work function for the billet listed is required at each site. It must also be confirmed that the support activity cannot be done by the station's public works department or other Navy support activities because of the site's distance from a regular Navy base.

[14] See the next section for an analysis of these jobs.

Table 9.3 *Support Primary Duty Billets*

Master Billet List	Honolulu	Guam	Norfolk
OIC office			
2. Clerk (Typing)	Same	Clerk (typist)	
3. Military clerk	Personnel petty officer		
4. Communications specialist			Same (80%)
5. Administrative clerk	Administrative assistant (50%)		Same
6. CMAA	CMAA/1st Lt. division chief (25%)	Same (50%)	
		(T) MAA (90%)	
		(W) MAA force	
		(U) Guard mail orderly (2)	
		(V) Security force (2)	
Supply division			
8. Supervisory supply clerk	Supply officer (50%)	Supply clerk—50 Dept.	PO inc ready supply store (50%)
9. Supply clerk	Same		Same
10. Storekeeper	Assistant supply officer		
11. Galley chief	Food services petty officer		
12. Galley captain	Provisions storekeeper		
13. Watch captain			
14. Galley watch			
15. Mess attendant			Food service worker
16. Cook	Same (2)		Same (90%)
	(H) Exchange operations supervisor (40%)		(I) Asst. resident asst. navy exchange officer
	(I) Exchange operator (2)		(J) Sales clerk
	(D) ATCU supply clerk		

- Determine how many full-time equivalent workers are required for this work function at each site. This depends on the size and layout of each site and whether the function is (or can be) provided to any extent by the main station or by other Navy support services (such as regional medical services).

In this way judgment can be systematically applied in allocating these billets.

9.5.4.2 Support Collateral Duty Factors

Table 9.4 is a composite summary of support collateral duty jobs now being done at the four sites and constitutes this support planning factor. Column 1 briefly describes the type of job involved, such as cleaning. This is followed by a list of support jobs, by number, as a cross reference to the data submitted by each site, and the total man-hours per year required to do each job clustered in that job category. A more detailed description of those collateral support jobs, including the method for calculating support, was obtained from each site and used as the basis of Table 9.4. This detailed description of each job included the work unit measure, the hours needed by one man to complete one work unit, and the number of work units done per week by all the men involved. Thus 52 times the product of the number of times each man does a work unit per week and the number of men doing them simultaneously is the total man-hours per year required for the job.

Based on the data submitted, the command could review these lists and decide

- Which collateral jobs must be done, and how often.
- Which jobs are really part of service diversions or off-hours' activities and not counted as productive work.
- How many man-hours are needed for each job. (Navy headquarters stressed that requirements can include only working time; for "on-call" duty, only actual working time can be counted.)
- Who should do the work—operational or maintenance personnel (or both), primary support duty, or outside personnel.

Requirements for collateral support stated by the sites absorb a substantial part of the division's total direct labor. Further analysis and validation of these requirements by the command is, therefore, very important.

9.5.4.3 Supervisory Factors

The third support planning factor involved determining the number of supervisors required for each of the organizational units at a site. Since

Table 9.4 *Support Collateral Duty Jobs*

Job Type	Honolulu		Guam		Norfolk		Italy	
	Job Number	Total Man-hours	Job Number	Total Man-hours	Job Number	Total Man-hours	Job Number	Total Man-hours
On-the-job training	—	9788					5	2208
Technical (acceptance testing)			20	1750				
Test equipment			21	480				
Cleaning	4,5,6,7	9360	1–5,15,26, 32,35,36	6679	3	3458	2–3	873
Military watch (security tours, fire tours, telephone watch, etc.)	1–3	11830	23,24,25	—	1–2	3252		
Inspections (fire, material, etc.)	8,9	84						
Pickup and deliveries	21,22	1599	10,11,28	720	5	1248		
Committee meetings	14–20	865						
Counseling	13	780						
TAD (except cleaning duties)			13–15,33,34	3520	6,7	3312		
Vehicle, equipment, and facility care			7,9,37,38	1068	9–11,14	7199		
Record keeping			22,30	2379	13	1378		
Storm condition			6	540				
Equipment removal			8	852				
Technical control coordination					8	1875		
Librarian					12	546		
Various service diversions and training (nonavailable time)	10–12	1957	12,16–19,29	3163	4	520		
Power outages							1	76.6

insufficient analytical resources were available to analyze the various supervisory jobs involved, it was decided to calculate a supervisory overhead rate for each unit, to compare these among sites, and to determine an acceptable standard. This supervisory overhead rate was defined as the ratio of the total number of full-time equivalent supervisors in each organizational unit being analyzed to the full-time equivalent nonsupervisory personnel currently assigned to that unit.

Data for these calculations were provided by each site as their estimate of what proportion of each person's time was devoted to supervision versus "direct labor." Thus an assistant watch supervisor was a working supervisor and estimated as devoting, say, 20% of his time to supervision.

Supervisory overhead calculations were then made for each of these organizational components:

- Total site overhead.
- General management (percent of total direct labor).
- Watch operations (percent of total direct labor, including maintenance personnel).
- Total operations division (percent of total watch and day operations personnel).
- Maintenance division (percent of maintenance personnel, excluding those on watch).

The results of these supervisors' overhead calculations are shown in Table 9.5. Comparing the total supervisory and general management overhead rates among sites shows that they are reasonably close to one another, but much higher than industry (say 5 to 15%). Further analysis of the individual overhead rates for the individual organizational components shows little consistency among sites; moreover, overhead rates are

Table 9.5 *Supervisory Overhead Analysis Results (Percent of Direct Labor)*

	Honolulu	Guam	Norfolk	Italy
Total supervisory overhead	24.9	25.8	22.5	20.0
Watch operations	19.7	23.7	46.2	8.1
Total operations division (including day operations)	24.1	41.0	67.7	11.1
Maintenance division	40.0	16.5	4.8	9.4
General management	1.4	1.4	3.1	10.0

abnormally higher than expected. Discussions with supervisors at the transmitter and receiver sites regarding the division of work between the supervisors and workers revealed that:

- The supervisor works side by side with the workers, doing some of the operating work load previously described, particularly during busy hours.
- The only operating work load not listed and done by the supervisor consists of (1) providing on-the-job training; (2) spot-checking the quality of work of his personnel; (3) being available as the senior person for any problem that arises during the watch; and (4) evaluating personnel.
- Although the supervisor has overall responsibility for proper operations during the watch, he delegates this responsibility among all watch personnel. Thus the only man-hours this ultimate responsibility really costs are in performing the tasks described in the preceding item.

Since the overhead ratios calculated in Table 9.5 were obtained from site estimates based on job titles and not on an analysis of work function, their accuracy was questionable. Experience indicates that the overhead ratios are probably smaller than those shown in the table. A satisfactory estimate of the supervisory overhead planning factor may be obtained in one of these ways:

- For supervisors who do not perform direct labor, determine their work functions to substantiate the need for a full-time position, given the size of site and the number of direct-labor personnel. For example, large sites might require an assistant officer in charge; small sites might require only a chief in charge.
- For working supervisors estimate the amount of supervisory tasks not already being counted under direct labor (or listed among the support jobs) and estimate the time required to do them. Recalculate the supervisory overhead ratio as before. Navy headquarters concurred that, as in industry, the supervisory overhead ratio probably should be 5 to 15%.

9.5.5 Standard Workweek Planning Factor

To convert the total number of man-hours required to the number of billets required, the characteristics of a standard workweek must be specified.

9.5.5.1 Standard Workweek for Military Personnel Ashore

The standard workweek for military personnel at CONUS activities and oveseas bases where dependents are authorized is 40 hours. Included in the workweek is an allowance for service diversions; this allowance provides for quarters, sick call, personal business, and so on. The 40-hour standard workweek for the military consists of:

	Hours/Week
Service diversion training	4.83
Leave	1.85
Holidays	1.38
Time available for work	31.94
Total	40.00

9.5.5.2 Standard Workweek for Civilians

The standard workweek for civilians is 40 hours. Training includes classroom lectures, on-the-job instruction, and safety indoctrination. Diversions include minor unavoidable delays, such as fire drills, chest X-rays, blood donations, and so on. The 40-hour standard workweek for civilians consists of:

	Hours/Week
Leave	4.60
Holidays	1.38
Training	0.22
Diversions	0.44
Time available for work	33.38
Total	40.00

Similar data were also obtained for the standard workweek for other categories, such as firefighters, and for military ashore where dependents are not authorized.

9.5.6 Summary of Planning Factors

The previously described analysis provided a set of planning factors for each transmitter site that can be arranged into the 16 categories listed in Table 9.6. Most of these factors represent the man-hours required to do one unit of work (such as CM or PM or tuning one transmitter). To obtain the total man-hours per year associated with this job, one must take the product of the planning factor, the frequency in which the job must be

Table 9.6 *Transmitter Planning Factors*

Maintenance
1. Conventional operator planned maintenance subsystem (PMS) factors
2. Conventional technician PMS factors
3. Make-ready, put-away time factor
4. Other noncorrective maintenance (non-CM) factors
5. CM factors

Operations
6. Operational usage factors
7. Tunings and retunings to usage factors
8. Tuning and retuning unit time factors
9. Quality control (QC) checks factors
10. Other operational activities factors

Support
11. Support primary duty factors
12. Support collateral duty factors
13. Supervisory factors

Navy headquarters work standards
14. Personal fatigue and delay (PF&D) factor
15. Standard workweek

done, and the number of units of equipment involved (if applicable). The frequency with which the job must be done is sometimes specified by standard operating procedures (e.g., PM schedule, as required for CM). Or it is determined by the equipment usage (e.g., the number of tunings depends on the number of hours that a transmitter is operating). In the latter case, planning factors were developed showing the usage of transmitter systems last year at each site.

Two types of planning factors were developed, each for a different purpose. The planning factors derived from the original data were called localized planning factors. They represent the manpower required at each site to do the same job. However, this manpower may differ from site to site because of certain site differences, such as physical environment, personnel quality in terms of training and experience, and age of equipment. These localized factors were to be used when a specific site (or one similar to it) was undergoing realignment, such as reducing its amount of equipment or equipment usage.

If the environment of an individual site cannot be related to one of the four sites analyzed, its manpower planning factors can be estimated in another way. Each set of four localized manpower factors was also converted into one command planning factor by taking a weighted average of each of the four localized manpower planning factors. The resulting factor relates to an "average environment" within the command rather than a specific site. Such command planning factors were to be used for manpower calculations for all other sites, excluding these four. Thus if a total annual command budget for all sites was to be prepared and a particular site's environment could not be assumed to approximate one of the four sites for which data were available, the best data to use for these sites would be those comprising the weighted average command planning factor. Since a large number of different sites are included in such calculations, individual deviations will tend to compensate for one another.

Individual estimates of equipment usage should be made for each site. However, last year's data are readily available at the site.

9.5.7 Developing the Planning System

Figure 9.7 illustrates the procedure developed by which a communications system planner from the command may calculate the number of billets required for a particular communications system. A detailed pro-

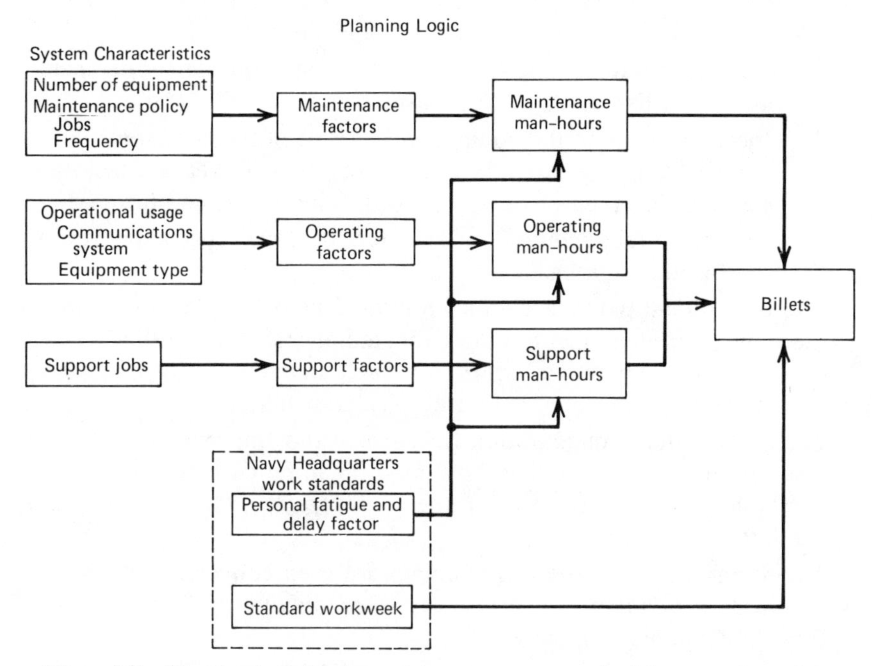

Figure 9.7. Planning logic for determining manpower required for transmitter sites.

cedure (contained in Appendix 9.A to this chapter) was developed as a user's manual for use within the command. This planning procedure may be summarized as follows.

The communications systems planner performs a set of preliminary analyses, examining the need for communications services of various types. Factors he considers include geographical coverage, number of messages per unit time to be handled by each communications system (such as full-period termination, broadcast, etc.), division of responsibilities among worldwide sites, operating loads to be accommodated for both peak operations and the entire year, and the division of these loads between satellite and HF equipment. Further system design considerations are then made, culminating in the configuration of alternative communications system designs.

1. For each alternative being considered, the communications planner must specify the following as inputs to the manpower planning system:
   a. The set of equipment (both numbers and types) to be in inventory at the station being considered.
   b. Total maintenance policy to be followed (e.g., whether the prescribed PM schedule is being followed for each unit of equipment, frequency of equipment overhaul, and the like).
   c. Specific operating procedures, as selected from the set of operational jobs listed in the data base.
   d. Operational use of the equipment in terms of the communications systems being operated, the number of hours per year each system operates, the type of operating jobs being done, and how often.
   e. All support jobs required, as selected from the set of support jobs listed in the data base.
2. The operational planning factors provided help the planner estimate the number of equipment-hours expected. These include the number of transmitter operating hours for each communications system–transmitter type of combination at each transmitter site.
3. For each system configuration being analyzed, the planner uses the planning factors to calculate how many man-hours of what types of personnel are required at each site for operation, maintenance, and support.
4. Man-hours of each type of personnel are then converted into billets required by dividing by the number of productive hours in one year (1661 for military personnel).
5. Additional constraints are added, such as having a minimum of two persons always "on watch" at a transmitter site.

The procedure followed is similar to that used by personnel specialists at Navy headquarters to calculate billets required as a function of the average weekly work load at the site. Peak work loads that deviate from the average are accommodated by:

- Assigning additional manpower for those shifts that have predictable peak work loads.
- Using the electronic technician to help the operator when needed.
- Having the maintenance man do CM work before he does PM work.
- Bringing support personnel into operations and maintenance (O&M) activities if they can be trained to take on some of the simpler jobs during a peak.
- Working longer than the average standard shift or workweek. Overtime should be repaid with compensatory time off. This policy is implicitly included in calculating billets based on the total annual work load because peaks are included in that total. All other assumptions have been previously described.

The detailed planning procedure has been included as Appendix 9.A to demonstrate an important principle of systems planning. A planner must explain how to use the results he has derived in appropriate language that his client understands. Although the planning logic may appear to be simple algebra, some of the personnel who would be using the procedure may refer to these algebraic equations as "higher mathematics." Hence, in addition to providing the equations, an English narrative of this logic was also provided as part of the procedures manual. Appendix 9.A illustrates the type of detail that must be provided.

In addition, forms useful in performing the calculations were also constructed. All of these aided in the acceptance of the procedure developed.

9.6 APPLICATION OF THIS ANALYSIS TO OTHER SITES

While the development of the planning system for the transmitter sites was progressing and discussions of the jobs being done at the other sites were begun, it became apparent that the same analytical approach developed for the transmitter sites would also apply for other sites. For example, all the work functions of the receiver sites exactly paralleled the transmitter sites. Hence a similar set of planning factors was calculated using the data from the receiver sites, and the planning logic of the transmitter sites was reworded and used for the receiver sites.

In addition, the work of the centralized electronics maintenance division (EMD) exactly paralleled the work done at the decentralized transmitter and receiver sites. Only the equipment was different. In addition, maintenance men from the EMD had to travel to the various remote buildings where the equipment was located and hence additional credit was given for their travel time, based on the trip records showing the number of trips they took during the past year and the mileage and travel time involved in each trip.

Analysis of the fleet center, quite different from the other type of operations, since this involved message processing, is described in Chapter 10.

9.7 ORGANIZATIONAL AND BEHAVIORAL ASPECTS OF GATHERING DATA

Most of the previous discussion was concerned with the technical aspects of gathering and analyzing data to produce the desired resource analysis system. It is equally important to describe some of the behavioral aspects involved in organizing the effort, gaining a correct understanding of the problem, gathering the data required, and validating the information obtained. This will further illustrate some of the points first addressed in Chapter 3.

9.7.1 Gaining an Understanding of the Problem

A planner should use all the resources available to him to gain a proper understanding of the problem and the operations involved. In this case the four volumes of the Delta, Inc., report proved to be a rapid way of doing so. Another way was through discussions with knowledgeable personnel at the command. Since "everyone is always busy," a good way of obtaining the needed access to the proper people was to organize for this. I asked the assistant commander for the names of the knowledgeable personnel in the different major areas of operations, plans, and so on, and asked him to issue a memorandum announcing the project and assigning these personnel as "contact points" in each of the key areas. Thus each person was an official part of the project and could justify some expenditure of time to it. If anyone did not have the information I was seeking, he could provide me with an introduction to someone he thought could provide it. It is important to stress this procedure, because in the initial phase of the effort the planner must establish his information network, and having "insiders" aid him in doing so is extremely helpful. This team is also valuable in validating the planner's understanding of the problem and his analytical approach as he develops it.

Another organizational aid used was to designate the Norfolk Communications Station as the "analytical test bed." Just as a laboratory is essential to the scientist, exposure to current operations as well as the operators is essential to planning. My first visit to Norfolk provided me with my first opportunity to see the various equipment, layouts, work areas, and operations and a chance to discuss them with the system operators. In addition, it was another opportunity to test the analytical concept and hear any objections.

9.7.2 Need for Motivating Participating Personnel

Although the admiral's directive to Norfolk to participate in the project was important, behavioral obstacles still needed to be overcome. There are a number of reasons why personnel in an operating organization are not pleased (and sometimes hostile) to participate in a planning project such as this one. The simplest reason is that inevitably they see this as more work for them, in terms of the time that must be spent in gathering and explaining the data. Also they invariably view a stranger, particularly a professional analyst coming from headquarters to analyze their operations, as a natural threat to their existing method of operations. In our case, they knew that the purpose of the planning effort was to reduce their number of personnel. They did not know exactly how this cut would be made, and since previous reductions had been perceived as arbitrary, they tended to view our problem as a "can't win" exercise.

For these reasons it was necessary to gain the cooperation of the participating personnel, particularly if we hoped to obtain a product of reasonable quality. Here are some of the behavioral steps that were taken to increase the chances of success and have general applicability to all planning projects. Note that they follow some of the principles discussed in Chapter 3.

1. Use the good offices of the patron.

After having the command telephone the Norfolk commander and send an official message announcing our forthcoming visit to Norfolk, I arranged to have a member of the headquarters task force from the operations area accompany me on the trip. He was an experienced civilian of middle management stature who also was known and respected by the Norfolk personnel. This was extremely beneficial, particularly at the beginning of the project, since he served as a bridge tying together headquarters, me, and the Norfolk personnel whom he knew. He also knew Navy protocol and the functions of each of the sites we were to visit. He was invaluable in reviewing with me what took place after each meeting and in aiding me in overcoming obstacles.

2. Orient the key personnel.

We followed the procedure described in Chapter 3: When entering a new area, always start by orienting the key personnel on (1) the objectives of the project, (2) what you expect from them, and (3) reasons they should help you. This same procedure was used for the initial visit to all sites:

Start with the commanding officer. It is a necessity in any organization to start the orientation with the head. We wanted to give the commanding officer an opportunity to understand what we were planning to do, how we were planning to do it, and what was expected from his organization. After explaining the purpose of the project, I asked for the following resources:

- The assignment of a project officer, the key point of contact for any arrangements or any problems that might arise.
- A brief visit to each of the decentralized operations to see firsthand the different operations and the flow of incoming and outgoing messages. This included visits to several outlying locations.
- A joint meeting with the department heads, the various officers in charge at each of the sites in the Norfolk area, and the key enlisted men who would actually be collecting the data required. Attendance of the latter personnel is important because if they are not there they would be getting their information secondhand, which could lead to inaccuracies.

Implement orientation of other personnel. This proceeded in accordance with the request previously described.

Motivate participants. The biggest problem in a project such as this is to motivate the participating personnel to collect the data accurately. Remember that the project was dependent on the accurate collection of data and that these data would have to be collected by the site personnel themselves, since I had no resources of my own to do so. Since we all knew that their own data would be used to reduce their personnel, I faced a challenge with two components: (1) getting the participants to gather data in accordance with a tight time schedule and (2) making certain that the data were accurate.

One of the best ways of obtaining participation is by appealing to self-interest. I explained that a reduction in manpower was inevitable. All that was in question was the degree of reduction. Since I was familiar with how budget people evaluate the needs for resources (through my experience and through my working with the manpower people at Navy headquarters), I could help establish their needs for personnel through the manpower analysis approach that I had structured. But what I needed to build their case was the data that they had in their files. Since they had been complaining that they were undermanned for many of the jobs that had to be done, challenging them to "show me and I'll build your case for you" proved to be an appropriate way of motivating them to provide the assistance required. Another way was to mention that Army and Air Force had already applied a similar approach and had been successful in using their analyses to obtain the manpower they felt they needed. Those familiar with Air Force communications operations agreed that the Air Force had indeed been successful in using this type of approach.

Several methods were used to safeguard the accuracy of the data they were to provide. First, I warned that both the budget people and I were professionals in this area and when reviewing the analysis could rapidly spot any inflated data. Second, I had built into this exercise certain checks and balances that enabled me to recognize incorrect data supplied. I did not elaborate what these were so as to retain some mystery, but these safeguards were available to me: First, I insisted that the sources of all data would be identified. Thus an audit of any of the original data could always be made if there was any doubt about accuracy. Second, I was planning to compare their data against any Navy standards I could locate. Such standards for maintenance functions were obtained and used. Perhaps the best incentive for preventing an inflation of manpower data was *pride*. I indicated that I would be comparing the manpower required at each of the sites to do the equivalent job. Ultimately, this proved to be the best check and balance.[15] In fact, in visiting Honolulu and Guam, I was especially impressed by the pride that these personnel exhibited in describing how they were able to do their jobs more efficiently than even the maintenance standards called for.

3. Aid personnel in interpreting instructions.

Several types of assistance are required on projects such as this one. As carefully as the instructions may be written out and explained to the

[15] Obviously, this assumes that a mass conspiracy among sites to inflate all data will not occur, a good assumption.

participants, it is only after they start collecting data that they realize that their data may not be in the form envisioned by the planner. Hence suitable "fall-back" positions must be generated. Several examples of these include:

- The Naples transmitter site accumulated corrective maintenance data on only a small number of transmitters. However, these corresponded to 81% of the total PMS standard for the site. Hence one can interpret the data available as an excellent "data sample."
- If other data (such as PM manpower required) were not recorded, the sites were asked to take work samples over a period of several weeks, and this constituted a reasonable work standard, since the results obtained were less than the official Navy PMS standard.
- Equipment usage times were recorded in the station log. However, a complete data collection effort would have been quite time-consuming for some sites. Hence, a 10% sample was deemed to satisfy the minimum accuracy requirement. Using an available table of random numbers,[16] 37 numbers were selected, converted to days of the year, and given to the sites if they wished to gather data for only these days.

4. Represent both clients honestly.

Another factor that was problematical was whether I could represent the command and the sites as clients at the same time. Although the command gave me the original assignment, I agreed to construct the sites' case in the best way possible. Operationally, this meant that I must identify all jobs and determine what manpower was required for each. This represented no conflict of interest. In addition, the Norfolk management insisted that they see all the data gathered from the sites before it was sent to the Washington headquarters, a normal process. However, the staffing review of the data not only would add an additional time lag in the analytical process but might tend to isolate me from the direct source of data. My suggestion for coping with this natural reluctance was as follows. I would work directly with each of the sites, assisting them in how they would gather their data (but they would gather the data themselves). I would take the data, review their calculations, and write their report. However, none of the actual data would be given to the Washington headquarters until the Norfolk management reviewed the report and approved it. This enabled me to work directly with the sites unencum-

[16]If such a table had not been available, the use of the last four numbers from a commercial telephone directory list would have been a fairly good substitute.

bered but still permitted Norfolk their full review prerogatives. This policy was followed for all stations surveyed.

5. Validating the data submitted.

Based on the approach described, all data submitted by each site were compared against all comparable data from the other site. Any discrepancies were called to the attention of the person who had submitted the data. During the site visits suggestions were made for overcoming any data collection problems.

I prepared a common report for each type of site, displaying and describing the data submitted by each site and the data sources used. This report was returned to each site for validation.

6. Report of site analysis.

The same format was used to prepare a common report comparing the planning factors derived for a particular type of site and the planning logic to be used. This was prepared in the form of a procedures manual so that the method of using the data would be clearly understood.

7. Staff review of site analysis.

Finally, the combined data for each site were reviewed by a team of representatives from the command.

9.8 SUMMARY OF PROBLEM-SOLVING APPROACH: Telecommunications System Planning Case

This section summarizes the key principles of planning as they were applied to this case.

There were a number of generic problems associated with this case and these needed to be related so that each of the planners associated with the total problem could determine which part of the overall problem he was responsible for (bounding the problem, part of the problem definition phase):

The decision maker's ultimate problem was to determine the specifications of the final, preferred system using the new satellite system and cutting back on the HF system. However, these specifications of the preferred system will depend greatly on the budget available for the telecommunications command at the time. This is because much of the cost of the system depends on the number of redundant units kept available. This number relates to the risk that the decision makers are willing to take. Hence the real problem is to provide the decision makers with a set of options consisting of the alternative systems that could be configured, a measure of the communications service each provides, and the total cost of each.

Thus the major problem was to develop a planning system that would show in a rational way communications services provided versus total system cost for each alternative being considered.

The first step was to analyze the current system so as to

- *Define the system demand function.* This function would be treated as a parameter, consisting of the following components:

  The maximum amount of equipment that would be kept available for peak operations as required.

  The average number of messages and channels of each type of communications that would actually be operated during the year.
- *Determine the work load required to implement this system demand function.* These would be based on standard operating procedures and the planning logic and planning factors derived.

Having specified each entity that could be used in the system, the next major task was to define

- The various work activities associated with each entity.
- How often each activity was required to be done (as a function of the system demand function assumed).
- The time (and manpower) required to do each activity.

Thus this generic problem may be defined as the development of a manpower planning system, the first module of an overall resources planning system. The same approach described here would be applicable in determining the other resources required, such as fuel, spare parts, supplies, and so on. This problem may also be defined as determining manpower standards for different types of operations based on operational measurements.

Another key generic problem in this case involved how to enter a complex organization and perform a functional analysis. Here we are analyzing all the different jobs that go on at different sites and trying to determine the common functions that take place so that common work standards can be developed. Obviously, it is also important to determine the differences among sites as they related to these manpower requirements.

The next step in developing the manpower model was to:

1. *Focus on several sites doing the same function* and gain an understanding of the current operational system, applying the same generic models as before.
2. *Identify the different types of jobs and operations* at each job site. Helping in this identification process is the table of organization, which provides job titles and locations. The various supervisors can also describe what work their personnel do. We may note that all jobs come under three major classifications:
   - Operations.
   - Maintenance.
   - Support.

   In addition, each of these three functions may be subdivided into direct labor (those doing the work function) versus supervision (those supervising the direct labor). Finally, there are general management personnel, those who are managing several work functions.
3. *As an aid in identifying all the work functions,* look not only at the personnel who are working, but at the equipment that personnel are associated with and its usage.

The specific data to be collected for the direct-labor model consisted of:

- The man-hours required to do the job once.
- The number of times per year each job needs to be done.

Total man-hours per year required for each job operation consists of the product of these two factors. The time required to do a complex job may be more accurately estimated by constructing the operational flow model showing the series of tasks involved and measuring the time required to do each task. The latter may be obtained by taking work samples or by using available records. The frequency of occurrence of each job is obtained in the same way.

Explain the approach to the personnel at the sites selected to provide such data so as to:

- Obtain their validation of the approach.
- Motivate them to collect data properly. This involves:

Reviewing their records to see their coindition (data source).

Advising them how to extract the data from their files properly (total data versus data samples).

Building in data validation techniques so that possible data errors may be detected. These include available standards, correlation of man-hours used with manpower available, actual work sampling, data correlation with other sites doing similar jobs.

- Reduce the work required by the site personnel collecting the data. Such aids can include

  Furnishing printed instructions on how to gather data.

  Reviewing such instructions with these personnel to validate their understanding.
- Have them furnish raw data only (including citing data sources), with the analyst doing the calculations of the raw data.
- Generate the final report for validation by the site.

Comparing "similar" data collected from different locations of different operators is important to validate that:

- The same set of tasks are done.
- The same procedure is used.
- The same frequency of doing tasks occurs.

Any differences in these factors should be identified and a more detailed analysis performed to identify why the differences occur. The results of this analysis will be either a common agreement about the jobs (by which all sites will use the preferred procedure, or at least allocate only those resources required to do the jobs in the preferred fashion) or a determination of whether there is justification for these differences among sites. Valid reasons for such differences are:

- Environmental differences.
- Layout of equipment.
- Age of equipment.
- Personnel training.

Such differences can be compensated for by allocating different amounts of manpower in accordance with the differences involved.

Once the man-hour standards for each operation have been determined, the total number of positions required can be determined from:

- The product of the total set of operations and their annual frequency of occurrence.
- The average number of productive hours per year expected from each operator.
- Any additional constraints, such as safety or layout, that require an individual to be continuously on duty at a location, even though there may not be work for him at all times.

A number of behavioral problems are also encountered in a project such as this. These mainly involve problems of organizing and motivating people to provide accurate data.

It is important in data-gathering projects such as this that "fall-back" methods be identified if the data desired either are unavailable or require too much manpower to collect.

APPENDIX 9.A. PLANNING LOGIC FOR TRANSMITTER SITES

Having described how to calculate the various planning factors associated with a transmitter site, the next step was to generate a procedure for calculating the number of billets needed to operate, maintain, and support the equipment for any alternative being proposed. This section describes this procedure in the form that it was given to the client, thus enabling him to perform these calculations.

Maintenance Manpower Requirements

Equipment Needs

Decide on the total number and types of equipment, including spares, that must be kept operationally ready for peak operations, such as major fleet exercises or contingencies. This information can be obtained from the users. List the equipment type in column 1 and the total number required in column 2 of Worktable 9.A.1.

Such needs should then be validated by comparing them with former usage under similar conditions. Such data are not now part of the planning data base; these data should be collected as exercises are conducted.

Planning Factors

Decide which set of planning factors is to be used for the realignment alternative under consideration: either the command-wide planning factors

Worktable 9.A.1 *Maintenance Man-hour Requirements*

(1) Equipment required	(2)	(3)	(4)	(5)	(6)	(7)	(8)
Type	Number	Total Operator PMS Factors	Total Technician PMS Factors	PMS Operator Man-hours	PMS Technician Man-hours	CM Factors	CM Man-hours

Worktable 9.A.2 *Tuning/Retuning Operating Man-hour Requirements*

(1) Communications System	(2) Equipment Type	(3) Operating Hours	(4) Number of Tuning/Retuning Required	(5) Average Tuning/ Retuning Times Required	(6) Tuning/Retuning Man-hours

or the set of planning factors related to a particular geographical zone as represented by one of the four sites.

Equipment Inventory

Decide on the equipment inventory to be maintained at full readiness. Also decide what PMS schedule to follow, including all non-CM actions such as overhaul and appropriate work schedules.[17]

PM Man-Hours

Based on the PMS schedule to be followed, calculate the total PM man-hours required for each equipment type. First, calculate the sum of the unit PM man-hours[18] needed for the total PM schedule over the full year (from the list of all PM jobs and their unit manpower requirements as included among the maintenance planning factors). List the unit PMS factors for operating personnel in column 3 and the PMS factors for maintenance personnel in column 4. The product of columns 2 and 3 gives the PMS man-hours required of operators; this number is listed in column 5. The product of columns 2 and 4 gives the PMS man-hours required of technicians and is listed in column 6. Find the total operator PMS man-hours (sum of column 5 entries) and total technician PMS man-hours (sum of column 6 entries).

The total operator and technician man-hours required (columns 5 and 6) should also include the appropriate "make-ready and put-away" and PF&D factors. The Navy requirements for these two factors are 30 and 17%, respectively. Thus the Navy headquarters requirement for operator and technician PM man-hours would be 1.47 times each of the totals shown in columns 5 and 6. These totals should be listed as the last lines of columns 5 and 6.

CM Man-hours

Calculate the CM man-hours required for each equipment type and list the total in column 8. This number consists of the product of the number of equipment units in inventory (column 2) and the CM planning factors listed in column 7. Find the total CM man-hours required (the sum of column 8 entries).

[17] According to current policy, all site equipment is to be fully maintained for both CM and PM. However, manpower may be saved (at the cost of more time to reach full operational readiness) when all equipment is not fully maintained all year, and greater use is made of strategic warning in starting the readiness process early enough. Further analysis of such a proposed policy change is required. If current policy were changed, the calculations of PM and CM man-hours would be modified accordingly.

[18] Unit PM man-hours is the annual man-hours needed to do PM for one piece of this equipment.

Calculating the Navy CM requirement is a simpler process, since the CM requirement is defined to be equal to the total PM requirement (1.47 times the PMS standard). Thus the separate CM factors do not have to be listed in column 7, and the total of column 8 is equal to the total of the last line of column 5 plus the last line of column 6.

Tuning and Retuning Manpower Needs

Equipment Needs

List in columns 1 and 2 of Worktable 9.A.2 each communications system and the types of equipment to be operated during the coming year. List first all continuously operated systems and then all intermittently operated systems.

Operating Hours

Estimate the number of operating hours for each equipment type during the coming year and enter the estimate in column 3. In this estimate you may wish to consider operational usage facts at particular sites as a baseline, adjusting it up or down to reflect the proposed operation. Find the total of operating hours for the continuously operated systems and for the intermittently operated systems.

Number of Tunings and Retunings Required

List in column 4 the number of tunings and retunings required for the continuously operated systems and for the intermittently operated systems, as obtained from Figure 9.3.

Tuning and Retuning Unit Times Required

Calculate the average time required for one tuning and retuning of the continuously operated transmitters, based on a weighted average of the number of transmitters of each type operating continuously, as previously described. In a similar way, calculate the average time required for one tuning and retuning of the intermittently operated transmitters. List each of these times in column 5.

Tuning and Retuning Man-hour Requirements

Calculate the total tuning and retuning man-hours required for continuous systems and for intermittent systems as the product of columns 4 and 5, and list them in column 6. Find the total operating man-hours for tuning and retuning as the sum of the entries in column 6.

Additional Man-hour Requirements

Quality Control Checks
Decide on what QC checks are to be made and how often.

Manpower for QC Checks
Estimate the total annual man-hours needed for QC checks in the following way. Review the list of QC checks and decide which ones are to be done, how often, and the time required for each. List this in columns 1, 2, and 3 of Worktable 9.A.3. Then calculate the annual man-hours required for each check by multiplying column 2 times column 3 times 52. List the man-hours required for each QC check in column 4. The sum of the entries in column 4 is the total QC man-hours required.

Power Failures
Calculate the total man-hours needed to cope with power failures the same way as QC requirements. First, list in column 1 all operational activities that must be done following each power disturbance (such as retuning and readjustment). Next, list in columns 2 and 3 the average number of work units expected each week (annual estimate divided by 52) and the man-hours associated with each disturbance. The total man-hours required will then again be the product of columns 2 and 3. Record this in column 4.

Direct Labor Support

Support Needs
Decide which support jobs are needed at the site by reviewing the data base on support jobs and determining which of these the site has to do for itself, thus requiring site billets. In column 1 of Worktable 9.A.4, list the direct-labor support, primary duty functions, such as medical services, in which billets are to be provided by the NavCommSta rather than by outside organizations. The number of direct labor support billets required for these functions is listed in column 2. The support primary duty planning factors may be used in deciding how many billets should be allocated to these functions. List those support jobs being done as collateral duty in Worktable 9.A.3, along with the average number of work units done per week and the unit man-hours required for each work unit (columns 1, 2, and 3). Calculate the total man-hours per year required for each job and list this total in column 4.

Worktable 9.A.3 *Man-hour Requirements for Additional Jobs*

(1) Job Description	(2) Average Work Units per Week	(3) Support Planning Factor	(4) Total Man-hours per Year	(5) Watch Allocation	(6) Maintenance Technician Allocation	(7) Primary Duty/Support Allocation	(8) Supervisor Allocation

Worktable 9.A.4 *Support Primary Duty Requirements*

(1) Support Primary Duty Functions Required	(2) Billets Required

Support Man-hours

Determine who will do each job in terms of these categories:

- On watch.
- Maintenance technicians on day shift.
- Primary duty support personnel.
- Supervisors.

Allocate the total support man-hours required among these billet categories and list them in columns 5, 6, 7, and 8 of Worktable 9.A.3. Although using O&M personnel for this purpose may not seem efficient, it does offer the advantage of having extra O&M workers available for peak operations. Add the total man-hours required for each category.

Total Billet Requirements

The remainder of this section explains how to calculate billet requirements for each class of personnel. The characteristic being calculated is given in column 1 of Worktable 9.A.5 and is called an item of this column. The data for each calculation should be listed in column 5.

Work Elements

In column 1 list the various work elements done by the operator watch personnel. These elements are:

- Tuning and retuning operations.
- QC checks.
- Power failures.
- Operator PM action.
- Support collateral duty work load done by operator watch personnel.

Man-hours per Work Element

In column 2 list the man-hours required for each work element. In all appropiate cases the working man-hours must be converted into total man-hours by applying the PF&D factor appearing in column 3. Thus the total number of man-hours for each work element is

$$\text{TMH} = (1 + \text{PF\&D})(\text{WMH})$$

where TMH = total man-hours
WMH = working man-hours
PF&D = personal fatigue and delay factor

Worktable 9.A.5 *Calculating Total Billet Requirements*

(1) Characteristic Being Analyzed	(2) Working Man-hours Required	(3) PF&D Factor	(4) Total Man-hours Required	(5) Numerical Factor
1. Tuning/retuning operations		1.17		
2. QC checks		1.17		
3. Power failures		1.17		
4. Operator PMS actions		Included		
5. Support collateral duty work load done by watch personnel		Included		
6. Total operating man-hours required				
7. Standard workweek (for labor mix)				
8. Number operating billets required				
9. Watch supervisory overhead ratio				
10. Number watch supervisors required				
11. Additional maintenance workers added to watch				
12. Additional supervisors added to watch				
13. Total maintenance technician PM work load				
14. Total maintenance technician CM work load				
15. Total maintenance technician work load				
16. Total maintenance watch man-hours available				
17. Percent time watch technician does peak operating load				
18. Total maintenance watch man-hours available for maintenance				
19. Maintenance work load done by day shift				
20. Maintenance billets required for day shift				
21. Maintenance supervisory overhead ratio				
22. Number maintenance supervisors required				
23. Number support primary duty personnel				
24. Support supervisory overhead ratio				
25. Number support primary duty supervisors required				

The PF&D factor should have been included in the operator PM requirements calculated in Worktable 9.A.1. Obtain the total operating man-hours required (row 6 of the table) by adding the man-hours of the five work elements and listing the total in column 4.

Number of Watch Standers[19]

The next step is to calculate the total number of operator watch standers required (row 8 of the table). There are three major factors to consider in this determination:

- Average work load.
- Peak work load the system is designed for and how flexible the system is in sharing operating work load with other watch standers (such as maintenance and supervisory personnel).
- Constraints, such as safety.

Each factor is considered in greater detail here. The number of operator billets B_o based on average work load is determined first:

$$B_o = \frac{\text{TOW}}{52}\,(\text{TAW})$$

where TOW = total operator work load per year
TAW = time available for work per week

According to the standard work week of 40 hours (where dependents are authorized), TAW equals 31.94 hours per week for military and 33.98 hours per week for civilian personnel. An assumption here is that a watch stander assigned to a five-man-for-four-section watch also has about 32 hours per week available for work because of time out for meals.

TAW thus is based on a weighted average of these two factors and depends on the civilian-to-military mix at the site. For example, if there were 10 civilian to 40 military direct-labor personnel at a site, TAW, the weighted average would be:

$$\text{TAW} = \frac{10(33.98) + 40(31.94)}{50} = 32.35 \text{ hours/week}$$

Enter this weighted average of TAW in row 7. Enter the results of the

[19]There are two categories of workers in the Navy. "Day workers," such as maintenance men, work only during the day shift. The position is not required to be occupied on a 24-hour basis. "Watch standers," such as radiomen, work at positions that are manned 24 hours a day.

calculation of B_o in row 8, column 5. Carry the billet calculations to the nearest hundredth of a billet until all calculations are completed and a final "round off" of fractional billets is made.

Determine the number of watch supervisors B_{ws} assigned to the watch:

$$B_{ws} = B_{wo}S_{rw}$$

where B_{ws} = number of watch supervisor billets required (row 10)
B_{wo} = number of watch operator billets required (row 8)
S_{rw} = watch supervisor overhead ratio (row 9)

Enter the values for these characteristics in worktable 9.A.5, column 5, in the appropriate rows.

Allocate the watch operators and supervisor among the four watches and transmitter buildings and see that anticipated peak loads during the week are accommodated. Note that watches do not have to be manned equally, and "peak loaders" can be used. After the allocation is made, check to see that the safety constraint is satisfied (minimum of two men per watch). When either of these factors is a problem, it can be alleviated by adding maintenance technicians to the watch (plus the proportional amount of supervisors). Insert this information in rows 11 and 12. This strategy may yield two benefits simultaneously. First, the technician can satisfy the safety constraint; second, because of his flexibility, the technician can be always gainfully employed either doing CM or PM actions or aiding the operator(s) during a peak.

Finally, since the total operator work load includes PM work, and since the PM work can be dropped during a peak, some extra manpower is available for peak demands for tuning and retuning.

Additional Direct-Labor Maintenance Personnel

Determine the total number of additional direct-labor maintenance personnel required during the day shift by following the items listed in column 1, entering the data requested in column 5.

First, enter the PM and CM work loads to be done by technicians (either on watch or day shift) in rows 13 and 14. Enter the total in row 15. Enter the total maintenance watch man-hours available in row 16:

$$\text{TMWM} = 52B_{mw}\text{TAW}$$

where TMWM = total maintenance watch man-hours available
B_{mw} = number of assigned maintenance watch billets
TAW = time available for work per week, as already described

Then enter, in row 17, an estimated percentage of time to be spent by the maintenance man doing the peak operating load. As discussed, operating peaks, when they occur, are handled by a maintenance watch stander (when such an assignment exists) or watch supervisor. In either case the individual drops his normal work and responds to the peak operating request. Thus this time is used in operations and is not available for maintenance or supervision.

A working supervisor's time is already properly allocated between direct labor and supervision. For a maintenance technician on watch, including day shift, some fractional part of a billet needs to be added to this operating function to account for that fraction of time when he is taken off his maintenance work to keep the operator during a peak:

$$\text{TMWMA} = (\text{TMWM})\left(1 - \frac{p}{100}\right)$$

where TMWMA = time available for maintenance work by the watch maintenance technician

p = percentage time on peak operating load

Enter TMWMA in row 18.

Next, determine the resulting maintenance work load to be done by the day shift (row 19). This is equal to the total PM required of technicians plus the CM to be done (as previously calculated) minus the maintenance man-hours spent by maintenance technician watchstanders. In calculating the total maintenance man-hours, the CM planning factors have non-productive time built in, whereas the PM planning factors do not. Hence only the latter time must consider the PF&D factor as well as the make-ready, put-away factor; these were included in Worktable 9A.1. Finally the number of maintenance billets B_m required on the day shift (row 20) is:

$$B_m = \frac{\text{TMW}}{52}(\text{TAW})$$

where B_m = direct-labor maintenance billets required (row 20)

TMW = total maintenance work load to be performed by maintenance personnel on day shift (row 19)

TAW = time available for work per week, as previously described

Maintenance Supervisors. Determine the number of maintenance supervisors required (row 22):

$$B_{\text{ms}} = B_m S_{\text{rm}}$$

where B_{ms} = maintenance supervisor billets (row 22)

B_m = maintenance billets on day shift (row 20)

S_{rm} = maintenance supervisor overhead ratio (row 21)

Support Primary Duty Supervisors. Determine the number of support primary duty supervisors required:

$$B_{ss} = B_{sp}S_{rs}$$

where B_{ss} = support primary duty supervisors (row 25)
B_{sp} = support primary duty billets (row 23)
S_{rs} = support primary duty supervisor overhead ratio (row 24)

The service diversion work load should be examined as part of the entire service diversion requirement to ensure that the total does not exceed an average of 8 hours per week. When it does, an appropriate number of additional billets may be added.

Fractional Manning

After the number of billets for each function has been calculated to the nearest 100th of a billet, fractional manning problems may arise. In the past, this was solved by arbitrarily selecting the equivalent of one-half (0.5) as the cutoff point. Any work load that earned at least one-half space was awarded the next whole number without regard to work center size. Those that earned less than one-half did not get the extra manpower.

Overload factors are established based on the premise that separate criteria should be applied to small and large work centers. A maximum individual work overload is established at ½ hour per working day and is cumulative until reaching a maximum of one-half billet. The cutoff point is the highest value the fractional manpower can reach before the manpower requirement is rounded to the next higher integer. Table 9.A.6 reflects

Table 9.A.6 *Fractional Manpower Cutoffs for Computing Standards*

Manpower Authorized	Fractional Manpower Cutoff	
	Military	Civilian
1	1.081	1.078
2	2.162	2.155
3	3.243	3.233
4	4.324	4.310
5	5.405	5.388
6	6.486	6.466
7	7.500	7.500
Over 7	Authorized manpower + 0.500	Authorized manpower + 0.500

fractional manpower cutoff points for both military and civilian manpower.

Qualitative Requirements

Next, determine the qualitative requirements of each position in terms of designator, grade, rate, and series. This should be done uniformly, based on the total number of people required in each functional unit.

10

Analyzing the Message-Processing Operations

In Chapter 9 we were concerned with the development of a procedure for determining the manpower requirements of sites involved primarily with communications transmissions (transmitter and receiver sites). Manpower is also required for various message-processing functions. This chapter describes the manpower analysis done for such functions.

Message processing occurs within the system in primarily two locations: telecommunications centers (TTC) and fleet centers. The beginning of Chapter 9 described the overall method of transferring a message from a message originator through his TCC to a ship-based addressee. All messages are automatically routed to the fleet center serving the ship addressee for manual handling by a fleet center operator, if necessary. The manpower required at a TCC does not depend on whether satellite or HF communications are used. However, if the satellite was used, most of the fleet center operations could be automated. Hence because of the constraint on analytical resources, it was decided to limit the message-processing analysis to operations in the fleet center.

10.1 SUMMARY OF ANALYTICAL APPROACH

The same approach used in analyzing the transmitter sites was also used for the fleet centers as shown in Figure 10.1. Specifically, the data listed under system characteristics were requested from each of the four fleet centers analyzed. In particular, the operational usage of the equipment in the fleet center was quantified in terms of the number of messages of a certain type handled each year, the number of circuit-days of operational use for each type of circuit, and the maximum number of circuits each operator can handle in parallel because of layout constraints.

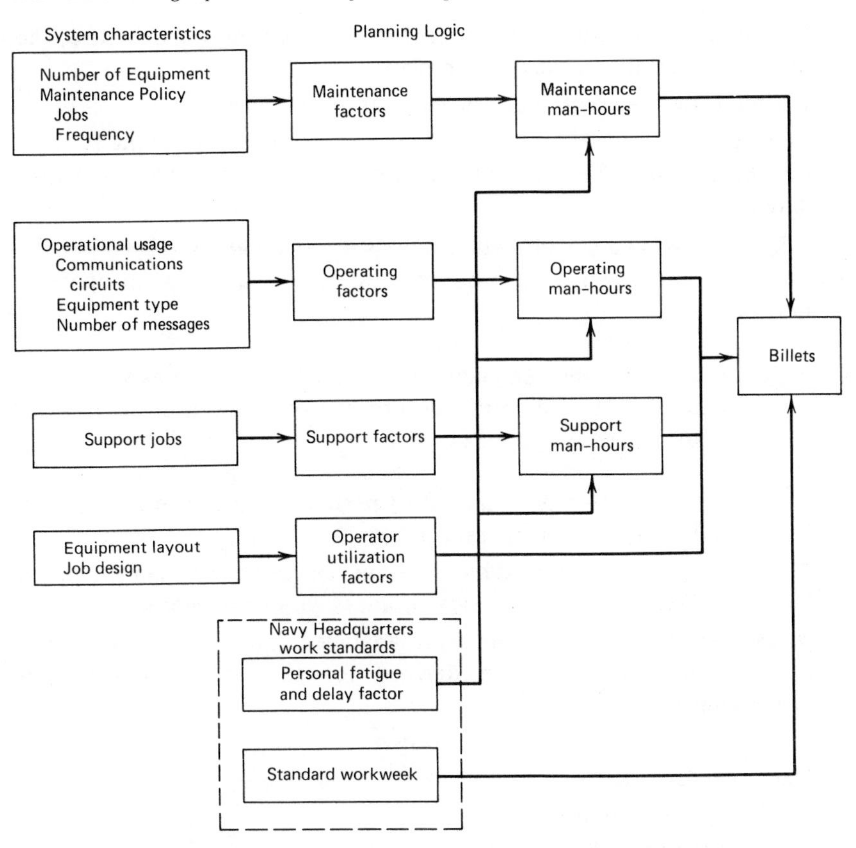

Figure 10.1. Manpower planning logic for determining manpower required for fleet centers.

Since the analysis of the maintenance and support manpower requirements was the same as that described in Chapter 9, we now concentrate on the analysis done in determining the operational manpower requirements. The detailed planning logic developed as user's manual for use within the command is contained in Appendix 10.A.

10.2 DETERMINING OPERATIONAL MANPOWER REQUIREMENTS

These were determined by means of the following steps:

- Identify all message-processing jobs to be analyzed.

- Calculate the man-hours required for each unit job (in this case, the unit consisted of the message processed).
- Estimate the message load for each job next year.
- Find the total man-hours required per year for each job as the product of the man-hours per unit job and the number of message units estimated.
- Cluster together jobs that can be done by the same operator.
- Convert the number of man-hours required for each type of operator into billets.

The operational manpower-planning factors derived to make the man-hour calculations are based on this description of fleet center operations, as validated by the sites:

- The entire message-processing effort performed at a fleet center[1] can be divided into a set of operations.
- The operations planning factors relate the operator man-hours required to the work load for various message-processing jobs.
- Message-processing jobs that were analyzed and for which quantitative planning factors were derived for the following message-processing jobs:

  Full-period termination (FPT), receive.

  Data speed reader.

  Full-period termination, send.

  Allied/NATO/SEATO, receive.

  Allied/NATO, SEATO, send.

  CW (continuous wave) broadcast.

  PG (patrol gunboat) broadcast.

  Encrypt message.

  Decrypt message.

  Service center.

  Data base operation.
- Each job is done at a work station consisting of a set of equipment and manned by an operator.
- Each operator does one or more jobs. For example, the full-period termination operator handles both receive and send circuits, as does the Allied/NATO/SEATO operator. In addition, an operator may also

[1]Some message-processing jobs are also done at some receiver sites.

have certain collateral support duties, such as cleaning his area at the end of his shift.

- Each job can be described as a sequence of activities and illustrated as an operational flow diagram, similar to that describing any production job, as in Case 1.
- The basic planning factors derived for each job consist of

  The *sequence* of activities associated with the job. For manpower planning, this sequence should be fairly standard within the command.

  The *average time* required to do each activity associated with the job. This should also be standard within the command.

  The *average proportion of time each activity is done* during the processing of a single message. With some activities, such as transmitting a send message or logging in the message, the activity is done exactly once for each message. The proportion equals one. With other activities, however, such as retransmissions or piecing (i.e., rework) of a message, the proportion may be less than one. With others it may be greater than one. This factor often depends heavily on HF propagation characteristics or other local conditions and is generally unique to a particular site.

  The *average message lengths;* this can be treated as unique to each site if large differences exist among sites, and it will influence certain activity times.

  The *number of messages processed* by this job last year. This is unique to each site and may be used as an indicator of next year's traffic load.

- From these basic planning factors, two higher-level planning factors unique to each site have been calculated:

  The *average total time to complete all activities satisfactorily* (both operator and machine) involved in one work unit of each job (such as one FPT-receive message). This time was obtained from an expected value calculation, taking into account all the activities involved in the job and the proportion of time each activity is performed during an average message.

  The *average total time the operator is occupied* during completion of the work unit; this time is calculated the same as total time, but in this case only those activities during which the operator is working on this job are considered.

- Thus the *annual operator man-hours required* at a site to process the given number of messages per year can be calculated as the product of the average operator time required to process one message times the total number of messages to be processed by the particular site. However, to obtain the total operator work load per year, these factors must also be considered:

All the jobs handled by this operator.

Any additional, unavoidable operator idle time not already included.

Personnel fatigue and delay (PF&D) (17%).

These factors and how they are combined are described in that part of the planning logic dealing with calculating total billet requirements.

In the following analysis the derivation and results of the basic planning factors associated with the "full-period termination, receive" job is presented, plus the calculations for determining the operator man-hours required per year for this job. The analysis of all other operations jobs was done in the same way.

10.3 OPERATIONAL MANPOWER REQUIREMENTS: ANALYSIS OF FULL-PERIOD TERMINATION, RECEIVE, JOB

Data describing this job, submitted by all four sites, consisted of the following parts, which were needed to construct the operational flow model for each site:

- A qualitative description of the way the job was being performed at each site.
- Quantitative data describing the time required to perform each activity in an acceptable manner. These times were obtained from work samples taken by each site.
- Quantitative data describing various nontime characteristics regarding the job as it is done at each site. These data were taken from available logs as well as work samples.

The first step in correlating the data was to compare the operational descriptions and flow diagrams submitted by each site with one another to make certain there was a common basis for comparison. To aid in this correlation (and to reduce the data gathering done by the sites, all sites were supplied with Norfolk's operational floor diagram and their description of the activities associated with this job. Each site could then use this as a reference and describe which of their activities differed from the reference. Each of the sites confirmed that the process they followed in doing this job was basically the same as that shown in the reference.

Figure 10.1 was therefore constructed as a flow diagram for use as a standard.

10.3.1 Derivation of Time Standards

Data describing each activity shown in Figure 10.2 was then listed in column 1 of Table 10.1 so that the data submitted by each site could be readily compared and a standard time for each activity calculated:

- The activity title and designation, as given in Figure 10.1, is listed in columns 1 and 2. Underlining this designation in Column 2 indicates that the operator is completely occupied during this time. The designations submitted by the four sites are listed in columns 3, 6, 9, and 12. Related activities were clustered into six main groups, as is described in the next section.
- The times required to do each activity, as submitted by each site, are in columns 4, 7, 10, and 13.

Although we wished to obtain a standard time for each activity as the mean of the times submitted, this could not be done when

- An apparent arithmetic error was made in the time submitted by a site.
- Certain conditions peculiar to a site made the activity somewhat different from its counterpart at another site (one example of this: when the time required for punching tape is a function of average message length, which varies from site to site).

In both cases the time submitted was translated into a "standardized" time (i.e., appropriate to the same standard set of conditions assumed); this standardized time was used in calculating the mean. The reason for making these changes is described in this analysis.

The standard time recommended for each activity is listed in column 15. Each value is derived as the mean (weighted by the number of messages per year at each site) for the four sites. (Throughout this analysis the word *mean* indicates the weighted mean.) If a site did not include a time estimate, the mean was derived as the weighted mean of the data submitted by the other sites.

10.3.2 Description of Activities

This section describes each activity, using the designations in Figure 10.1. Operator activities are again indicated by underlining the designation.

Table 10.1 *Activity Time for Full-Period Termination, Receive*

(1) Activity	(2) Std. activity designation	Honolulu (694)[b] (3) Activity designation	Honolulu (694)[b] (4) Required time (sec)	Honolulu (694)[b] (5) Standard-ized time (sec)	Guam (2794) (6) Activity designation	Guam (2794) (7) Required time (sec)	Guam (2794) (8) Standard-ized time (sec)
Network control							
Total operator time	—	—	—	—	—	—	—
Total time	—	—	—	—	—	—	—
Transmission							
Punch tape and print copy	a	a	95	144	a	168	144
Tear message off machine	b	b	5	2	b	2	2
Inspect message	c	b + g + r	20	23	c	10	10
Total operator time				25			12
Total time				169			156
Administrative							
Receipt for message	e	c	30	30	g	25	25
Log-in message	g	e	10	10	k	5	5
Carry message	h	f	5	5	l	14	14
Total operator time				45			44
Total time				45			44
Total time if no retransmission is required							
Total operator time[a]				70			56
Total time[a]				214			200
Retransmission							
Request retransmission	j	g	20	20	i	30	30
Total operator time (j + b + c)				45			42
Total time (j + a + b + c)				189			186
Copies disposal							
Total operator time	m	m	5	5	f	20	20
Total time				5			20
Piecing							
Piece message	l	c	180	180	j	172	172
Total operator time				180			172
Total time				180			172

Table 10.1 *(Continued)*

	Norfolk (489)			Italy (604)			
(1) Activity	(9) Activity designation	(10) Required time (sec)	(11) Standardized time (sec)	(12) Activity designation	(13) Required time (sec)	(14) Standardized time (sec)	(15) Standard time (sec)
Network control							
Total operator time	—	—	—	—	—	—	—
Total time	—	—	—	—	—	—	—
Transmission							
Punch tape and print copy	a	144	144	a	150	144	144
Tear message off machine	b	3	3	b	5	5	3
Inspect message	c	5′	5	c	15	15	12
Total operator time			8			20	15
Total time			152			164	159
Administrative							
Receipt for message	g	20	20	e	60[a]	60[a]	25
Log-in message	l	20	20	f	10	10	8
Carry message	m	5	5	g	15	15	12
Total operator time			45			85	45
Total time			45			85	45
Total time if no retransmission is required							
Total operator time[a]			53			105	60
Total time[a]			197			249	204
Retransmission							
Request retransmission	l	20	20	d	60[a]	60[a]	27
Total operator time (j + b + c)			28			80	42
Total time (j + a + b + c)			172			224	186
Copies disposal							
Total operator time	j	20	20		—	—	17
Total time			20			—	17
Piecing							
Piece message	k	180	180		—	—	174
Total operator time			180			—	174
Total time			180			—	174

[a]Unexplained "outlier" and not used in calculating standard time (weighted mean) of column 15.
[b]Messages handled per week.

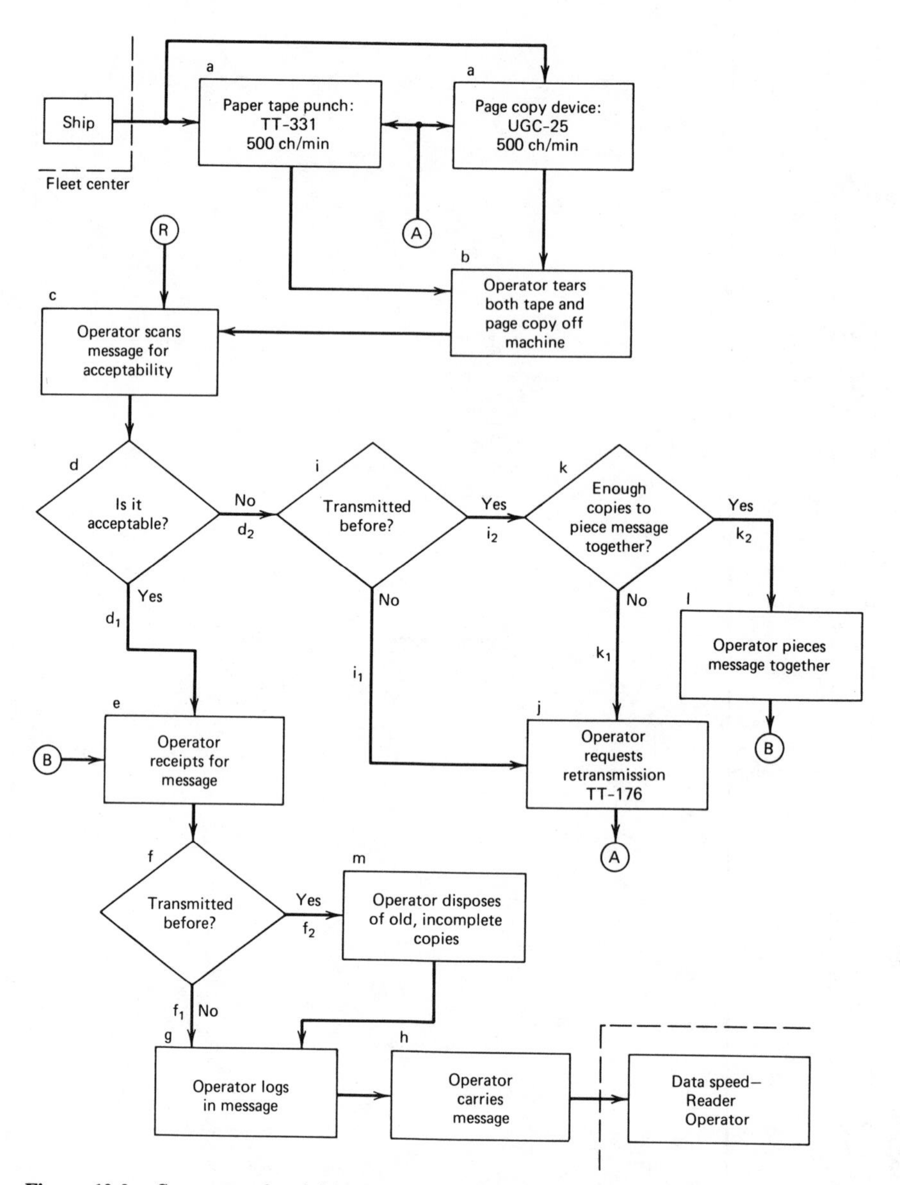

Figure 10.2. Sequence of activities to complete one work unit of full-period termination, receive.

10.3.2.1 Network Control Activities

No activity associated with setting up a full-period termination, receive, circuit was mentioned by any site. It seems that some fleet center operator man-hours are required for this function. The sites were then asked to describe these activities and estimate the times required and the frequency with which they are done so that appropriate planning factors can be developed.

10.3.2.2 Transmission and Administrative Activities

- *Activity a—Punch tape and print copy*. The message is received from the ship in page copy and paper tape (via UGC-25 and TT-331 equipment). Honolulu submitted an initial machine time of 95 seconds, compared with 144 to 168 seconds, as submitted by the other sites. Since all the machine rates are the same (500 characters per minute), the time difference is reflected in all calculations involving message length. The standardized time, 144 seconds, was calculated on the basis of a standard message length of 1200 characters.
- *Activity b—Tear message off machine*. The operator tears both page copy and tape off the machines. Honolulu submitted one time of 5 seconds for the operator to tear a message off the printer and inspect it (activities b and c). To conform with the standard structure, the 5-second period was subdivided into two times—2 seconds for b and 3 seconds for c.
- *Activity c—Inspect message*. The operator scans the message to determine whether it is acceptable or whether it requires retransmission. For simplicity, it was assumed that the time required for message inspection in activity c also includes the time required for decision making in activities d, f, i, and k. Thus node d is treated as a no time-loss decision node, allowing us to separate the flow of activities for received messages that are acceptable from those that are unacceptable. If a message is acceptable on first transmission, proceed down path d_1 to activity e.

Honolulu also included two times not reported by the other sites; these seem to be part of activity c. These activities are labeled g and r on Honolulu's flowchart. Activity g consists of 10 seconds to determine that a retransmission is required. The other sites presumably included their times in activity c. Activity r requires 10 seconds for the full-period termination operator to request the command's video display terminal (VDT) operator to hold the outgoing line to the ship so that a retransmission request may be made. The need for this function is not clear and should be reexamined. However, these 20 seconds were added to the 3 seconds carried over from activity b to yield a total of 23 seconds for Honolulu. It is assumed that the times listed are the average working times for both original messages as well as those returned by the data speed-reader operator.

- *Activity e—Receipt for message*. The operator receipts for an acceptable message on the TT-176. Italy indicates a requirement of 1 minute to receipt for a message, compared with 20 to 30 seconds for the other sites. Since this outlier may result from a ''round-off error'' in the time unit used, it was decided not to consider Italy's estimate in calculating the mean.
- *Activity g—Log in message*. The operator logs the required information on the log card, then delivers the message for further processing.
- *Activity h—Carry message to data speed-reader*. The operator carries the acceptable message to the data speed-reader operator, who continues subsequent operations at his station.[2] Honolulu included 10 seconds to carry the message and add the format line 1 (FL/1) header. The time was divided into two parts—5 seconds for carrying, 5 seconds for adding the header. At decision node e, if the acceptable message was transmitted before the flow of activities, follow path e_1 to activity f.

10.3.2.3 Retransmission Activities

- *Activity j—Request retransmission*. Given that a message has been received in unacceptable form, the operator must complete two decision nodes (whose times were included previously) before he requests a retransmission. The first, node h, is a decision node enabling him to separate the unacceptable messages into two subsets. When this transmission is the first unacceptable transmission of a message received, the operator proceeds directly through path i to activity j and requests a retransmission. But when this is not the first unacceptable transmission of a message, he proceeds through path i_2 to node k to determine whether the set containing the current, retransmitted message and all past transmissions of the same message permits the operator to piece together an acceptable message on his UGC-6. If not,

[2]As described, messages judged unacceptable by the data speed-reader operator and that cannot be corrected by him are returned to the full-period termination operator at point *R* of Figure 10.2.

he proceeds through path k_1 back to activity j, requesting another retransmission. When he can piece together an acceptable message, he proceeds through path k_2 to activity l. In activity j the operator requests retransmission of message on the TT-176 teletype. The process continues with repeated activities a, b, and c.

- *Activity m—Disposal of old copies.* After an acceptable copy is obtained (through retransmissions or piecing—see the next function), the operator discards all old, imperfect copies of the message, then proceeds to activity e. Italy did not include a time estimate for discarding old messages, but it is assumed that the job must be done. Therefore, the mean time of the other three sites was calculated as the standard.

10.3.2.4 Piecing Activity (Activity l)

When the operator has enough copies to piece together an acceptable message, he does so using the UGC-6. After he completes the job, he disposes of all the used and incomplete copies (activity m). Guam included disposal time of 20 seconds as part of its piecing activity l. Thus its time, 192 seconds, was reduced to 172 seconds to conform to the standard flow diagram.

10.3.3 Derivation of Nontime Planning Factors

Figure 10.3 shows, and Table 10.2 lists, the set of data submitted by each site and used to derive the planning factors unique to each site's environment. The next step was to derive the proportion of the time it takes each site to do each activity. To simplify these calculations all the activities of Figure 10.2 were clustered into a set of major work functions, as shown in Figure 10.3. The flow of messages through these major functions is described here as part of the derivation of nontime planning factors, using the data provided thus for illustration.

The first section of Table 10.2 contains the message flow characteristics for a job, as submitted by each site, using the designations shown in Figure 10.2. The second part of the table consists of the planning factor proportions associated with the job and derived from (or associated with) the data in the first part. The calculations used in deriving these planning factors are also included with each line.

The sites submitted this data in one or two acceptable ways. Certain sites submitted data from part one (total number of messages passing through each function), from which we estimated the proportions of messages with respect to the input number of messages. Honolulu invariably submitted the proportions directly on its flow diagram (presumably based on work samples taken), from which we derived the number of

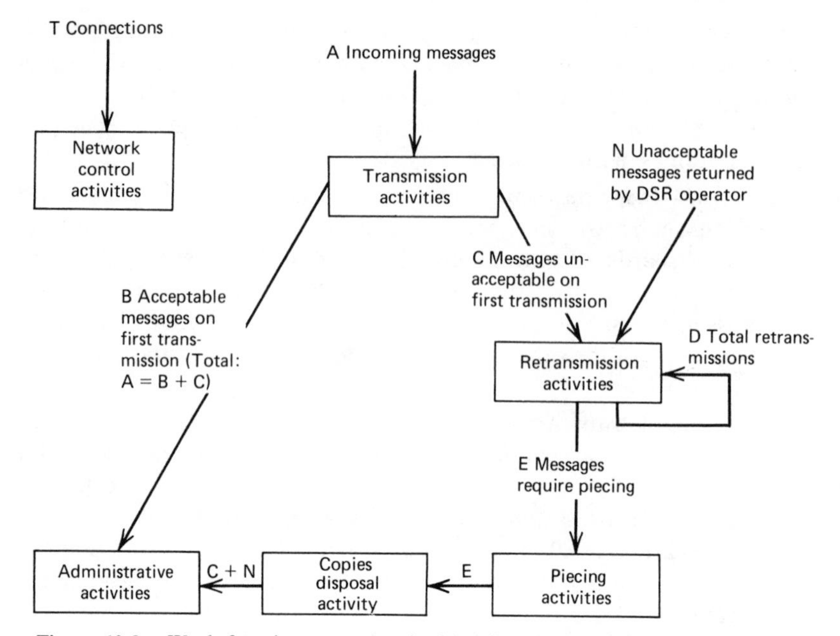

Figure 10.3. Work functions associated with full-period termination, receive, job.

messages involved. When both types of data were missing, we derived the required data by using the mean of the pertinent data submitted by the other sites.

In all cases the data we derived are explained in the text and are shown in parentheses in the table. Each site was asked to reexamine its operations and submit the data it omitted so that the most accurate planning factors could be derived. But the analysis that follows shows how "average" planning factors can be derived based on the entire set of data originally provided for by the sites.

In establishing a satisfactory communications link between ship and shore before any messages can be processed, *T* connections per week need to be made for the number of circuits operated (see Figure 10.3). Line 1 of Table 10.2 indicates that no site submitted the appropriate data.

The most important planning factor is the number of messages received per week (*A*) listed in line 2 of the table, since many of the other planning factors and the basic manpower calculation are derived from this number. Honolulu omitted this important data. We estimated the number of messages as 694,[3] based on the percentages supplied with their flow diagram.[4]

[3]All units labeled as *messages* in this volume imply a quantity of messages in one week.

[4]This estimate was based on the following data they supplied: (1) there were 1070 transmis-

Unfortunately, the estimated total is not consistent with the other data Honolulu submitted, and Honolulu was asked to reexamine the set of data and the percentages they initially submitted.

Norfolk submitted its data on a per work station basis. These had to be converted to a total to allow us to calculate billets properly.

All A messages undergo the initial transmission activities. Of these A messages, B are received in acceptable form on the first transmission (line 3 of Table 10.2). The proportion of acceptable messages is thus B/A (line 9). The value of B was not submitted by Guam and Italy. Its calculation was based on the combined proportion of line 3 to line 1 for Honolulu and Norfolk:

$$P = \frac{538 + 392}{694 + 489} = \frac{930}{1183} = 0.79$$

For Guam,

$$B = (0.79)(2794) = 2207 \text{ messages/week}$$

acceptable on the first transmission, and for Italy,

$$B = (0.79)(604) = 477$$

The remaining $C = A - B$ messages are received in unacceptable form on the first transmission. Not only do the C messages have to be retransmitted (at least once), but so do N messages (listed in line 4), which have been returned to the operator by the data speed-reader (DSR) operator. The proportion of unacceptable messages requiring transmission is $(C + N)/A = [1 - (B - N)]/A$, listed in line 10. Because of $C + N$ unacceptable messages, a total of D messages are retransmitted.

The total number of transmissions therefore will be $A + D$, which is how the sites presented their data (line 5). The proportion of retransmissions (D/A) as derived is listed in line 11 of Table 10.2. Guam did not submit either set of data. So the proportion of transmissions Guam received per week (line 11) was derived as the weighted mean (the ratio of line 5 minus line 2 to line 1 for the other three sites):

$$P = \frac{(1070 - 694) + (588 - 489) + (785 - 604)}{694 + 489 + 604} = \frac{656}{1787} = 0.37$$

For Guam, $(A + D) = (1.37)(2794) = 3828$ transmissions/week.

---

sions per week, of which 67% were acceptable; (2) there were 13 pieced messages per week; (3) of the acceptable transmissions, 20% were transmitted before; (4) of the unacceptable transmissions, 5% can be acceptably pieced together.

Table 10.2 *Nontime Planning Factors for Full-Period Termination, Receive, Job*

	Honolulu	Guam	Norfolk	Italy
Flow characteristics submitted by sites				
1. Number of network connections per week (T)	—	—	—	—
2. Number of messages received per week (A)	(694)	2794	489	604
3. Number of messages acceptable on first transmission (B)	538	(2207)	392	(477)
4. Number of messages returned as unacceptable from the DSR operator (N)	—	—	—	—
5. Number of transmissions received ($A + D$)	1070	(3828)	588	785
6. Number of messages requiring piecing (E)	13	(335)	60	(72)
7. Average length of message	650	1400	1200	1261
8. Average length of retransmitted message	450	(1638)	1400	(1475)
Derived characteristics				
9. Proportion of messages acceptable on first transmission (B/A)	0.77	(0.79)	0.80	(0.79)
10. Proportion of messages requiring transmission $(C + N)/A = (A - B + N)/A = (B - N)/A$	0.23	(0.21)	0.20	(0.21)
11. Proportion of retransmissions required $(D/A) = (A + D)/A - 1$	0.54	(0.37)	0.20	0.30
12. Proportion of piecings required (E/A)	(0.02)	(0.12)	0.12	(0.02)
13. Message length deviation relative to 1200 characters standard (line 7 − 1200)/1200	−0.46	+0.17	0	+0.05
14. Retransmission message length deviation relative to 1200 characters standard (line 8 − 1200)/1200	−0.63	+0.36	+0.17	+0.23

Of the C total of unacceptable messages, a fraction E requires piecing, which is the set of activities combining the information on several messages into one acceptable message.

Guam and Italy did not report the number of messages pieced per week (line 6). The proportion of piecings required by the other two sites (line 12 of Honolulu and Norfolk) shows that they had a large difference between them (1:5) but the proportion of retransmissions required (line 10) was in the other direction much lower (4:1). These differences might be explained by different site policies concerning piecing, as opposed to retransmitting.

For example, Honolulu's data, compared with Norfolk's, indicate that Honolulu pieces fewer of its messages, relying instead on more retransmissions. Furthermore, it seems that Guam and Italy are more like Norfolk than Honolulu with respect to proportions of unacceptable messages and retransmissions (lines 2 and 5). Italy noted further that it requests a retransmission only twice. If these are still unacceptable, the message is "logged out to control," resulting in a small number of piecings by the full-period termination operator (but a higher work load for control). We therefore decided to use Honolulu's lower piecing proportion (0.02) for Italy and Norfolk's proportion (0.12) as the standard for Guam. The number of messages requiring piecing (E in line 6) was then calculated this way: Italy's piecings (E) = (0.02)(604) = 12; Guam's piecings (E) = (0.12)(931) = 112.

After the C number of original unacceptable messages are received in, or pieced into, acceptable form, the incomplete copies are disposed of. The planning factor derived as a proportion (C/A) is listed on line 10.

Once an acceptable message is obtained, the administrative activities are done. These are done only once following message processing. It is assumed here that all A messages are completed and that they all go through this step.

The average lengths of original and retransmitted messages are listed in lines 7 and 8; the ratios of their deviations in length relative to a 1200-character message are listed in lines 13 and 14.

Neither Guam nor Italy submitted the average length of a retransmitted message. But they did indicate that when a retransmission is required, the entire message is retransmitted. Because a Honolulu retransmitted message is shorter than the average message, it implies that Honolulu only retransmits that portion of a message that is unacceptable. Since longer messages are apt to require retransmission more often than short ones, we can infer from Norfolk's data that Norfolk also retransmits the entire message. We then decided to use only the Norfolk data (lines 7 and 8) as the basis for estimating line 14:

$$P = \frac{1400}{1200} = 1.17$$

Thus Guam's retransmitted message length = (1.17)(1400) = 1638 characters per retransmission (line 8), and Italy's = (1.17)(1261) = 1475.

There is another inconsistency in the Honolulu data. That station's shorter message length on retransmissions and its small proportion of piecing seem inconsistent. But the extrapolation is the best that could be done with the available data.

10.3.4 Calculating Manpower Requirements

Using Figure 10.3 and Tables 10.1 and 10.2, the manpower requirements for the full-period termination, receive, job at Guam were calculated (using Table 10.3) as described here.

10.3.4.1 Calculating Average Times Required per Message

List the major work functions in column 1 and the standard operator and total times they require (in seconds) in columns 2 and 3, as obtained from Tables 10.1 and 10.2.

Also list in column 1 any relative deviations from the standard times listed. The only deviations that occur relate to message length for a transmitted or retransmitted message. These relative deviations are listed in lines 13 and 14 of Table 10.2 as +0.17 and +0.36, respectively. Next, convert these relative deviations to time deviations by multiplying by 144 seconds (the machine time required for 1200-character message); list the value obtained in column 3 of Table 10.3.

List in column 4 the proportion that indicates how often each function is done for one message. These data and the formula used are obtained from the bottom half of Table 10.2.

Since no data concerning the network control function were provided by any of the sites, the calculation of the total time required for this job was made without considering this function. But the calculation for this function can be illustrated in algebraic form. Let M equal the number of man-minutes required each time a full-period termination circuit is to be terminated. Let N equal the number of such terminations per year. Thus the actual working man-hours per year (MH) required for network control are

$$\text{MH} = \frac{MN}{60}$$

Calculate the average operator time required for each function as the

Table 10.3 *Man-hours Required at Guam for Full-Period Termination, Receive, Job*

(1)	(2)	(3)	(4)		(5)	(6)
			Proportions		Average	Average
Work Functions	Operator Time	Total Time	Designation	Value	Operator Time	Total Time
Calculating average operator and total time per message						
1. Transmission	15	159	*A/A*	1.0	15	159
Message-length deviation (0.17) × 144		+24	*A/A*	1.0		+24
2. Administrative	45	45	*A/A*	1.0	45	45
3. Retransmission	42	186	*D/A*	0.37	16	69
Message-length deviation (0.36) × 144		+52	*D/A*	0.37		+19
4. Copies disposal	17	17	$(C + N)/A$	0.21	4	4
5. Piecing	174	174	*E/A*	0.12	21	21
6. Network control	—	—	—	—		
7. Average time required per message	293	581	—	—	101	341
8. Operator time ratio					0.30	
Calculating working man-hours required per year						
1. Total operator time (in seconds) required per message (column 5)					101	
2. Number of messages per year (supply)					145,288	
3. Operator time (in seconds) per year (lines 1 × 2)					14,674,088	
4. Convert time from seconds to hours per year (line 3/3600)					4,076	

product of columns 2 and 4 and list this in column 5. Calculate the average total time required for the function as the product of columns 3 and 4, and list in column 6. Find the average operator time required as the sum of column 5; find the average total time required as the sum of column 6.

10.3.4.2 Calculating the Operator Time Ratio

Calculate the operator time ratio (OTR) as the ratio of the average operator time to the average total time required per message. Thus for this job

$$\text{OTR} = \frac{101 \text{ seconds}}{341 \text{ seconds}} = 0.30$$

See Table 10.3, line 8. Use of this factor is explained under the calculation of billet requirements given in Appendix 10.A on planning logic.

10.3.4.3 Calculating the Working Man-hours Required

Once the average operator time per message has been calculated, the man-hours required for actual operator work time can be calculated as the product of the average operator time per message year and the number of messages per year, converted to man-hours (Table 10.3, second part). As discussed in Appendix 10.A, there may be some unavoidable operator idle time that must be added to these man-hours to determine the total man-hours per year required.

10.3.5 Piecing Versus Retransmission

Further analysis of the proportions listed in Table 10.2 (line 12) indicated that Norfolk pieced about 12% of its messages; Honolulu pieced only 6%. On the other hand, Norfolk required only 20% retransmissions (line 11), whereas Honolulu required 54%. This seemed to be a difference in policy regarding piecing, since the more retransmissions obtained (at a cost of extra time for the ship operator and the circuit), the fewer piecings required by the fleet center operator. Therefore an analysis was made to see which policy is better (based on the least time required).

To make this analysis, the average times required per message (Table 10.3) was reconstructed using Honolulu and Norfolk data (Table 10.4). The first half of the table shows the times (operator and total) and the proportions associated with the two main functions being analyzed (retransmission and piecing). The table shows that the average operator times required per message for the two functions are about the same—27 and 29 seconds. However, Norfolk's total time is much less than Honolulu's—58 versus 104 seconds. The times for the other two functions

Table 10.4 *Average Times Required per Message for Different Rework Policies—FPT Receive*

Work Function	Operator Time	Total Time	Honolulu Proportions	Norfolk Proportions	Average Operator Time Honolulu	Average Operator Time Norfolk	Average Total Time Honolulu	Average Total Time Norfolk
Retransmission	42	186	0.54	0.20	23	8	100	37
Piecing	174	174	0.02	0.12	4	21	4	21
				Subtotal	27	29	104	58
Transmission and administrative	60	204	1.00	1.00	60	60	204	204
Copies disposal	17	17	0.20	0.20	3	3	3	3
				Subtotal	63	63	207	207
				Total	90	92	311	265

were also calculated (under standard conditions) and added to the first two so that the relative figures could be shown.

The conclusions drawn from this analysis are

- There is only a 2% gain in operator time by doing less piecing at the cost of more retransmissions.
- The total time required by more retransmissions is 58 seconds per message (or 17% more). This becomes a problem only when the operator does not have access to enough circuits to keep busy, and the layout should be constructed so this will not happen.
- The major waste is in the extra ship operator and circuit time required for the extra retransmissions required by Honolulu. No calculation of this was made because the two policies gave roughly the same results in operator time without this additional factor.
- Honolulu's policy of higher retransmissions and less piecing seems inferior to Norfolk's.

10.3.6 Work Samples Taken

To obtain some validation of the time estimates provided by the sites, work samples of the full-period terminations send and receive and the data speed-reader jobs were taken at Norfolk; the results appear in Tables 10.5 and 10.6. Eight sets of measurements (columns 1 through 9) were taken for this operation, showing

- The ship circuits that each operator serviced during his test (lines 1 and 2).
- The total number of send and receive messages serviced by the operator during his test (lines 3 and 4).
- The operator working time (in minutes) required for the send messages (line 5), receive messages (line 6), idle time when no messages were being serviced (line 7), and total time elapsed—that is, the sum of these three times (line 8).

Column 10 shows

- The number of send and receive messages serviced during the eight tests (lines 3 and 4).
- The working times devoted to send (line 5) and receive (line 6) messages.

Table 10.5 *Summary of Data Collected on Full-Period Termination Operation (all times in minutes)*

(1) Sample Characteristics	(2) 1	(3) 2	(4) 3	(5) 4	(6) 5	(7) 6	(8) 7	(9) 8	(10) Total
1. Send terminal (No.)	Forrestal (1) JFK (2)	Independence (1)	Autec (1) Mitscher (1) Forrestal (1)	Autec (1) Mitscher (1) Forrestal (1)	Shenandoah (1) Independence (1) Nimitz (1)	Inchon (1) Shenandoah (1) Iwo Jima (1)	Shenandoah (1) Nimitz (1) Independence (1)	Forrestal (1) JFK (2)	
2. Receive terminal (No.)	Forrestal (1) JFK (2)	Independence (2)	Autec (1) Forrestal (1) Mitscher (1)	Autec (1) Mitscher (1) Forrestal (1)	Independence (2) Nimitz (1)	Inchon (1) Shenandoah (1) Iwo Jima (1)	Independence (2) Nimitz (1)	Forrestal (1) JFK (2)	
3. Number of send messages	23	15	5	6	7	5	26	21	111
4. Number of receive messages	23	5	14	10	7	4	6	19	88
5. Send time (min.)	15.18	22.78	4.35	5.28	12.06	6.48	33.81	37.50	137.34
6. Receive time (min.)	18.88	7.89	14.25	14.40	8.74	4.63	6.63	20.62	96.04
7. Idle time (min.)	20.11	11.88	10.80	5.64	11.70	22.89	7.96	15.23	106.21
8. Total time elapsed	54.29	42.55	30.52[a]	25.32	32.50	34.00	48.40	73.35	340.93
9. Retransmitted messages (send only)	0	0	0	0	1	0	2	3	6
10. Occupancy (U)	0.63	0.72	0.65	0.78	0.64	0.33	0.84	0.79	5.38

[a]The operator spent 1.12 minutes placing new tape on monitor machine.

Table 10.6 *Summary of Work Sample Data*

(1) Operation[a]	(2) Time per message[b] (seconds)	(3) Standard deviation (seconds)
FPT send	74 + 5	29
FPT receive	65 + 4	17
DSR operation	72 + 2	72

[a]The respective sample sizes and standard deviations are from top to bottom n_s = 111, n_r = 88, n_d = 65 + 2 rejects, s_s = 0.4816, s_r = 0.290, s_d = 1.206.
[b]Confidence intervals have been obtained at the 90% confidence level ($Z_{.05}$ = 1.96).

Before the actual sampling, all supervisors and participating operators were briefed as to why the exercise was being conducted, and the operators were asked to go about their work as usual. Cooperation was excellent.

There was some inconsistency in the procedure in that different operators usually followed several distinct sequences of activities, and they often varied their sequences. However, this may be viewed as a good thing, since the measurements obtained could be considered a good average of different operators and different procedures. Certainly the results represent an average of what actually takes place at a fleet center.

There was also some inconsistency in circuit layout. In some cases the existing layout or work station configuration was such that for one terminated ship, the send set of equipment was not adjacent to the receiver—in fact, it was relatively far away. Different operators were handling the different ends of the same termination in different and separated work stations. This arrangement was considered impractical in HF transmission; when send messages are received garbled by the ship, requests for retransmission come to the fleet center via the receive message end of the termination. This means that one operator is distracted from normal procedures, and either walks to the send side or shouts the necessary information to the send operator for the message to be retransmitted. The same problem arises when received messages are garbled. Thus the time measurements calculated should be higher than those that could be obtained if a proper layout were available.

Although a "normal" operator assignment might consist of a maximum of three send and three receive channels per work station, aircraft carriers were sometimes generating so much traffic that two operators were assigned to one work station. One operator handled send and the other receive, adjacent to one another. Again, the measurement could be viewed as representative of a mix of various conditions.

Since the basic objective of the work samples was to estimate the amount of time per message unit, times for individual activities were sacrificed. Total times were obtained by letting the clock run continuously and by identifying busy interval send, busy interval receive, and idle time, while counting the number of messages sent, received, and retransmitted. These data yielded average time per message for both send and receive, as well as utilization, or occupancy, as a function of message arrival rates and number of channels terminated.

From these data the average full-period termination operator time required for a send or receive message was calculated; this is listed in column 2 of Table 10.6. The standard deviations M for these measurements are also calculated for the sample sizes (column 3) and used to calculate confidence intervals at the 90% level (column 2).

The times measured can be compared with the average times calculated in Table 10.3 this way: Since none of the 88 receive messages required either retransmission or piecing, the measured time of 65 ± 4 seconds compares favorably with the standard operator time of 66 seconds for transmission and administrative activities only. Obviously, we have no check on the time required for other functions.

10.4 SUMMARY OF PROBLEM-SOLVING APPROACH FOR ANALYZING MESSAGE-PROCESSING OPERATIONS

The approach used in Chapter 9 also applies to the analysis of message-processing operations, as well as to the analysis of any "production type" of operation:

- Identifying all jobs to be done (processing one message of a certain type).
- Determining the average operator time required to do this job.
- Determining the total operator man-hours per year required for the job (the product of the average time per operator and the number of messages to be processed per year).

The main difference in the approach for message processing and the jobs analyzed in Chapter 9 is that one operator is able to do several jobs in parallel, since he is not totally occupied during the entire time required for a job (i.e., the machine does certain tasks unattended). Hence it is necessary to design the operation so that the operator can be utilized as much as possible. This involves problems of equipment layout, so that the operator has ready access to different machines during his available time.

The major steps involved in obtaining manpower standards for this type of operation were

- Agreement on a common procedure for the production process.
- Agreement on a suitable time required to do each work task.
- Calculation of an average operator time and total time required to do a job, based on the probabilities of requiring rework.

APPENDIX 10.A PLANNING LOGIC FOR FLEET CENTER

Having described how to calculate the man-hours required to do each job (as a function of the number of messages per year to be processed), the next step was to generate a procedure for calculating the number of billets needed to operate, maintain, and support the fleet center equipment for different proposed alternatives. This section describes this procedure in the form that it was given to the client, thus enabling him to perform these calculations.

Figure 10.1 is a diagram of the manpower planning process as envisioned for fleet center jobs. Inputs to the process are the characteristics describing a specific system configuration at each site being analyzed, as previously discussed. The system characteristics are then combined with the derived planning factors to give the man-hours needed for the various jobs.

As in Chapter 9, the procedure is to calculate the man-hours per year required for all maintenance, operations, and support jobs identified. These manpower requirements are then converted into billets by considering the standard workweek, equipment layout constraints, and the operator utilization resulting from such constraints.

Manpower requirements for maintenance and support were calculated as described in Chapter 9. The major difference in the approach was in the calculation of operator manpower required.

Operating Manpower Requirements for Each Job

This section shows how to calculate the man-hours required per year to operate each of the fleet center circuits and other jobs as analyzed. Refer to the analysis of the full-period termination, receive, job for an example of this calculation. The logic to be followed may be summarized this way:

- Use as a worktable format the particular table developed to calculate the man-hours required for each job being analyzed (Table 10.3).
- In column 1 list the major work functions involved for the job being analyzed.
- In columns 2 and 3 list the standard operator and total times required for each function.
- In columns 1, 2, and 3 also list any deviations from the standard times caused by the average message length at the particular time.
- In column 4 list the frequency factor associated with each function, following the formulas and planning factors listed for this job at this site.
- Calculate the average operator time required per message for each function as the product of the corresponding data of columns 2 and 4; list this in column 5.
- Calculate the average total time required per message for each function as the product of columns 2 and 4; list this in column 6.
- Calculate the average operator time and total time per message as the sum of columns 5 and 6, respectively.
- Estimate the total number of messages per year expected at the site for this (using the operational work load for factors as a guide); list this in the table as shown.
- Calculate the total operator man-hours required per year as the product of the operator time per message and the number of messages per year (converted to man-hours per year).

Adjusting for Operator Idle Time

In calculating the minimum number of man-hours required for each job, we assumed no operator (avoidable) idle time by using the average operator time per message rather than the average total time required per message. One major consideration affecting this idle time is the total number of circuits one operator can handle in parallel.

Referring to the FPT, receive, example, we see that operator time per message is 101 seconds and total time per message is 341 seconds. From this we calculated (in Table 10.3) the operator time ratio per circuit, defined as the ratio of average operator time to average total time per message (equal to 0.30 in the example).

Thus if the operator were assigned only 2 FPT receive circuits to handle, his working time would be 202 seconds, and the total time would still be 341 seconds; operator idle time would be 139 seconds. The operator time ratio would then be two times 0.3, or 0.6. If the operator were assigned three circuits, the operator working time would be 303 seconds, and the operator time ratio would be 0.9.

But if the operator were assigned four FPT receive circuits, the operator time ratio would be 4 (101)/341, or 1.18. This ratio, however, can never exceed 1. So the calculation really means that the operator is fully utilized, and there is no idle time.

Consider another example. The operator is assigned two of these circuits, as well as a third circuit whose operator time ratio is 0.2. Thus this operator's ratio is 2 (0.32) + 1 (0.2), or 0.84, and there is some idle time.

If the operator also had a fourth circuit of the second time, the ratio would be 2 (0.32) + (0.2), or 1.04; the operator would be fully occupied and there would be no idle time.

Thus in calculating the operator man-hours required for message processing, it is important that the operator time ratio be calculated as shown previously. If it is less than 1, the correct value of man-hours required is obtained (to include operator idle time) by dividing the first value obtained by the operator time ratio. Thus if the operator handles M circuits of one type of job:

$$\text{MH} = \frac{(\text{OT})\,(N)}{(\text{OTR})\,(M)}$$

where MH = man-hours required
OT = operator time per message
N = number of messages per year
OTR = operator time ratio per circuit
M = average number of circuits each operator handles in parallel.

When an operator handles more than one job, both the numerator and the denominator consist of the sum of the products indicated for each job. The denominator cannot exceed 1.

Total Billet Requirements

We now explain how to determine the billet requirements for each type of position analyzed—that is, full-period termination operator and allied operator. Unlike the other sites, where all operators do the same jobs, each fleet center operator is restricted to a specific set of jobs, and therefore a set of calculations must be made for each operator situation.

Two different methods were developed for making these calculations. The first method is to determine the total man-hours per year required to do all jobs associated with the position; using the standard workweek factor, the number of billets required is calculated. Then various constraints are introduced, each tending to increase the number of billets. These include

- The total number of billets required is, on the average, the number of circuits expected to be operating during the year divided by the maximum number of circuits one operator can handle in parallel because of the equipment layout at the site. That is, even if one operator could handle a given message rate, if these messages were distributed over a very large number of circuits, the extra walking time involved would probably require additional operators.
- A minimum of five billets per watch-standing position is required (one billet per watch).[5]

The operator productivity is then calculated (the minimum number of billets required to do the actual work load divided by the number of billets required by layout considerations). If this productivity is low, consideration must be given to changing the equipment layout or job design so that one operator can operate more circuits.

The second method essentially goes through the same calculations, but in a different order. The first determination is the total number of operator billets required (the average number of circuits expected to be operating during the year divided by the maximum number of circuits one operator can handle). Again, a minimum of five billets per watch-standing position is required. Next, the operator productivity is calculated (as described previously) leading to equipment layout changes and job redesign for those positions having low utilization.

The first method was adopted and is described in greater detail.

Work Elements

List the operator position being analyzed at the top of Table 10.7, and in column 1 list all work elements of that position that can be done in parallel by the operator while on watch:

- All operator message-processing jobs.

[5]Assigning four men for every watch position being manned continuously constitutes a four-duty section watch. This results in a 42-hour workweek (including mealtime). Assigning a fifth man for each watch position allows for service diversions, training, leave, and holidays, and results in 33.6 hours per week available for work (including mealtime).

Table 10.7 *Calculating Billet Requirements: Full-Period Termination Operator*

(1)	(2) Man-hours Required	(3) PF&D Factor	(4) Total Operator Work Load	(5) Numerical Factor
1. Work elements				
FPT receive	4076	1.17	4769	
FPT send	408	1.17	477	
Operator PMS	0	Included	0	
Other operating activities	1200	Included	1200	
Collateral support	0	Included	0	
2. Total operator man-hours required			6446	
3. Standard workweek for labor mix (hours)				32.35
4. Minimum number billets required (B_o)				3.83
5. Number billets required by layout				35
6. Operator productivity				10.9%

- Any operator PM action (if applicable).
- All other operating activities.[6]
- Support collateral duty work done by operators.

Man-hours for All Jobs

In column 2 list the operator man-hours per year required for each job in column 1. In the example shown in Table 10.7:

- A total of 4076 man-hours (minimum) was calculated as being required for the full-period termination, receive, circuits.
- Assume that 408 man-hours are required to operate the full-period termination, send, automated circuits.
- No operator PMS is required for this position.
- Other jobs require 1200 man-hours.

In all appropriate cases the working man-hours must be converted into total man-hours by applying the PF&D factor (in column 3). Thus the total number of man-hours for each work element is

$$\text{TOW} = (1 + \text{PF\&D})\text{WMH}$$

where TOW = the total operator work load (listed in column 4)
WMH = the working man-hours
PF&D = the personal fatigue and delay factor

Generally, the PF&D factor is already included in the operator PMS and collateral support man-hours requirements; it therefore should not be listed in column 3 for these jobs. Obtain the total operator man-hours required (row 2 of the table) by adding the man-hours of all work elements; list the total in column 4.

Number of Watch Standers

The next step is to calculate the total number of billets required for each operator watch-stander position considered. Because of layout and operational constraints, the planner must make a series of iterative calculations to arrive at the final value for watch-stander billets.

[6]Any job done by an operator that takes him away from his primary operating location may require a relief operator during the time the operator is away from his primary operating jobs. The relief operator could be the watch supervisor. If these additional jobs cannot be performed in parallel by the original operator, they must not be included in this operator billet calculation, but must be calculated separately in the same way as described here.

Minimum Billets Required

The minimum number of operator billets B_o based on average work load is determined first:

$$B_o = \frac{\text{TOW}}{52}\,(\text{TAW})$$

where TOW = the total operator work load per year (as previously calculated)

TAW = the time available for work per week

According to the standard workweek of 40 hours (where dependents are authorized), TAW should equal 31.94 hours per week for military and 33.98 hours per week for civilian personnel. But a watch stander assigned to a five-man-for-four-section watch is at his station 33.6 hours per week, less time out for meals. TAW therefore is based on a weighted average of these two factors, and it depends on the civilian-to-military mix at the fleet center.

For example, if there were 10 civilian to 40 military direct-labor personnel at a site, the weighted average would be

$$\text{TAW} = \frac{10\,(33.98) + 40\,(31.94)}{50} = 32.35 \text{ hours/week}$$

Enter this value in row 3 and the results of the calculations B_o in row 4, column 5. Carry the billet calculations to the nearest hundredth of a billet until all calculations are completed and round off fractional billets.

Equipment Layout

Another important consideration is equipment layout. Therefore the average number of circuits expected to be used must be considered. Suppose that our illustration is for Guam, which used an average of 45.8 multichannel, single-channel, and dedicated circuits for the work load calculated. For the work load being considered, one man per watch could not handle all these circuits (even though he could handle the message work load if it were at three circuits) because of their extended layout.[7] In

[7]Our analysis assumes that the activity times submitted by the sites from which the standard activity times and man-hours required are derived include delays incurred in the operator's moving from one circuit to another during normal operation. If this assumption is incorrect, an additional average walking time would have to be added to the total time required for those operators whose work stations are separated by a substantial distance. Furthermore our method of calculating man-hours would also show the extra man-hours required by the large number of circuits because we would add "walking from circuit to circuit"; thus the average operator time required per message would increase.

fact, Guam recommends a manning of seven billets per watch for its work load (considering both number of messages and circuits).

Additional billets per watch must be added, based on the maximum number of circuits one operator can handle (as opposed to message load, which has already been satisfied by the work load calculation). Guam's figure, seven billets per watch (or a total of 35 billets), is the recommended manning for the 45.8 average circuits it states were active during the past year. This means one operator can handle an average of 6.5 circuits in parallel under relatively light loads.

List the total number of billets required in row 5, column 5.

Additional Constraint on Minimum Billets per Watch

Check to ensure that there are at least 5 billets for each position (including supervisors). Also determine whether the safety requirement is satisfied (minimum of two men per watch in an isolated area).

Calculating Operator Productivity

Because Navy headquarters bases its billet allowance mainly on operator working time, and the compartmentalizing of jobs invariably leads to lower operator utilization because of all the factors described, it is up to each site to defend its billet recommendations for all positions in which there is considerable operator idle time. This may be done by calculating the operator productivity for each position (defined as that proportion of time the operator is actually working). For those positions whose operator productivity is low, consideration must be given by the site to

- Changing the layout of the equipment so that the operator can handle more circuits in parallel.
- Redesigning the total set of positions to include the possibility of combining low productivity positions, thereby reducing the total number of positions required.
- Using working supervisors to accomplish the same effect.

If none of these can be done, the site must show that each was considered and give reasons why it cannot be done.

Operator productivity (OP) for each position may be calculated this way:

$$\text{OP} = \frac{B_o}{B_r}$$

where OP = operator productivity

B_o = minimum operator billets as originally calculated and based on the total operator per work load

B_r = operator billets required, as finally calculated, based on the other constraints

Fractional Manpower Cutoffs

After the number of billets for each function has been calculated to the nearest hundredth of a billet, fractional manning problems may arise. These are handled the same way as previously described in Chapter 9.

Other Personnel Required

Determine the number of other nonsupervisory watch standers (such as computer operators and programmers) and watch supervisors required. Since no quantitative data regarding the work done by people in these positions were supplied by the sites, judgment must be used when allocating these positions. References describing these positions and recommending a billet allocation are available. These references can be used as a guide. Allocation should be made uniformly unless environmental conditions at the different sites vary the work load for these positions.

Dealing with Periodic Peak Loads

Since the operator work load (number of messages or circuits, whichever is the limiting factor) is generally distributed unequally among the four shift or watch sections) the total billets required should not be distributed equally throughout the watch sections. This, of course, is the peak loader concept, in which the total number of billets is distributed according to the work load on each of the four watch sections. The following example shows how to calculate total billet requirements from an operational viewpoint:

- Three watch standers on each of the first and second watches over the five weekdays only; 30 watch positions.
- One watch stander on each of the first and second watches, weekdays; five watch positions.
- One watch stander on each of the three watches, weekends; six watch positions.
- The total is 41 watch positions.

Since 5 billets equal 21 watch positions, a total of 9.8 billets is required.

Dealing with Peak Loads from Fleet Exercises and Contingencies

In accordance with Navy billeting policies, peak work loads can only be handled by:

- Reducing the number of watch sections.
- Transferring men to the site from reserve components or from other sites not affected.
- Cross training individuals for less complex jobs in the fleet center and deferring some of the less critical jobs. These men thus may be used in the fleet center to handle some of the work load.

Dealing with Random Peak Loads

Random peaks, which do not occur periodically, may be handled this way:

- The precedence system allows higher-precedence traffic to be handled first within allowable time lags at the cost of delays in lower-precedence traffic. If the higher-precedence traffic constitutes such a large proportion of the total traffic that the time delay standards are not met (as shown by work samples), a case can be made for an increase in billets.
- Any backlog at the end of each shift should be eliminated by extending the shift, thus doubling the manpower available. In a "worst case" situation asking some watch personnel to come in an hour early may be possible sometimes. Obviously the extra time spent by an operator should be compensated for when the work load is down.

The annual work load includes at least one fleet exercise containing a peak work load; therefore the number of billets required already exceeds the normal average, excluding fleet exercise work loads.

Qualitative Requirements

Next determine the qualitative requirements of each position in terms of designator, grade, rate, and series. This should be done uniformly, based on the total number of people required in each functional unit.

Part V

DESIGNING TRAINING SYSTEMS

CASE 4: COMPUTER-MANAGED INSTRUCTION VIA SATELLITE

Case 4 has three major objectives. First, Chapter 11 expands the reader's knowledge of the system design process by ilustrating the use of functional analysis in arriving at the system design that provides the best interface with other parts of the total system. Our focus is on the design of a training system that uses a satellite communications system to connect an existing computer-managed instruction (CMI) training system located in the United States with remote sites so that CMI can be conducted at these sites. The part of the case described in Chapter 11 builds on the knowledge of the communications system developed in Case 3. Chapter 12 continues the case with the second objective, showing how to do a more complex systems evaluation than is done in Case 1. Here we compare the CMI via satellite system with alternative approaches for conducting training. Finally, meeting our third objective, Chapter 13 illustrates the analytical simplifications that can be made when time and analytical resources are extremely limited. We show how a "best efforts" type of analysis can be made when all the desired data are not available.

COMPUTER-MANAGED INSTRUCTION VIA SATELLITE

Problem as Given

You are a systems planner with the Information Services Company and have been asked to meet with Tom Carlson, project manager of the company's COMISAT project. Tom hands you a newspaper article and says:

Here is an article from *Navy Campus Magazine* describing the Navy's current CMI (computer-managed-instruction) training system. The Navy is sold on CMI as a way of providing individualized, self-paced instruction. It has been decided to use this technique for as many courses as possible at their various U.S. training centers. Since over 20% of the Navy is in training at U.S. training centers at any one time, the next logical step of development is to explore the use of satellites for delivering certain training courses to students right on their ships or at remote job sites on land, rather than at U.S. learning centers.

As you may know, we have recently been awarded a 29-month contract to evaluate this method of training. The first phase of this contract is a 9-month feasibility study to put together a preliminary plan for conducting a satellite demonstration. It involves a preliminary design of the demonstration, including a description of how it would be operated, the data to be gathered, and the equipment required. It also includes the development of an economic model whose objective is to determine whether satellite-delivered CMI to shipboard or remote land sites is as economical as CMI in the learning center environment. The actual evaluation of the operational COMISAT system will be conducted following the demonstration phase, using the economic model and the performance and cost data gathered during the demonstration.

We are currently doing a preliminary analysis and design of a number of communications systems, each of which would utilize one of a number of possible satellites available to us during the 6-month demonstration phase scheduled to begin 18 months from now. These satellites include NASA's Application Technology Satellite series (ATS-6, 1, and 3). However, all these designs require that a satellite terminal be obtained and installed on board the ship, which is quite expensive.

We originally approached the Navy about using the Naval telecommunications system for our demonstration, since there are shipboard satellite terminals already installed on larger ships. The Navy would not allow us to have exclusive use of their Gapfiller satellite for 1 hour a day as NASA would. Accordingly, we went ahead with our NASA plans. Now we find that there is a reasonable chance that neither NASA nor the Navy will provide the funds for the demonstration shipboard satellite terminal. We think we should reconsider the Navy's offer to transmit our training messages to the Memphis computer, using the standard Naval telecommunications system. The main disadvantage of this is that the total delay time will be several hours rather than the several minutes delay expected with the ATS-6.

I'd like you to take on these problems: Find out if we can use the Navy communications system for our demonstration. Then if we can, design an information-handling system for CMI training, using the Navy Gapfiller satellite.

If such a system will meet our needs, evaluate that system against the other information-handling systems being considered.

Second, I'd like you to tackle the demonstration objective of determining whether satellite-delivered CMI to shipboard or remote land sites is as economical as providing CMI in the learning center environment. Describe the factors you would consider, the data you would gather, and the approach you would use in answering this question.

Here is the article describing the current Navy CMI system.

COMPUTER MANAGED INSTRUCTION—NAVY STYLE[1]

The Navy has placed into operation, following several years of research and development, a computer application to instruction. It is the purpose of this article to describe this system and its major characteristics. But first we will speak to some instructional philosophy by way of introducing CMI, Navy style.

Because we will be discussing one aspect of instructional technology, we should have a commonly acceptable definition of that term. In 1970 a Commission on Instructional Technology made its report to the President and the Congress, in which the following definition was offered:

"Instructional technology is a systematic way of designing, carrying out and evaluating the total process of learning and teaching in terms of specific objectives based on research in human learning and communications, and employing a combination of human and nonhuman resources to bring about more effective instruction."

There are two major dimensions to the application of instructional technology.

The first of these is the technology involved in the structuring of a course or program of instruction . . . in the military training commands this is by common agreement called *"instructional systems development."*

The second dimension is that activity which makes available to the student the media for effecting the desired behavioral changes, or put another way, allows him to achieve his learning objectives. We might refer to this dimension as the *"delivery system."* There are, broadly speaking, two major categories of these delivery systems. One we could call "conventional group" describes the kind of instructional delivery system with which most of us associate when we recall our own personal school experiences. This, of course, is the system where thirty or so students sit in a classroom and listen to the facts which the teacher has decided are most helpful for them to know. The other we can call "individualized instruction."

[1]From Worth Scanland (Chief, Naval Education and Training, Pensacola, Florida), *Navy Campus Magazine*, December 1975, pp. 25–27, reproduced by permission.

Delivery sub-systems that could fall within the general classification of methods by which we deliver instruction would include textbooks, lectures, discussion, tutoring, demonstration, laboratory, experimentation, programmed instruction, and computer assisted instruction. These subsystems fit into any major system of instructional delivery, and that includes both conventional group and individualized instruction. Individualized instruction uses the same subsystems as other kinds of instruction, but is different because it attends to a certain very, very important human characteristic, called individual differences.

In the Navy, individualized instruction attends to individual goals, to individual entry skill levels, to individual learning rates, to idiosyncratic learning methods, and to criterion rather than normative referenced measures for determining mastery of objectives by the student. If we attend to all these individual differences we make instruction truly humanistic but we also make it very difficult for the instructor to manage. He no longer has thirty or so students to whom he can give out a common assignment, or lecture to, or administer a test for normative discrimination. Now he must attend to thirty (or some number of) students each of whom is, in effect, a class by himself.

An example of the kinds of communications and data exchanges required to cope with a well individualized instructional program can be seen in the flow diagram which is intended to show a student entering an individualized course and proceeding down through the network of activities. The solid lines represent the path of the student through the program, while the dotted lines represent the information and data flow. At every step of the process there is some kind of information or data flowing back and forth between various elements of the system. If an instructor is managing this operation, one can appreciate the kind of demand which is placed upon his ability to listen, diagnose, prescribe, test, counsel and administer. In fact, it can quickly become overwhelming if the number of the students passes some critical level, probably between ten and fifteen. If the number of students exceeds ten to fifteen, the instructor must get an assistant instructor, and so on.

Now we can make the case for the Navy concept of individualized instructional management. There are three basic axioms upon which the Navy CMI system has been built:

- that well delivered individualized instruction demands a very high level of information and data flow and processing.
- that in these times of automated data processing technology, large quantities of data justify the application of ADP equipments and techniques, and
- that thus viewed, the ADP process is intended for the *management* of instruction, *not* for its delivery, and there *is* a significant difference. We have implacably resisted the many well-intentioned efforts to corrupt our management system with "just a *little* delivery, please," and will continue to resist until our established objectives have been fully met . . . then we may reconsider.

When one is contemplating the problem of how much data can be manually handled before the high costs of ADP equipments are seriously considered, he usually turns to others experiences for useful information. There was, however, at the time when we were faced with these questions, no other experience upon which we could draw. No one had previously thought of managing hundreds, let alone thousands, of students in an instructional program through automated data handling techniques, so the alternative to real world experience was simulation. With the assistance of a federal group known as FEDSIM, we developed a computer program to reasonably simulate our CMI system as it was envisioned and tested in a research mode. By altering the principal variables of student loading and instructor to student ratios and costs, we arrived at the indicated conclusion that somewhere in the neighborhood of 1,300 students on the system at one time caused the *"instructor managed instruction"* curve to cross the computer managed instruction curve. Thereafter, as the number of students increased, the savings attributable to CMI also increased as in the graph shown.

This argument won for us approval and funding for the acquisition of our new hardware and the purchase was consummated last spring.

In the meanwhile we have been renting computer time on a machine located twenty miles from our instructional site.

The third illustration is a fairly simplified diagram intended to show the principal features of the Navy CMI system. As is usually the custom in such systems, the student is registered on the computer, which acknowledges this action by giving him a student number and directing him to a pretest to determine his entry level into the program. The student takes this pretest and feeds to an optical scanner the answer sheet he has just completed. The scanner reads the sheet and transmits the information to the computer, returning the answer sheet to the student. In an average turn-around time of about 12 seconds the computer diagnoses the answers, prescribes a learning module for the student and delivers this to him at the

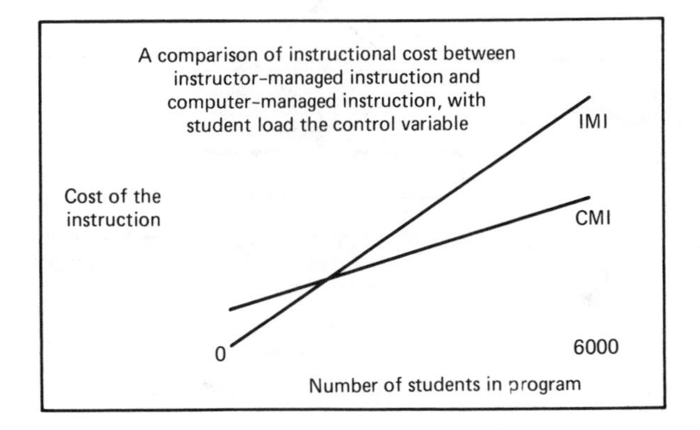

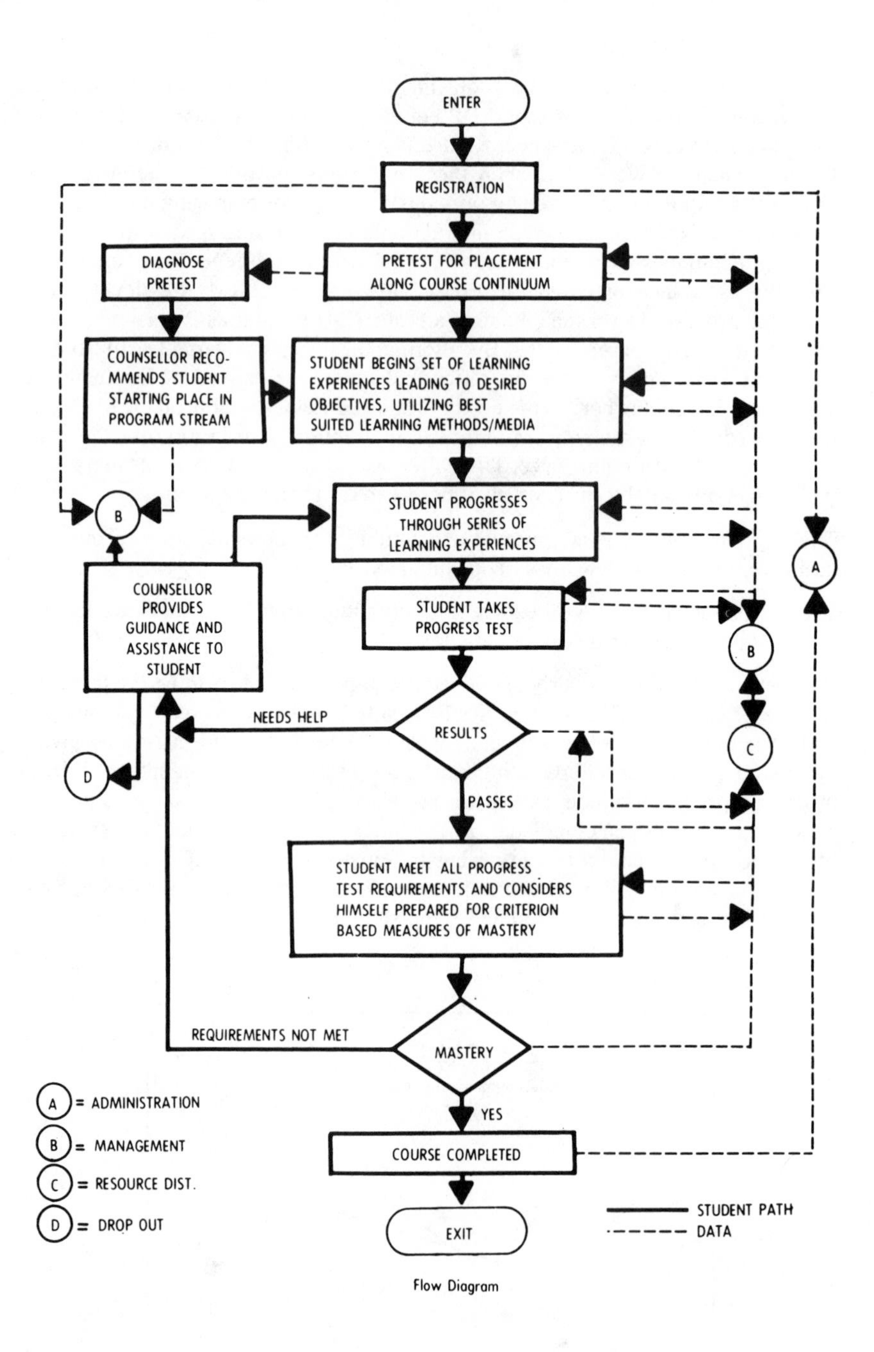

Flow Diagram

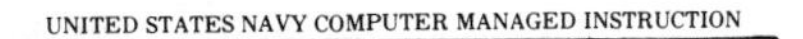

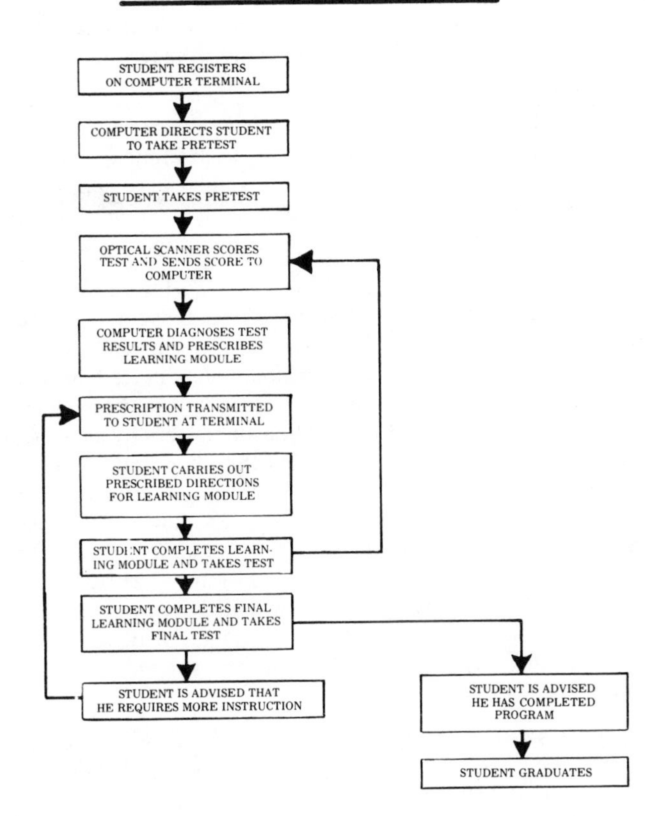

terminal by way of a small, high speed printer. The student then tears off his prescription and carries out the directions offered to him.

At the end of that learning experience he takes a progress test and repeats the procedure with the optical scanner, iterating this process until the computer informs him that he has successfully indicated mastery of all the objectives set for him and has therefore completed the program. This is very similar to the first flow diagram which describes the well developed individualized instructional program. What we have done here is substitute a computer program for the counselor and the administrator. The student does all his learning from materials provided from the learning center, not from the computer. If he runs into trouble and needs help, there is someone (the Learning Center Supervisor) on hand to provide it, although usually he is able to move along entirely on his own.

The Learning Center Supervisor also uses the computer to assist him in the administration and management of his students. For example, he may wish to know whether student Jones is falling behind the predicted rate of advance, or what the current average time for course completion may be, or where student Smith is today in his progress through the program. These and other kinds of information are available to him from the computer upon inquiry.

The Navy recently purchased a large Honeywell Series 60 computer for installation at the training facility near Memphis. Learning centers are now in operation at both Memphis and San Diego, both sites using leased time on a Xerox Sigma 9 machine located at the Memphis State University. Plans are in process to install additional learning centers at the Great Lakes Naval Training Center and at Pensacola, to be tied in with the new computer under installation at Memphis.

There have been some encouraging achievements during the early life of the Navy CMI system. There are currently about 3,000 students on the system. This is a growth from 25 in the summer of 1973. Because of shortened student time to mastery, we have saved an estimated ten million dollars in student salaries, or at least the charging of those students' time to the training command. This has been done through the reduction of mean student time to mastery by an average of over 42 percent. We need to make it clear here that this shortening is not *because* of CMI—rather it is because of the good design of the programs and the individualization of the instruction. The latter would have been impossible for the numbers of students we are dealing with *were it not for the availability of CMI.* In addition, we have seen a mean reduction of student to staff ratio of 23 percent and that is truly attributable to CMI. It gives a partial answer to the restraints the Congress is placing upon us for reductions in the costs of training and the size of training staffs. And finally, this has all happened while the students' performance has improved and the attrition rate has reduced.

It is hoped that these data will give you some idea of the reasons for the enthusiasm we have for CMI, Navy Style, and the reasons we believe it is the way of the future. Speaking of the future, here are our plans for the years ahead. By 1978, barring budget cuts which remove our capability to expand, we will have 17,000 students on the system. Were it not for the availability of CMI, we would need 1,700 more instructors to manage that number of students in an individualized mode, or we would be forced to put the students back into a lockstep mode, and that would add some 5,000 more students to the pipeline, at an annual cost of 25 million dollars. By 1980 we expect to substitute satellite communications techniques for our currently expensive land lines for the transmission of data between the central computer and outlying stations. Perhaps most exciting of all, we will be managing the instruction of students on ships at sea and remote overseas bases by our CMI system delivered by these same satellites. That will truly be CMI, Navy Style!

DEVELOPING THE ANALYTICAL APPROACH

Understanding the Problem

The following statements characterize a current statement of this problem:

- There exists a CMI system that is currently operational for the Navy.
- This system consists of

  Audiovisual training materials for a number of CMI courses the students can use at various U.S. learning centers located at San Diego, Great Lakes, Orlando, Memphis, and Pensacola.

  A set of course instructions that starts each student in an assigned course.

  An information-handling system that connects the student from his learning center to a computer located in Memphis and programmed to interface with students for these CMI courses. The primary function of this information-handling system is to accept each student's test answers, feed them into the computer which scores the test, and then inform the student of his test results and his next assignment in the course. The computer also informs the student's learning supervisor (LS) of the progress of each student.
- The Navy is pleased with the results achieved so far. They find CMI to be superior to traditional instruction, which is not self-paced, or instructor-managed instruction, which is self-paced but requires the instructors to do the same functions done by the computer. There are several reasons for this conclusion:

  The students complete the course in shorter times than they do in traditional courses.

  The amount of instructors and support personnel[2] required at a learning center for a given volume of students has decreased appreciably.
- For these reasons the Navy has decided to expand its CMI development program at the shore training centers. In addition, if some form of information-handling system can be designed to operate between a student located *at a remote site* such as a Naval ship and the Memphis computer, part-time students will be able to take available CMI

[2]The CMI computer also performs many administrative functions that previously required support personnel. These functions included keeping track of student progress, thus determining when a student is expected to complete the course, and arranging for the assignment of new students to learning center carrels when they become available.

courses right on the ship rather than having to be transferred to a U.S. training center. Having such an information system available, the concept of operation would be similar to the present CMI system operating at the U.S. training centers.

Understanding the Type of Problem

Two different problems need to be addressed in this case:

- Design of an information-handling system using the existing Naval telecommunications system and satisfying the needs of the COMISAT demonstration.
- Development of a systems evaluation approach[3] for determining whether satellite-delivered CMI to ships and other remote sites is more economical than CMI as currently used in the training center environment.

The first problem is described in Chapter 11, the second in Chapter 12.

[3]The evaluation approach was defined as an "economics model."

11

Designing the Satellite Data-Handling System

11.1 DEVELOPING THE ANALYTICAL APPROACH

Since the first two problems relate to the design and evaluation of alternative information-handling systems, the analytical approach developed in Parts II and III can also be used for these problems.

11.1.1 Analysis of the Current CMI System

One of the first steps was to obtain a better understanding of the operation of the current CMI system, particularly the current data-handling system. This was obtained by a review of available CMI reports describing the system and by meeting with personnel who were

- Operating and continuing to develop the current CMI computer and information delivery system in Memphis. This was important because we had to understand all the constraints that the satellite system would have to meet in interfacing with the CMI computer and the type of data currently being transmitted through the system.
- Operating various training centers using CMI. This gave us an opportunity to see the current system in operation and meet with students, learning supervisors, and officers in charge of each course to discuss our problem with them.

The information flow within the current CMI system can be modeled as in Figure 11.1:

- After the student completes the audiovisual material in a lesson, he takes the examination (a multiple-choice test) and also inserts his identification (Social Security) number and lesson number on the test sheet.

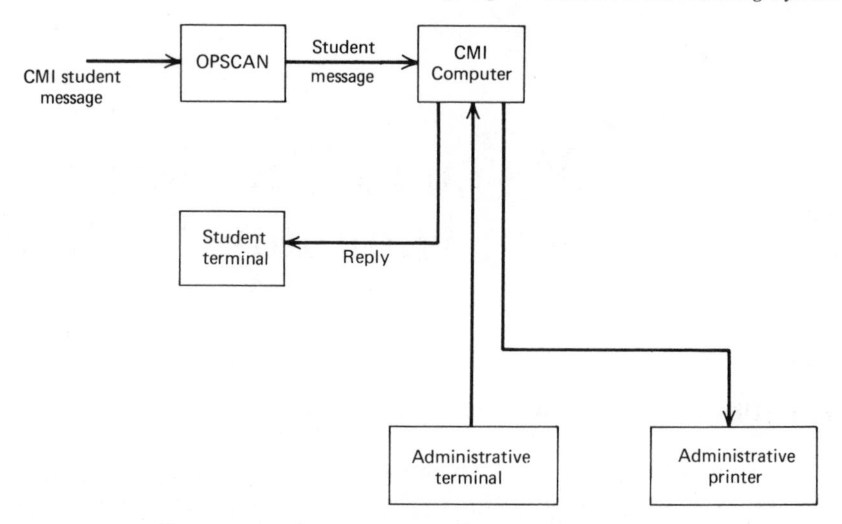

Figure 11.1. Current system information flow diagram.

- He inserts the sheet into an optical scanner (OpScan) device, which senses the answers and converts these to an electrical coded signal that is transmitted over telephone lines to the time-sharing, CMI computer located near Memphis.
- The answers are then compared with the correct answers for this lesson, a score generated, and a reply transmitted back to the training center originating the message and typed on the student terminal. The total process requires 12 to 20 seconds, depending on the total number of students requesting service at that time.
- The reply appears in English text, phrased in the same way that an instructor would communicate with the student. Replies are of two types. If the student's answers met all the criteria associated with the lesson, the student is congratulated and told to advance to a given page, his next lesson. If the criteria are not all met, he is told to go back to a particular remedial section for further study.
- The entire process is repeated for each lesson.

In this manner the CMI system manages the student's instruction. In addition, it performs two other functions:

- Assists the learning supervisor in managing student progress.
- Manages administrative support functions.

Each of these functions is now described.

11.1.2 Managing Student Progress

The primary functions of the LS are to provide technical assistance to the students on subject matter as requested and to monitor the progress of each student by determining whether he is meeting the expectations of the school. The CMI computer functions as a management control system.

- The milestones for completing the various lessons are constructed for each student, based on a predictor model using the results of a battery of standard tests taken by that student while in recruit training.
- Each time the student completes a lesson, the completion time is recorded.
- At any time the LS can compare each student's last completion time with his predicted time to see whether these times are within acceptable limits. If the student is not keeping up, he is counseled to see what the trouble is. In fact, one of the main reasons stated for the reduction in time to complete the course under CMI is the amount of monitoring and counseling an LS does with lagging students. For example, after the first week of training, if a student has not satisfactorily completed the number of lessons predicted, he is directed to evening study hours in a quiet facility during the second week. If he still has not caught up by the end of the second week, he meets the academic board for possible dismissal from the school.

11.1.3 Managing Administrative Support Functions

Another function of the CMI computer is to handle a number of routine administrative functions such as

- Keeping track of free learning center carrels.
- Assigning the next student on the class waiting list to a carrel.
- Registering the student.

This relieves an administrative clerk of these tasks.

11.1.4 Information Flow to Other Sites

In addition to serving the learning centers at Memphis, the CMI computer functions as a time-shared computer for a number of other locations around the country, including Great Lakes, Illinois; Orlando, Florida; and San Diego, California. Dedicated phone lines interconnect the CMI computer with the terminals at these sites (see Figure 11.2). In fact, by having

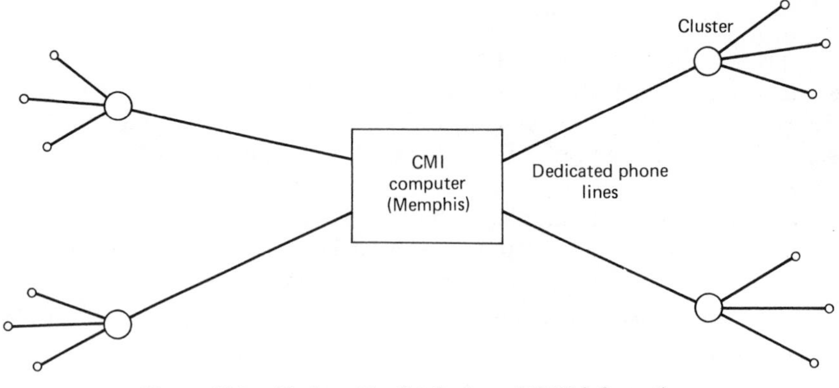

Figure 11.2. Nationwide distribution of CMI information.

each main line from Memphis go to a "cluster" node, it may service a number of learning centers in the vicinity, thus minimizing the cost of long-line communications links to Memphis.

11.2 ANALYSIS OF A SATELLITE-DELIVERED CMI SYSTEM

Another system design effort under way deals with a satellite system to connect a ship to the CMI computer.

Figure 11.3 illustrates the information flow through a satellite-delivered CMI system and serves as an operational concept for the system. Basically, the major change in the satellite system is to substitute a combination land-line and satellite path for the all-land-line path of the current CMI system. Figure 11.3 illustrates that a signal may be sent from a ship to the CMI computer in Memphis by using a path composed of

- A ship satellite terminal (SST).
- The satellite (the NASA ATS-6).
- The NASA earth terminal located in Rosman, North Carolina.
- Land communications (probably a dedicated phone line from the terminal to the Memphis computer).

In this case all CMI messages are inserted into the OpScan equipment during that time of the day when the satellite is assigned to this demonstration and the ship's satellite antenna is aligned to the satellite. Once the path is established, signals may be sent in either direction, just as in the

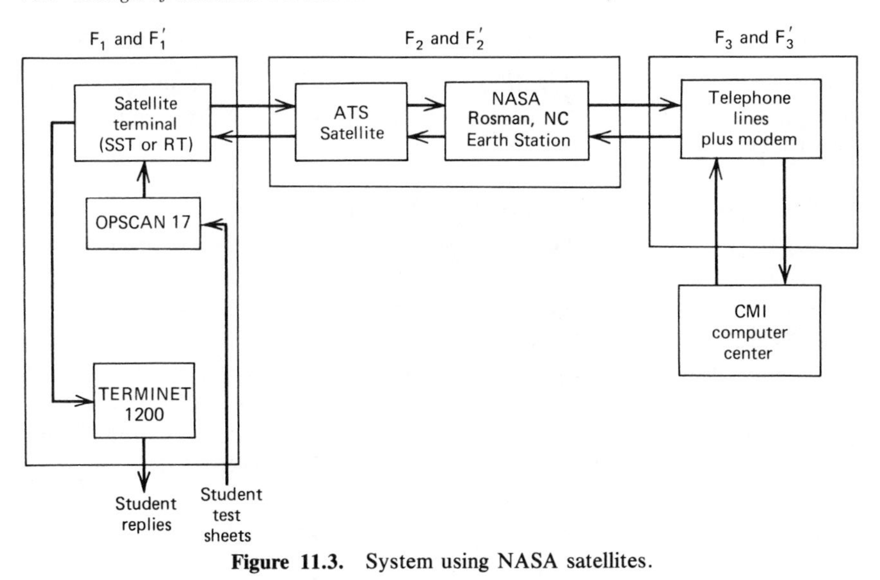

Figure 11.3. System using NASA satellites.

current CMI system. Since the ATS-6 would be available for our use only once or twice a day, all CMI messages would have to be sent in batch during the available period. If only two transmission periods were available each day, a CMI message might be delayed as long as 12 hours (6 hours on the average). It was felt this was not an insurmountable problem, as is described in the analysis presented later in this chapter.

11.3 DESIGN OF ALTERNATIVE DEMONSTRATION COMMUNICATIONS SYSTEMS

The remainder of this chapter describes the analysis performed in determining the preferred communications system to be used for the demonstration. This analysis included these parts:

- Analysis of the communications system alternatives considered.
- Evaluation of these alternatives in terms of performance provided and costs incurred for the demonstration.

In implementing these two objectives it was important that the systems planner make certain that (1) all feasible system alternatives were considered, and (2) the preferred system was selected on some objective basis.

11.3.1 Structuring the Analysis

To implement the first objective the entire information delivery system was structured (see Figure 11.3) and then generalized (see Figure 11.4) into these major functions:

1. The *transmission function* (F_2 and F_2^1), which can transmit the student test data from the remote site (ship) to the CMI computer (F_2), and transmit the reply message back to the site (F_2^1). These functions can be implemented by a number of alternatives listed in Figure 11.4. The objective of the system planner was to use the Naval communications system, whereas the other design efforts were to use all other available communication systems as alternatives. As an example, if an ATS satellite system were used, F_2 and F_2^1 would include the satellite and the NASA earth station at Rosman, North Carolina (as shown in Figure 11.3).
2. The *remote site interface* (F_1 and F_1'), consisting of all ways of interfacing the ship with F_2 and F_2'. Thus the ship-to-shore interface of F_1 consists of all equipment we must add to insert a CMI test message into F_2, whereas the shore-to-ship interface (called F_1') consists of all equipment we must add to convert the signal output from F_2 to a readable CMI reply message. For the NASA satellite systems F_1 would require an OpScan 17 terminal and a satellite terminal, whereas F_1' would require the satellite terminal and a printer such as the Terminet 1200 (Figure 11.3).

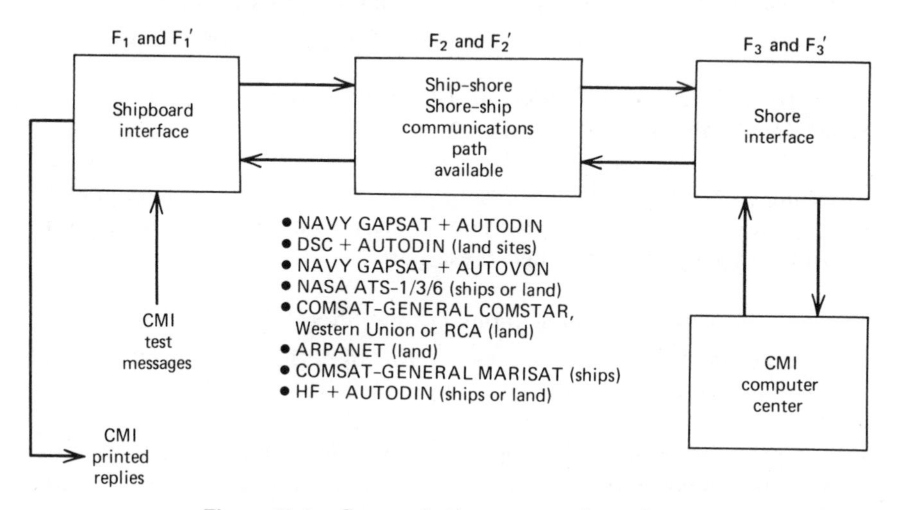

Figure 11.4. Communications system alternatives.

3. The *CMI computer interface* (F_3 and F_3'), consisting of ways of interfacing F_2 and F_2' with the CMI computer. Thus the ship-to-shore interface (F_3) consists of all equipment we must add to transport the signal output of F_2 to the CMI computer in Memphis and convert it into a form that can be accepted for processing by the computer. Conversely, the shore-to-ship interface (called F_3') consists of all equipment we must add to transport the signal output of the computer to F_2' and convert it into a form that can be accepted by F_2'. In the case of the NASA satellite systems, F_3 and F_3' consist of telephone lines from Rosman, North Carolina, to the CMI computer center and a MODEM, an electronic device needed to connect the telephone lines to the computer.

This first approach to an operational flow diagram is useful to the system designer in two ways:

- It indicates the input and output entities with which each part of the system must interface.
- It provides a useful way of partitioning the entire system so that those functions that must be treated as constraints can be separated from those that can be treated as design variables.

Having described the design structure to be followed, the detailed analysis of one of these options using the Navy telecommunications system is now presented. All other system alternatives using the satellite options shown in Figure 11.4 were also synthesized in the same way as that described for the Navy system. The performance and cost characteristics of all alternatives were then compared, as described later in this chapter.

11.3.2 Navy Communications System Options

Two fundamentally different communications system design concepts, each using the standard Navy communications system, were identified and analyzed. The first concept analyzed uses standard Navy voice circuits in the same way as the current CMI system operates or the NASA satellite system would operate. This system alternative was rejected for two main reasons. First, there are very few Navy satellite voice circuits and these are used mainly for tactical purposes. Second, operating this system would have required a drastic departure from standard operating procedures on the part of Navy communications personnel. In the second system design the CMI message is converted into a standard Navy message and transmitted to the Naval Air Station (NAS), where the CMI computer is located. This system alternative required very little additional effort from the Navy communications personnel and no departure from standard operating procedures. Based on the systems evaluation described later, this was the system selected for the demonstration. The analysis of this system was conducted in this way.[1]

11.3.3 Analysis of System Using Navy Message System

Figure 11.5 is a flow diagram showing the flow of data from the ship to the CMI computer and return using the standard Navy message distribution system. The three major functions involved in transmitting the CMI messages from the ship to the CMI computer are

- *Transmission function* (F_2). F_2 is the standard Navy telecommunications system, which can transmit a naval message from the ship to the Memphis NAS TCC, resulting in a magnetic or paper tape output.
- *Remote site interface* (F_1). F_1 is the function that converts the set of CMI messages into paper tape (the only message input the NAVMACS A-Plus system will accept) and transports it to the ship's telecommunications center (TCC).
- *CMI computer interface* (F_3). F_3 is the function of transporting the tape to the CMI computer.

In addition, there are three major functions involved in transmitting CMI reply messages from the CMI computer to the ship over this path:

- *Transmission function* (F'_2). F'_2 is the standard Navy telecommunications system, which transmits the message from the Memphis NAS TCC to the ship's TCC.
- *CMI computer interface* (F'_3). F'_3 is the function that converts the set of reply messages from the computer to the Memphis NAS TCC.
- *Remote site interface* (F'_1). F'_1 is the function of transporting the reply messages to the students or instructor.

Each of these functions is now described indicating the various design options available for implementation. The current Navy telecommunications system, or F_2, is discussed first.

[1]The analysis of the system alternative using the Navy voice system is presented in Appendix 11.A to this chapter.

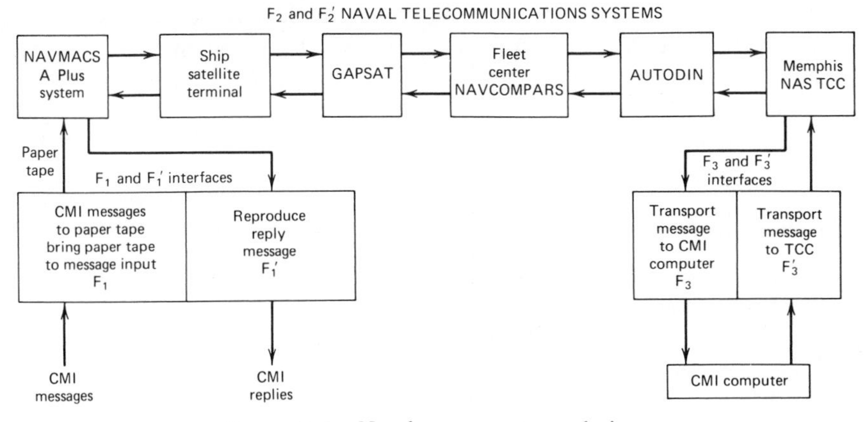

Figure 11.5. Naval message system design.

11.3.3.1 Analysis of Transmission Function (F_2): Naval Telecommunications System

F_2 is described first since it is the given element with which F_1 and F_3, of numerous variations, must interface. The current Navy system accepts a message in hard copy form at the ship's TCC and logs it in. A radioman, using a teletype, converts this message into paper tape form (the only form in which the NAVMACS A-Plus accepts messages), including a message "address"[2] preceding the message text and a message trailer following the text.

A radioman then feeds the paper tape into the NAVMACS (see Figure 11.3), where it is stored in the message-processing computer and placed in line to await transmission in accordance with its precedence. If the total message is longer than 40 lines, the message is automatically dissected into several sections with the same header used for each. After the message reaches the head of its line, it is transmitted via the Gapfiller satellite to the fleet center designated for the particular communications area in which the ship is located. In the East Pacific this is NAVCAMS (Navy Communications Area Master Stations) Honolulu. In the Atlantic it is NAVCAMS Norfolk. The message then proceeds over the AUTODIN system and is automatically routed to the Memphis NAS TCC as addressed.

[2]Technically this is called the "message header," which contains special symbols for telling the AUTODIN system how to route the message and indicating the date and time of the message.

At the Memphis NAS TCC the digital message is converted into either of three possible output forms as previously specified in the message heading:

- Paper tape.
- Punched cards.
- Magnetic tape.[3]

11.3.3.2 Analysis of Remote Site Interface (F_1): CMI Message Input to NAVMACS

The object of function F_1 is to convert the information on the CMI message into paper tape form, suitable for entry into NAVMACS, and transport the tape to the ship's TCC.

There are two types of CMI messages that will be transmitted from the ship to shore:

1. The first is called a student message and consists of a student test to be evaluated by the computer. Here the student completes his day's study and meets with the learning supervisor (LS) (on appointment) at the learning center. There the student takes and receives his test on an OpScan 17 test form provided by the instructor. This procedure avoids any collusion and gives the student an opportunity to see the instructor for any last-minute aid. To minimize student errors, each student is given a test paper that has been premarked with his identification number (social security number and any other designator desired for error checks). The average data content of a student message is 81 characters. Currently, a student message occurs once per student hour of instruction.
2. The second type message is called a learning supervisor message. If the LS desires a management report from the computer, he designates this request on his OpScan message form, especially designed for him to indicate which management report he desires the computer to send him.

Since the Navy message header containing the routing indicator and trailer require 267 characters,[4] communication efficiency is increased by batching together as many CMI messages to the Memphis computer as is possible. However, we must also consider the maximum delay time the educational process can tolerate (to be determined by the demonstration).

[3]It should be noted that currently the Memphis TCC can only provide a paper tape output.
[4]The data requirements of the communications system are described in a later section.

This may dictate that the CMI messages should be sent more often than once a day. It would seem that an LS message could be batched with student messages since each message has a separate identifier, and the computer could separate the two types of messages on that basis. Obviously, if for some (unlikely) reason the supervisor needed a faster response for a request, he would not use batching.

Five design alternatives for implementing F_1 are shown in Figure 11.6. *Alternative* F_{11} consists of the student inserting his completed test paper into a CMI "mailbox," locked to preserve privacy and holding all messages to be transmitted at the next scheduled time. The first step in the transmission process is to check the operability of the encoding equipment (the OpScan 17 and the UGC-6 teletype). To do so, the LS stacks a set of specially prepared test messages into the OpScan 17, which is connected to a UGC-6 teletype paper tape punch. This results in the generation of one paper tape containing the entire set of test messages in teletype coded form. The UGC-6 also provides a printout of the test messages. This permits the LS to perform a total system check of both the OpScan 17 and the teletype by verifying the hard copy of the test messages. He can even check the first few symbols of the paper tape to see if this was punched properly. After this system check, the LS stacks the real set of messages to be transmitted onto the OpScan 17 and repeats the process, obtaining the message tape he requires.

Three methods of obtaining the message header for the CMI textual messages were developed and reviewed with officers at the command:

1. The preferred way is for the LS to bring the tape and a signed, preprinted cover sheet (Figure 11.7) to the ship's TCC. A radioman (RM) would handle it as an ordinary message (i.e., log it in, type the message header, and then insert the text tape in the UGC-6 teletype reader). This would automatically create the addition to the message tape. The RM then types in the message trailer. The command estimated that this would require only 1 to 1½ minutes of operator typing. If only one to two messages per day were transmitted, the load on the radioman would be negligible.
2. In the second alternative analyzed the LS would compose the header. This is certainly feasible using his UGC-6 as follows. All of the preformatted message heading information such as routing indicator, precedence, from line, plain language address designator/identifier, standard subject identification code, and so on, could be prepunched onto a Mylar tape. The tape could then be notched at one edge at the place where the nonstandard information is to be inserted (date/time group and serial number, a pre-assigned block of numbers given to the LS).

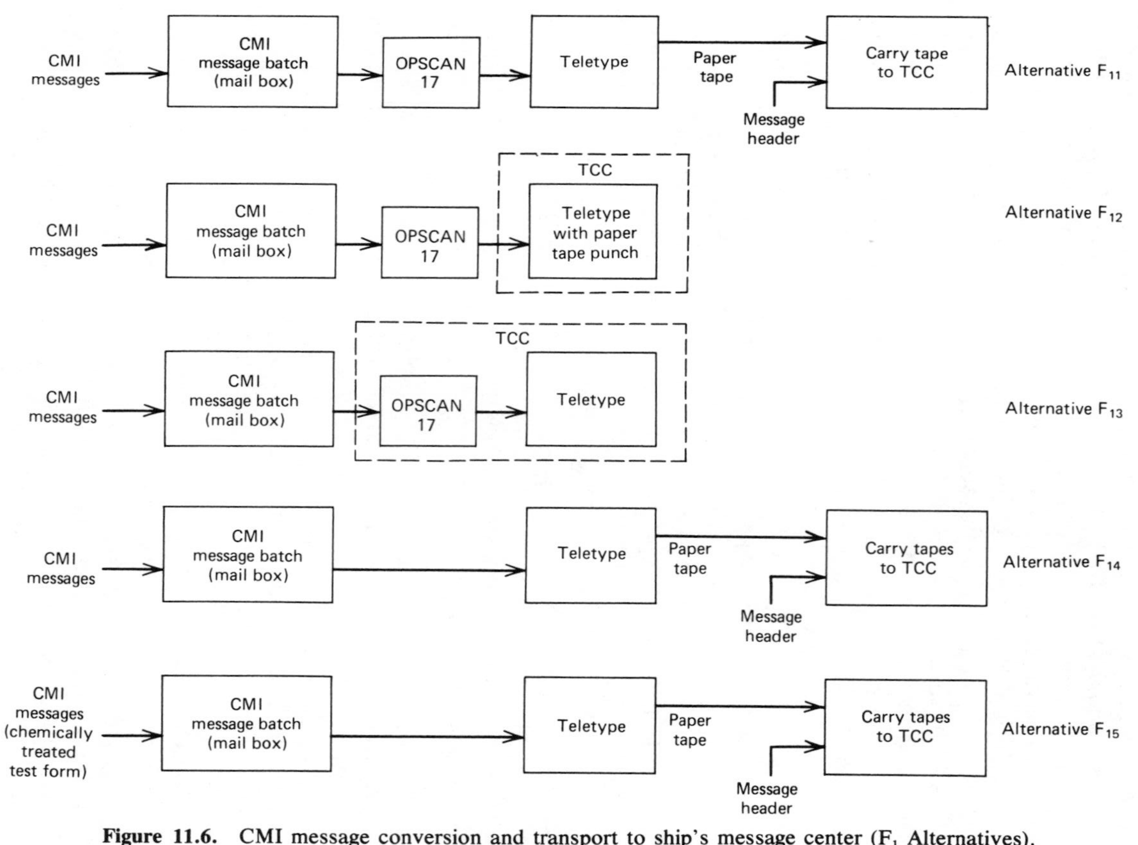

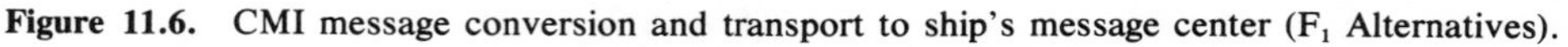

Figure 11.6. CMI message conversion and transport to ship's message center (F_1 Alternatives).

ROUTINE
FROM: U.S.S. FRANKLIN ROOSEVELT
TO: CMI COMPUTER CENTER
MEMPHIS, NAS
MEMPHIS, TENNESSEE

UNCLASSIFIED
(TEXT ON TAPE)
DATE: ____________________
TIME: ____________________
RELEASED BY: ____________________

Figure 11.7 Message release form.

The LS would insert the Mylar tape into the tape reader portion of the UGC-6, causing it to punch a new paper tape containing the preformatted information. The process would then stop at the first notch. The LS would then turn the transmitter distributor switch to OFF, move one space, and type in the unique information. He would then turn the T/D switch to ON and continue the process. The tape would stop for the next message, at which time the LS would start the OpScan 17 to activate the teletype. In this way, a complete paper tape could be generated. The LS would again fill out the message cover sheet and deposit both at the TCC. One Mylar tape would probably last for the entire demonstration. With this alternative, the command felt there was a chance of the LS making some error. Hence they felt that the slight extra work for the radioman was not worth taking this risk, and preferred alternative 1 to alternative 2.

3. The third alternative considered was to give the Mylar tape to the radioman, cutting down on his typing work. Here the problem was to keep track of all the preformatted tapes in the TCC. The command felt it would be easier to type the entire header in each time than to have to retrieve the Mylar tape and work from that.

The OpScan 17 and the teletype must operate at the same speed. The former can operate at 110, 300, 600, 1200, 1800 or 2400 baud, or 75 baud on special order. The NAVMACS system will use any of these tape outputs. Its tape reader is an optical head that counts the holes at any speed. For reliability considerations the OpScan 17 must be located in an air-conditioned room.

Alternative F_{12} consists of the same approach as Alternative F_{11}, except that the OpScan 17 is connected to a teletype located in the ship's TCC through a secure, shielded wire (to comply with Tempest Clearance security constraints). The teletype does not always have to be connected to the OpScan 17; the LS can call the TCC and request that the OpScan wire be connected to a spare teletype for transmission of his batch of messages. After the system is tested, the CMI messages are transmitted to this teletype and the paper tape is generated. Then the radioman generates the header tape and the date and time group, as described before, and inserts the total message into the communications system. The main disadvantage of this alternative is that the LS needs a radioman to assist him during the operation, since two locations are involved.

Alternative F_{13} is the same as Alternative F_{11}, except that both the OpScan 17 and the teletype would be located inside the communications room. This satisfies the requirement of the air-conditioned space and also may eliminate the need for an extra teletype unit. The spare teletype located in this room would satisfy the short time use required. Having both units of equipment in the same room would permit the LS to operate the system in the same way as described under Alternative F_{11}.[5]

Two other F_1 alternatives were also designed as backup alternatives if the OpScan 17 fails and cannot be repaired in time.

Alternative F_{14} was considered because of possible reliability problems with the OpScan 17. The data on the test papers could be manually converted to paper tape by operating a teletype. Here the LS would type the identification number and test answers of each student using the UGC-6 teletype, which has a monitor roll for reading the answers, thus verifying the typing accuracy. Again, the teletype could be located either outside or inside the communications room.

In *Alternative* F_{15} the student records his answers on a chemically treated paper. This gives immediate feedback of the test results to the student. The main benefit of this alternative is that the message would consist of information on only incorrect questions, thereby reducing both communications transmission time and transmission errors.

An evaluation of these F_1 design alternatives was made and F_{13} was selected as the preferred alternative for the reasons given in section 11.3.4.

11.3.3.3 Analysis of CMI Computer Interface (F_3): Getting the Navy Message from AUTODIN to the CMI Computer

As the CMI message leaves the ship, it travels over GapSat to shore whereupon the AUTODIN system routes the message to the Memphis

[5]If the LS was not permitted in the communications room, the radioman would have to operate the system under this alternative.

NAS TCC. Here the message is converted back to the alphanumeric code and transported to the CMI computer in one of six possible ways considered (see Figure 11.8).

The first alternative considered, *Alternative* F_{31}, uses the standard Navy delivery system for handling data messages. The decrypted message is routed from the AUTODIN switch at Albany, Georgia, to the Memphis NAS TCC, which is less than one mile from the CMI computer. There it is punched out on paper tape, the only data output the TCC can provide currently. The CMI computer center is notified by telephone that the ship's message tape is available. The center picks up both the tape and the page copy, which the teletype also provides, and delivers them to the center. There a check of the hard copy is made for any unauthorized characters. Following verification, the data are placed on a paper tape reader and read into the buffer storage (a Honeywell 6000 disk pack). The data then enter the CMI computer and are translated from Baudot teletype code to the original CMI ASCII data format. This alternative would require

- A paper tape reader at the computer center.
- Buffer storage, only if the DataNet now at Memphis cannot be used.
- Computer software to translate the teletype signal from Baudot code to the ASCII code used by the CMI computer.

Code translation was not a problem. NETISA said that the same translation approach developed by another site might be applied. Alternatively, a microprocessor could be designed to perform this Baudot-to-ASCII translation for under $1000.

Alternative F_{32} differs from Alternative F_{31} only in the use of magnetic tape instead of paper tape. This offers the advantage of using the computer center's magnetic tape readout device. The TCC would still need a magnetic tape recorder. Since they would have no other use for a magnetic tape recorder, this alternative was considered inferior to F_{31} on the basis of cost and the added maintenance work load imposed on the TCC.

Alternative F_{33} is the same as Alternative F_{31} except the message is automatically transmitted to the computer center by having a TCC operator insert the paper tape into a paper tape reader connected by phone lines to the CMI center buffer store. This alternative would require

- A paper tape reader at the TCC.
- A dedicated land line to the center.

We originally had been informed that special security safeguards might

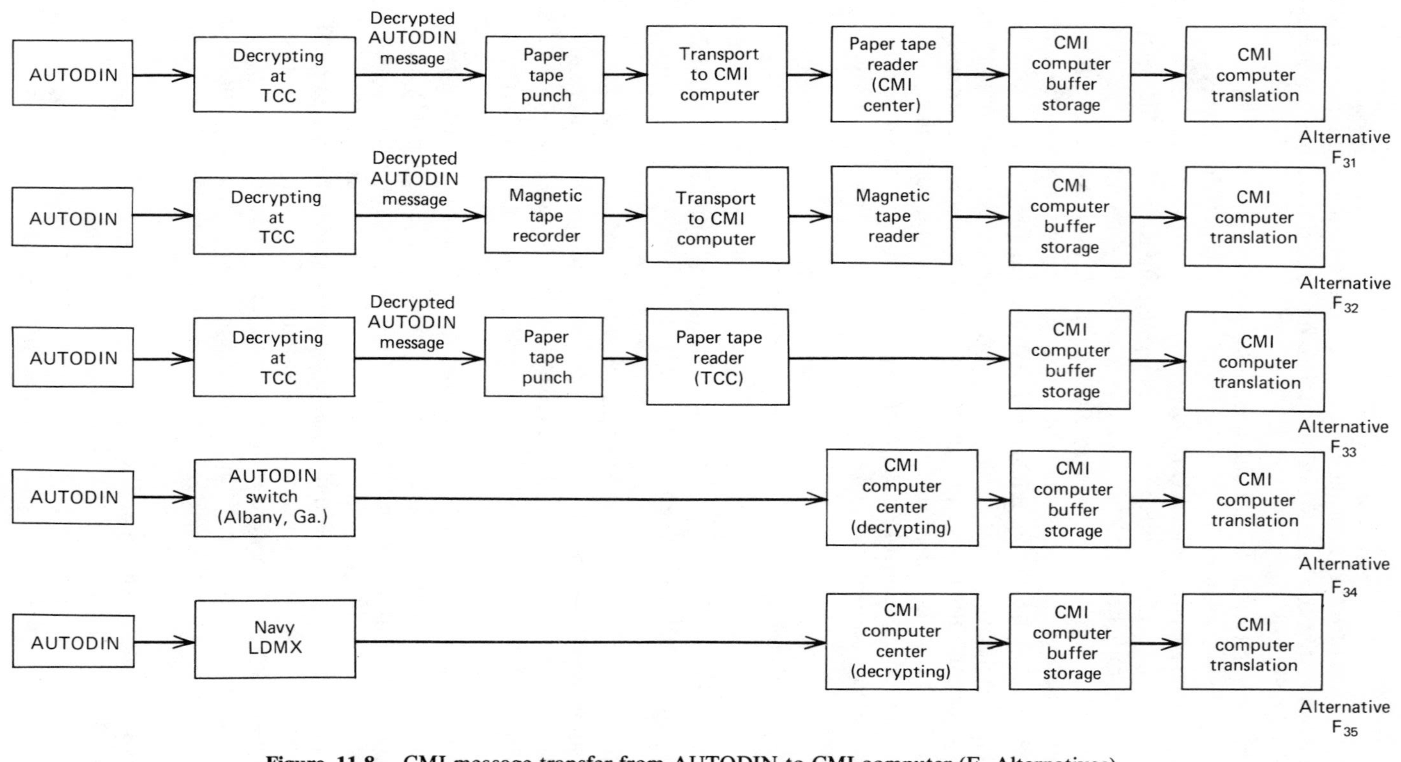

Figure 11.8. CMI message transfer from AUTODIN to CMI computer (F_3 Alternatives)

have to be satisfied in connecting the secret TCC facility to the computer center, even though the CMI message is unclassified. Later we were informed that this would not be a problem if only unclassified messages were transmitted to the computer center. Their paper tape reader is not now in constant use. It could be used if Western Union disconnected the reader from the AUTODIN line and replaced the connection with a patch circuit. Thus the reader would normally be connected to AUTODIN, but when a CMI message to the computer center came in, the tape reader could be switched to the land line going to the center. The tape would be read in, and the reader then reconnected to the AUTODIN line.

Alternative F_{34} essentially establishes the CMI computer center as a TCC, completely bypassing the Memphis TCC. This is done by running a land line from the nearest AUTODIN switch (Albany, Georgia) directly to the CMI computer center, establishing a new AUTODIN address and routing indicator. The AUTODIN signal then goes directly to the CMI Center. There it is automatically stored in buffer storage, as previously described.

Alternative F_{35} is the same as Alternative F_{34}, except that it connects the CMI computer center by land line with the Navy's nearest local digital message exchange (LDMX) instead of the AUTODIN switch. This would save some cost of the land line, and perhaps AUTODIN overhead costs.

An evaluation of these F_3 design alternatives was made. F_{31} was selected as the preferred alternative for the demonstration, with F_{33}, using its own paper tape reader, the preferred alternative for the operational system. The rationale for this choice is given in Section 11.3.4.

11.3.3.4 Analysis of Transmission Function (F'_2): Return Communications Path

After the different CMI messages are processed by the Memphis computer, the reply for each CMI message is stored in buffer storage so that one Navy message consisting of the set of narrative replies is automatically sent by the CMI computer to the ship over the reverse path, shown in Figure 11.9. Again, the three subsystems of F_1, F_2, and F_3 are involved, but in this case we define the return path functions as F'_3, F'_2, and F'_1:

- F'_3 passes the Navy narrative reply message from the CMI computer to the Memphis NAS TCC and into the AUTODIN system.
- F'_2 passes the Navy message through the entire Navy message communications system (through AUTODIN to the appropriate fleet center through a GapSat full period termination channel[6] to the ship's TCC).

[6]The command recommends that our CMI messages from Memphis to the ship go via full

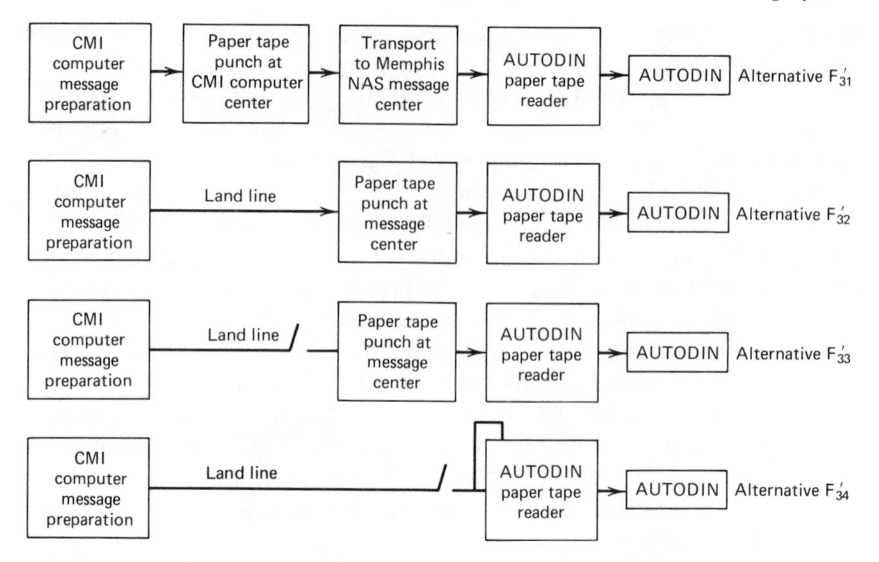

Figure 11.9. CMI message transfer from CMI computer to AUTODIN (F'_3 Alternatives)

- F'_1 converts the narrative batch message into separate CMI messages, each addressed to the originating student, then reproduces and distributes the individual messages.

Since F^1_2 is the standard Navy message system and operates essentially as previously described, no further description will be given.

11.3.3.5 Analysis of Return CMI Interface (F'_3): Getting the Message from the CMI Computer to AUTODIN

Four alternatives were considered for this function, as illustrated in Figure 11.9.

Alternative F'_{31}, the first alternative considered, uses an improved version of the standard Navy delivery system for handling the narrative message. First, the set of CMI narrative messages would be translated by the computer from the ASCII code to a code compatible with the AUTODIN system. Baudot software would be required to make such a conversion at the CMI computer. Next, the standard message header, stored in the computer, and the date and time group would be added to the message

---

period termination rather than via the fleet broadcast channels since they feel the time delay using the latter may be excessive. Permission for this negligible load will have to be requested from CNTC.

by the computer. The total message would then be provided as output from the CMI computer in the form of a paper tape that would then be manually delivered to the Memphis NAS TCC for transmission. The TCC operator would then insert the paper tape into the AUTODIN paper tape reader where it would be transmitted to the appropriate fleet center and then over the fleet broadcast via GapSat.

Alternative F'_{32} is the same as F'_{31} but eliminates the manual transport function by connecting the CMI computer to a teletype at the TCC through a dedicated land line. Thus the same complete message would be punched out on paper tape at the TCC. When the operator saw the paper tape, he would know this is a message to be transmitted, so he would log in the message and feed it into the AUTODIN tape reader.

Alternative F'_{33} is the same as Alternative F'_{32} except that it connects a TCC teletype to the land line only when needed. In this case when the total reply message has been completely generated by the computer, it is stored in intermediate storage. The computer notifies the computer operator that the reply message is ready for transmission to the ship. The computer operator telephones the TCC operator, saying he has a message to go out. The TCC operator then connects the land line to a spare teletype and notifies the computer center operator that the circuit is ready and that the message can now be transmitted from the computer's intermediate storage. After the paper tape is generated at the TCC, the remaining operations described previously continue.

Alternative F'_{34} automates the procedure even more than F'_{33}. Now the land line from the CMI computer intermediate storage goes through a TCC patching panel to the appropriate connection inside the paper tape reader supplying the signal for the AUTODIN line. In this alternative when the CMI computer operator calls the TCC operator for message transfer, the operator completes the paper tape message that may be feeding the tape reader at that time. He then switches the computer line to the inside of the reader and notifies the computer operator to transmit the reply message from the computer, thus feeding it directly into AUTODIN.

An evaluation of these F'_3 design alternatives was made and F'_{31} selected as the preferred alternative. The rationale for this choice is given in Section 11.3.4.

11.3.3.6 Analysis of CMI Message Printing and Distribution (F'_1): Converting Navy Message to Printed Copy

After the ship receives the reply message transmission, it is stored in the NAVMACS system for subsequent conversion to readable form. Two alternatives were examined for performing this function (Figure 11.10).

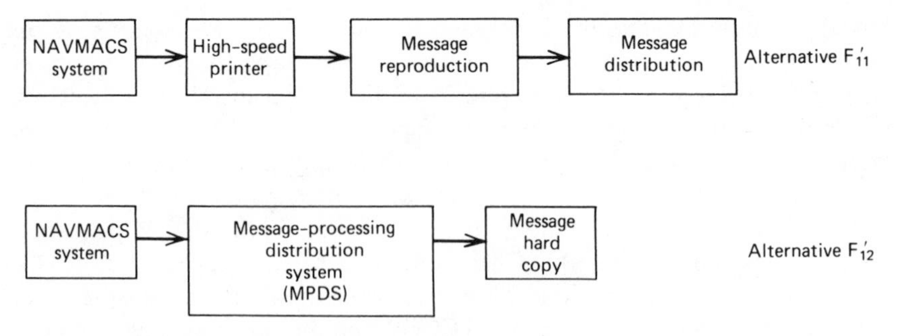

Figure 11.10. CMI message printing and distribution (F'_1 alternatives).

Alternative F'_{11} uses standard Navy telecommunications procedure. Each of the CMI messages making up the total Navy message is converted into hard copy form over the high-speed TT-624 printer at a speed of 800 lines per minute. The separate CMI reply messages for the student addresses are separated for distribution. Reproductions of the messages for the LS, or any other person, could also be made at the ship's TCC. Each message would be placed in the appropriate distribution box of the addressee's work station, where it would be carried to the addressee's incoming mail point for his pickup.

Alternative F'_{12}, an alternative method of distributing the message from the ship's message center to the addressee involves the use of the message-processing distribution system (MPDS) now installed on the U.S.S. *Nimitz* and to be installed on other ships in the future. This system uses a VDT display at a user location so that messages can be sent directly from the ship's message center, thus avoiding manual distribution.

An evaluation of these F'_1 design alternatives was made and F'_{11} was selected as the preferred alternative. The rationale for this choice is given in the following section.

11.3.4 Evaluation of Navy Message System Design Alternatives

Figures 11.6, 11.8, 11.9, and 11.10 illustrate the various design options considered for each function of the Navy message communications system. Since each function is independent, we can choose the preferred alternative for each function separately, and thus arrive at the preferred solution in a simpler fashion than having to consider all possible system combinations.

The preferred alternative for each function was selected by reviewing the following set of performance and cost criteria. However, the basic criteria used in designing the information-handling system for the demonstration, in order of importance, were

- Feasibility of operation (all alternatives presented were designed to be feasible).
- Ease of operation.
- Minimize any changes in the current operational procedure used by Navy communications personnel.
- Minimize the time spent by Navy communications personnel on the CMI messages.
- Minimize total cost to the Navy for the demonstration.

There was no simple way of selecting the preferred alternative (such as selecting for the lowest total cost to do the job). Rather the impact of each alternative on the evaluation criteria was considered.

The preferred system design alternative recommended for the demonstration is the standard Navy message system using the functions:

- F_{13}, OpScan 17, and teletype located in the message center.
- F_2, standard NAVMACS A-Plus system, and AUTODIN.
- F_{31}, carry paper tape from Memphis message center to CMI computer center, and read into computer intermediate storage.
- F'_{31}, carry paper tape from CMI computer, and read into AUTODIN reader.
- F'_2, AUTODIN, and standard Navy message system.
- F'_{11}, standard Navy printing and distribution system.

The evaluation used to select each of these alternatives now follows.

11.3.4.1 Evaluation of Function F_1: CMI Message Conversion and Transport to Ship's Message Center

The preferred alternative is F_{13}, locating the OpScan and the teletype in the message center. Reexamining the five design alternatives for F_1 (Fig. 11.6) revealed that there were three basic design variables involved:

- Do we use the OpScan 17 plus teletype or just the teletype?
- Where should the OpScan 17 and the teletype be located?
- What is our backup approach for inputting the CMI message into the communications system if the OpScan 17 malfunctions and cannot be repaired in time?

These issues were addressed in the following manner:

OpScan 17 Plus Teletype Versus Teletype Alone. For ease of operation we decided to use the OpScan 17 and a teletype as the primary means of generating a paper tape for the message input. To deal with equipment malfunction problems it is essential that sufficient spare parts be available to balance the equipment failure rate with the response time of the ship's logistics system. If we can afford the cost and off-line space requirements, we would like to have a spare OpScan 17.

We also know that if the OpScan is not located in a temperature- and humidity-controlled environment, reliability problems may occur. To increase reliability, a system checkout test prior to transmission is also recommended. A complete system check of F_1 could be made by inserting a set of standard pattern test sheets and checking the results on the teletype monitor roll. Thus a standard teletype (UGC-6) containing such a display should be used rather than just a paper tape punch. In addition, both pieces of equipment should be located next to each other so that the system check can be readily conducted (see next discussion).

Equipment Location. Because the teletype would be used only a short time each day, the ideal approach would be to use a teletype that is already part of the ship's equipment (and hence a "free" resource). Probably the spare teletype would be located in the TCC. However, this would require that the OpScan 17 be installed alongside for the system check. If room for it is not available, there are two other possibilities. Both pieces could be located elsewhere if the spare teletype could be moved to the TCC when the spare is needed there. If this was not possible, a separate (nonspare) teletype and the OpScan 17 could be located in any other environmentally controlled area.

Backup to OpScan 17. At certain times the desired response time for a CMI reply message would not be achieved because

- The OpScan 17 did not function properly and hence there is a delay in transmitting the original message.
- The CMI computer (or any other part of the rest of the system) malfunctioned and was unable to respond on time.

In these cases the following backup procedures would be employed (Figure 11.11).

- If the OpScan 17 failed and it appeared that it would not be repaired in sufficient time to obtain the CMI reply in time, the LS might choose to type each of the test results on the UGC-6, in narrative form: (student identification; Question 1: B; Question 2: D; etc.), generating a paper

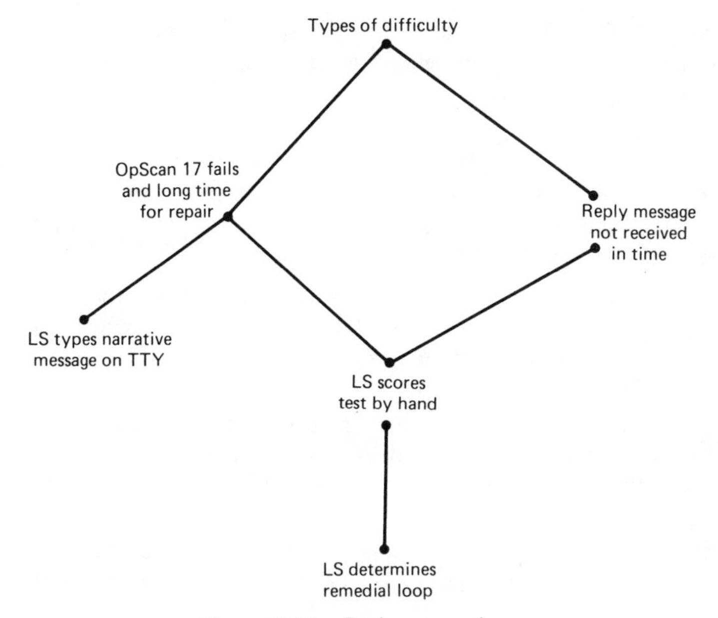

Figure 11.11. Backup procedures.

tape for transmission. The message cover sheet would also indicate that the message is in narrative form. Thus the radioman would so indicate this on the header. When the narrative message arrived at Memphis NAS TCC, it would be printed only in narrative page form and delivered to the CMI computer. There it would be keypunched, fed into the computer, and processed like a normal CMI message, using existing software. The message reply would be as previously described.

The LS has a second procedure available. He could score the test by hand, using an answer book available to him. He could then determine any remedial loop necessary, based on the set of student test answers, by using a second answer book developed for such purpose. This latter procedure is recommended only for the demonstration when just 30 students are involved. During the demonstration we shall gather performance characteristics regarding the magnitude of the system reliability problem. Later we shall determine how to cope with this problem for the operational system.

This method would be used if the reply message is not received by a certain time before the students' training period is to begin.

The use of chemically treated test paper[7] would alleviate this backup problem by eliminating the job of scoring each test. The ideal system would be designed as follows:

- Design the chemically treated paper so that not only will the correct answer appear but it will be readable by the OpScan 17.
- Have all students take their test using this chemically treated paper.
- If the OpScan 17 is available for transmitting the test results to Memphis, use it. If not, the LS scores each test using the crayon and manually determines the remedial loops using the answer book. When the OpScan 17 does become available, the LS transmits the stack of tests so that the computer will maintain the complete record of student progress. However, he adds a preprogrammed symbol that no replies are needed, since they have already been given to the students. This reduces unnecessary reply communications.

If the OpScan 17 malfunctions and the LS decides to teletype the test results in narrative form, the use of the chemically treated paper would

- Reduce the typing manpower required by the LS.
- Reduce the typing error rate.
- Reduce message length, although this may not be very valuable, since our total daily requirements are reasonably small now.

Since the chemical paper is more expensive than the standard CMI test paper, it should be used only if absolutely required by excessive OpScan unavailability or other system malfunctions. We should also determine how rapidly these papers can be designed and procured, if needed.

11.3.4.2 Evaluation of Function F_3: CMI Message Transfer from AUTODIN to CMI Computer

The preferred alternative is F_{31}, using the standard Navy delivery system. With this option the messages are converted to paper tape, with the Memphis NAS TCC telephoning the CMI computer center to notify them of the tape's arrival and the computer center picking it up and reading it into their intermediate storage device, through their own tape reader.

There are three key issues involved in the five design alternatives for F_3, illustrated in Figure 11.8.

[7]This is specially treated test paper that discloses the correct answer symbol when a special crayon is rubbed on the alternative answer locations.

- Should the AUTODIN message come into the Memphis NAS TCC or be routed into the CMI computer center directly from the AUTODIN switch in Albany, Georgia, or through the nearest LDMX to Memphis?
- Given that the message comes into the Memphis TCC, will it be hand-carried to the CMI computer center or will it be transmitted over a land line?
- If the message is hand-carried from the message center, should it be on paper tape or on magnetic tape?

These issues were addressed in the following manner, thereby bringing us to the preferred design alternative for F_3.

Does the Message Go to TCC or Directly to the Computer Center? Although the possibility of automatically routing the message directly to the computer center is attractive, there are two major problems connected with this alternative. The primary problem is the security aspect. If the AUTODIN signal went directly to the computer center, security regulations would require that the computing center be treated as a secret facility. There would always be some possibility that a classified message might inadvertently leave the AUTODIN line and be transmitted to the computer center termination. It is questionable whether the computer center would go along with such a restriction. The second disadvantage is cost. Although an automated delivery system is very attractive for the operational system with a high volume of messages coming in at different times, the small volume of messages for the demonstration system (one or two a day) does not justify the associated costs. The main savings would be in time (perhaps 10 to 20 minutes for the one-mile trip each way) and the associated manpower savings.

How to Transmit the Message from the TCC to the Computer Center. Here the same advantage of faster delivery time (perhaps 20 minutes) must be balanced against the costs of automatic delivery. These costs would include a leased line of one mile ($81 per month plus a one-time installation cost of $50 per device at each end[8]), an additional piece of equipment (paper tape or magnetic tape machine at one of the sites), and a trivial special-handling effort on the part of the TCC.

Hand-carried Message on Paper Tape or Magnetic Tape? Here we compared the operations and maintenance costs of a new piece of equipment (magnetic tape machine) for the Memphis TCC against similar costs of a paper tape reader at the computer center plus the cost of any interface efforts (and software changes) that the computer center might have to

[8]These figures do not assume the use of GSA Telpak rates, as do the long-distance figures.

bear.[9] It was felt that any additional effort required should be borne by the computer center rather than the TCC.

11.3.5 Summary of Resources Required

Here is a summary of the operational steps required to operate the recommended system. From this "scenario" may be obtained the various resources required during the demonstration (listed in Table 11.1). If any part of the recommended system cannot be implemented for any reason, one of the other design alternatives will be implemented.

1. *Construct the Navy ship-to-shore message* once or twice a day. This requires
   a. OpScan 17. CNTECHTRA (Chief of Naval Technical Training) provides one prime unit of equipment that can interface with UGC-6 teletype, one spare unit, set of spare parts; Navy installs in ship's TCC and maintains.
   b. Standard teletype (UGC-6). Use spare installed in TCC; Navy maintains.
   c. LS constructs paper tape of message text. Less than one half-hour of instructor time per message, plus time for him to walk to and from TCC.
2. *Navy transmits message.*
   a. Radioman logs in message and sets up and removes message from tape reader, after adding header and trailer. Ten minutes per message.
   b. Satellite transmission—5454 to 6522 characters per day.
3. *Transmission* through rest of Navy/DCS system. No additional cost.
4. *Memphis NAS TCC handling.* No additional effort required over their usual services.
5. *Computer center picks up message.* Ten minutes per message of computer center personnel.
6. *Computer center inserts message* into the computer.
   a. Paper tape reader. NETISA (Naval Education and Training Information Systems Activity) procures; Memphis central electronic maintenance personnel maintain.

[9]It could be assumed that either piece of government equipment could be purchased and then used elsewhere after the demonstration. Hence the major equipment cost is for operations and maintenance.

Table 11.1 *Resources Required for Navy Message System*

Equipment	Cost ($1000)	O&M Personnel	Man-Hours/Week
Ship			
1. SST	Navy provides service	Navy provides service	
a. Investment			
b. Installation			
c. Spares			
d. Tear Down and Restore			
2. OpScan 17 (2)			
a. Investment	2 @ 9.3 = 18.6[a]	Learning supervisor operates	Full time (7 mo)
b. Spare parts	6.5	Navy ET maintains	2 hr/wk (7 mo) = 61 hours
c. Interface	2.0[a]		
Total 2	27.1		61 man-hours + learning supervisor (7 mo)
3. Memphis NAS TCC	Navy provides service	Navy provides service	
4. CMI Computer Center		Courier picks up tape and returns reply	3 @ 20 min/day = 5 hr/wk (7 mo) = 152 man-hours
a. Software Changes	NETISA estimated 70		
b. Tape Reader/Punch-Investment	Navy supplies	Operator operates	3 @ 20 min/day = 5 hr/wk = 152 hours
Total 4	70		304 man-hours
Grand total—	97.1		365 man-hours + learning supervisor (7 mo)

[a]This equipment will have high residual value after the demonstration.

   b. Modification to computer to accept message. Software modification; extent to be determined by NETISA.

7. *Computer translates message* teletype format into CMI format.
   a. Computer software modification or microprocessor for Baudot to ASCII conversion; extent to be determined by NETISA.
8. *Computer handles message.* Nondifferentiating cost for all design alternatives.
9. *Computer reply message configured* and translated into standard Navy message format. This requires computer software modification, the extent of which is to be determined by NETISA.
10. *Computer center delivers message* to message center. Ten minutes of computer center personnel per message.
11. *Message center handling.* No additional effort required over their usual service.
12. *Transmission* through rest of Navy/DCS system. No additional cost.
13. *Navy transmits message* over full-period termination channel. 107,330 to 113,852 characters per day.[10]
14. *Reproduce and print message.* No additional cost over usual service.

11.3.6 Data Requirements of the Demonstration System

The previous sections of this chapter showed how to design a communications system such as the Navy message system alternative. This analysis would not be complete without a more detailed account of the requirements to be met by the communications system during the demonstration (the system demand function), including

- The amount of data to be transmitted from the ship to the CMI computer.
- The amount of data to be transmitted from the CMI computer to the ship.
- The time response required for the two-way communication.

It is important that these requirements be analyzed for two reasons:

- To make certain that there is compatibility between our data needs and the capability of the communications system to meet these needs.
- Although we have analyzed the manpower and monetary resources required by the system during the demonstrations, the providers of the

[10]See Appendix 11.B on data requirements for derivation.

communications system (in this case, the Navy) were always concerned about how much of their satellite we planned to use both during the demonstration and during operations. Hence to gain their participation, it was necessary to estimate our data requirements in advance.

The results of this analysis were as follows:

1. The total characters per day required to be transmitted over the satellite during the demonstration would be between 113,852 and 118,974. This corresponds to 95 average 1200-character Navy messages to be transmitted each day, a negligible part of the ship's total communications work load.
2. The training schedule could be designed to accommodate the communications response time expected.

Details of the analysis performed in determining the requirements of the demonstration system are contained in Appendix 11.B. The analysis of the data requirements of the operational system is contained in Appendix 11.C.

11.3.7 Cost Performance Analysis—Demonstration Communications Alternatives

This section describes the cost-performance analysis made in evaluating the various communications system alternative designs for the demonstration, including

- A comparison of the key performance characteristics obtainable from each of the alternatives considered.
- An analysis of all costs that would have to be expended in using each system during the demonstration.

11.3.7.1 Performance Analysis

Table 11.2 is a summary of the results of the analysis of the performance expected from each of the demonstration system alternatives analyzed. Column 1 contains the 11 characteristics considered in the performance evaluation. Columns 2 and 3 list two of the nine systems considered and the performance these would achieve. Each of the characteristics is now described.

1. *Installation requirements.* The antennas and other equipment required for the two Navy GapSat systems are already installed on the

ship. Except for the OpScan 17 and the teletype, no additional space is required. The same is true for a ship system using HF and AUTODIN, as well as a remote land site using AUTODIN. For the remaining satellite systems, their equipment requirements and the time required for installation are listed.

2. *Frequency band.* The frequency band employed by each system is listed.
3. *Operational availability*
   a. *Geographical constraints.* All DOD systems have worldwide coverage. The geographical coverage of the remaining systems is included under the description of each system.
   b. *Time of availability.* All DOD systems plus those of Columns 7 and 9 are available on a 24-hour-per-day basis. However, the Memphis CMI computer is on a two-shift basis, which is the limiting item regarding availability of these systems. ATS-6 would be available to the project one or two times during the day for a total period of 1 to 2 hours, a NASA constraint.
4. *Maintainability.* The DOD and commercial systems provide the best maintainability, since they are basically operational systems. Good maintainability is expected from the NASA experimental systems.
5. *Transmission system usage.* As indicated in Appendix 11.B on data requirements, 30 students taking 2 hours of training per day will require the equivalent of five to six standard Navy messages (each containing 1200 characters) per day being sent from the ship and 89 standard messages in reply to the ship. All other systems except the HF/AUTODIN system would require approximately the same transmission times. HF/AUTODIN requires more because of the higher error rate involved (thus more reruns or redundancy for error correction techniques).
6. *Message response time.* The DOD systems have the longest turnaround time (in the order of 1 to 8 hours), with the HF system response time being of the same order but more highly variable because of variable propagation characteristics. The other systems offer the same response time as the current CMI system (12 seconds).
7. *Additional response time due to transmission error.* The demonstration system will not provide any error correction capability. Hence any detected errors will require a retransmitted service request,[11] requiring an additional response time as indicated previously.

[11]This service request will probably consist of repeating the particular CMI messages affected twice with the reply also being sent twice so that a comparison of the three replies can be made. Thus the error correction will be based on the "two out of three" decision rule.

Table 11.2 *Performance Comparison*

(1) Evaluation Criteria	(2) Navy Message System	(3) AUTODIN (Land)	(4) Navy Voice Channel System	(5) ATS-1/3
1. Installation requirements				
a. Space requirements	Exists	Minimal	Exists	2 antennas + below deck space
SST/RT antenna size[a]	Part of WSC-3		Part of WSC-3	20-ft diameter required
SST/RT pedestal size	Part of WSC-3	No pedestal	Part of WSC-3	2 ft
SST/RT below-deck cabinet size	Part of WSC-3	Comparable to WSC-3	Part of WSC-3	22 × 24 × 72 in.
b. Installation time	0	0	0	8 months development 4 months installation
2. Frequency band	240-310 MHz	—	240-310 MHz	136–138 MHz
3. Operational availability				
a. Geographical constraints	Worldwide	Worldwide	Worldwide	ATS-1/3 Footprint
b. Time of availability	Memphis schedule	Memphis schedule	Memphis schedule	1–2 hours/day
4. Maintainability	Excellent	Excellent	Excellent	Good
5. Transmission system				
a. Type	Half duplex	Half duplex	Half duplex	Half duplex
b. Usage	5–6 standard messages-send 89 standard messages-reply	Same as Navy message system	Same as Navy message system	Same as Navy message system
6. Message response time	Less than 8 hours	Less than 8 hours	Less than 1 hour	12 seconds
7. Additional response time for transmission errors	Less than 8 hours	Less than 8 hours	Less than 1 hour	12 seconds
8. Earth station characteristics				
a. Location	Norfolk, VA (Atlantic) Honolulu, HI (East Pacific) Guam (West Pacific) Naples (Indian Ocean)	—	Norfolk, VA (Atlantic) Honolulu, HI (Pacific) Guam (West Pacific)	Rosman, NC
b. Circuit to Memphis	Furnished by Navy and DCS	Furnished by Navy and DCS	Furnished by Navy and DCS	Leased line to Memphis, TN
9. Interruption of normal work	Computer center picks up, delivers message	Same as Navy message system	Fleet center records, forwards message	None
10. Technical problems				
a. EMC problems	Solved	Solved	Solved	Requires study
11. Ease of transitioning system to next phase	Excellent	Excellent	Excellent	On-line satellite required

[a]SST = shipboard satellite terminal, RT = remote (land site) terminal.
[b]Comstar also covers Puerto Rico.

Table 11.2 *(Continued)*

(1) Evaluation Criteria	(6) ATS-6	(7) Domestic Commercial (Land)	(8) ARPANET (Land)	(9) MARISAT (Ships)	(10) HF/AUTODIN
1. Installation requirements					
a. Space requirements	2 antennas + below deck space	1 antenna	Minimal	2 antennas + below deck space	Exists
SST/RT antenna size[a]	10 in. to 18 in. (PRN) 4 ft (clear mode)	30 ft	No antenna	4 ft	—
SST/RT pedestal size	2 to 5 ft	—	No pedestal	5 ft	No pedestal
SST/RT below-deck cabinet size	22 × 24 × 48 in.	—	—	22 × 24 × 48 in.	A variety of systems are available. Most are comparable in size to the WSC-3
b. Installation time	8 months development 4 months installation	No development 4 months installation	0	No development 4 months installation	0
2. Frequency band	3700–4200 MHz downlink 5925–6425 MHz uplink	3700–4200 MHz downlink 5925–6425 MHz uplink	—	1537–1541 MHz downlink to ship 1638.2–1642.5 MHz uplink to ship	2–30 MHz
3. Operational availability					
a. Geographical constraints	ATS-6 Footprint. PRN operation within 50 miles of shore requires FCC agreement	All 50 states[b]	48 states + Hawaii	World-wide	World-wide
b. Time of availability	1–2 hours/day	Memphis schedule	Memphis schedule	Memphis schedule	Memphis schedule
4. Maintainability	Good	Excellent	Excellent	Excellent	Excellent

5. Transmission system					
a. Type	Full duplex at reduced power Half duplex at full power	Full duplex	Half duplex	Full duplex	Half duplex
b. Usage	Same as Navy message system	Same as Navy message system	Same as Navy message system	Same as Navy message system	Higher usage than Navy message system (×3) because of higher error rate
6. Message response time	12 seconds	12 seconds	12 seconds	12 seconds	Highly variable
7. Additional response time for transmission errors	12 seconds	12 seconds	12 seconds	12 seconds	Highly variable
8. Earth station characteristics					
a. Location	Rosman, NC	Various, within the 50 states	—	Southbury, CN (Atlantic) Santa Paula, CA (Pacific)	Various Navy sites on the east and west coasts, Honolulu, Guam
b. Circuit to Memphis	Leased line to Memphis, TN PRN, if used, requires unit at Rosman	Furnished by carrier or user	Leased line from Montgomery, AL to Memphis, TN	Furnished by Comsat-General	Furnished by Navy and DCS
9. Interruption of normal work	None	None	None	None	Same as Navy Message System
10. Technical problems					
a. EMC problems	Requires study	Requires study	None	Requires study	Solved
11. Ease of transitioning system to next phase	On-line satellite required	Excellent	Excellent	Excellent	Excellent

[a]SST = shipboard satellite terminal, RT = remote (land site) terminal.
[b]Comstar also covers Puerto Rico.

8. *Earth station path to Memphis.* The location of the earth station used by the various satellite systems and the land lines required to complete the path to Memphis are listed.
9. *Interruption of normal work.* The Navy message system alternative requires the CMI computer center to pick up the messages from the Memphis TCC, a distance of one mile, and return the replies two to three times per day. The Navy voice circuit system requires the fleet center to record the message and forward it to Memphis once they set up the voice circuit. They must also repeat the process for the return path. All the other systems require no interruption of the normal work of the CMI computer center personnel.
10. *Technical problems.* Any electromagnetic compatibility (EMC) problems to be solved are indicated.
11. *Ease of transitioning system to next phase.* All the demonstration system alternatives, except those using the ATS-6 or ATS-1/3, can be readily expanded into an operational system up to the permissible limits of its communications capacity.[12] Thus their ease of transition is excellent. If the ATS-6 or ATS-1/3 approach were to be used as the operational system, an on-line satellite would be required, thus taking additional time before an operational system would be available.

11.3.7.2 Cost Analysis

Table 11.3 is a summary of the costs to be incurred at each element of the total system for two of the nine system alternatives considered. Columns 2 through 10 list the various systems whose costs are presented.[13] Row 1 gives the system name, row 2 the location of the classroom (ship or land), and row 3 the type of communications path required to get to Memphis. Costs are divided into two parts:

- Man-hours of work for the entire demonstration, including predemonstration training, for example.
- Dollars of nonmanpower expenditures.

[12] See Appendix 11.C for an estimate of the amount of student training each percent of FleetSatComm could provide. This same type of analysis could also apply to other satellites being considered.

[13] A separate column for HF/AUTODIN is not included because its basic costs are those of the AUTODIN system as shown in Column 1. However, the CMI computer center at Memphis would need to provide comparisons of the triplicate messages sent from the classroom (one set for every half hour) to keep the HF error problems to reasonable proportions. This added cost has not been determined because the use of HF is felt to be of low probability due to its error characteristics.

Table 11.3 *Cost Comparison*

(1) System	(2) Navy Message System		(3) Navy Voice Channel System		(4) ATS-1/3		(5) ATS-1/3		(6) ATS-6		(7) ATS-6		(8) Domestic Commercial Satellite[c]		(9) ARPA-NET		(10) MARISAT	
Classroom Location	Ship/Land		Ship/Land		Ship		Land		Ship		Land		Land		Land		Ship	
Connection to Memphis	AUTODIN		AUTOVON		Telephone Line		Telephone Line		Telephone Line		Telephone Line		Telephone Line		Telephone Line		COMISAT-Provided Telephone Line	
	$(K)	MHRS	$(K)	MHRS	$(K)	MHRS	$(K)	MHRS	$(K)	MHRS	$(K)	MHRS	$(K)	MHRS	$(K)	MHRS	$(K)	MHRS
Ship/land site	27.1	61	35.5	76	196.2–493.2	591–682	66.5–93.5	423–484	281.5–663.5	575–666	165.5–270.5	423–484	175.5–285.5	575–666	35.5	91	233.1–508.1	575–666
Satellite usage	—	—	—	—	—	—	—	—	—	—	—	—	7.7[a]	—	—	—	22.8[d]	—
Earth station and land line	—	—	—	—	2.2	40–76	2.2	40–76	2.2	40–76	2.2	40–76	1.3	—	2.5	—	—	—
NAVCAMS	—	—	[b]	160	—	—	—	—	—	—	—	—	—	—	—	—	—	—
CMI computer center	70.0 (Software)	304	40 (Software)	—	40 (Software)	—	40 (Software)	—	40 (Software)	—	40 (Software)	—	40 (Software)	—	40 (Software)	—	40 (Software)	—
Memphis TCC	—	—	—	—	—	—	—	—	—	—	—	—	—	—	—	—	—	—
Total cost	97.1	365	75.5	236	238.4 535.4	631–758	108.7–135.7	463–560	323.7–705.7	615–742	207.7–312.7	463–560	224.5–334.5	575–666	78.0	91	295.9–570.9	575–666

[a]Based on $1,110/month × 7 months.
[b]Requires magnetic tape recorder, available from NETISA.
[c]The Domestic/Commercial Satellites are Westar, (including American Satellite), RCA Satcom, and Comstar.
[d]Based on 15 minutes/day, 5 days/week for 7 months.

Details of the cost elements comprising each of the total costs are contained in the sections describing each system alternative.

As seen, the Navy message system is the least costly system[14] for shipboard operation because it uses the existing ship satellite terminal. Its only added cost element is the additional software required for the CMI computer, including translation of the Baudot code to ASCII and vice versa, as well as the addition of a standard header (including date and time group) and trailer to compose the AUTODIN message.

11.3.7.3 Final Recommendations

As a result of this analysis the Navy message communications system was selected for the demonstration because it

- Meets all requirements of the demonstration.
- Requires the least resources in personnel and dollars.
- Permits the use of a ship's existing message system and communications equipment, thus eliminating the requirements for the installation of antennas or other communications equipment except for a small optical scanning device below deck to read the CMI tests.

The demonstration will place a negligible load on the ship's communications system.

11.4 SUMMARY OF PROBLEM-SOLVING PROCESS FOR COMISAT INFORMATION-HANDLING SYSTEM DESIGN PROBLEM

This section summarizes the key principles of planning as they were applied to this case.

11.4.1 Problem Definition

The generic problem described involves the design and evaluation of alternative information-handling systems to meet the requirements of the demonstration. Many of the principles summarized in Cases 1 and 2 also apply to this case. However, the differences between these three system design problems are as follows:

[14]As noted elsewhere, the Navy voice circuits are not intended for data transmission and thus could not realistically be expected to be available for uses such as COMISAT.

- In Case 1 the system demand function was well defined. We merely had to (1) design all primary system alternatives to meet this demand, (2) design all support systems to meet the needs of its primary system, (3) select the preferred alternative on the basis of lowest total cost or highest net return to the organization.
- By contrast, Case 2 concerned the design of a support system (an information system). Here the information actually needed was ill defined. We had to assume different levels of information to be provided (the system demand function) and design alternatives for each level of demand, again selecting the preferred alternative on the basis of lowest total cost to the organization.
- The Case 3 information-handling system had to be designed to interface with other parts of the higher-level system, which constrained the design alternatives. However, since most system designs involve such constraints and require that the new system properly interface with the other elements, this case represents the system design context that most frequently occurs.

11.4.2 System Design

There were several aspects to the system design problem. We had to design an information-handling system that could interface between the following system elements, which had to be assumed as a given, namely, (1) the Naval telecommunications system (it could only receive data inputs in a particular format) and (2) the CMI computer (it operated in a particular data code, ASCII). Since the Naval telecommunications system was a scarce resource, the messages we originated had to be designed for highest communications efficiency. In addition, the Navy system required a transmission delay far greater than the current CMI training system. For both these reasons the training schedule had to be designed to accommodate these constraints.

The analysis of the current information-handling system, as well as of a proposed system using the NASA satellite, provided:

- The data to be transmitted during the operation of the system. This is the start of the definition of the system demand function.
- The functions of the data-handling system requiring implementation.
- Those elements of the current system in which an investment had already been made and which hence should probably be used in the new design. The new design must interface with these elements.

Each of the demonstration system alternatives was designed using the following process, as illustrated in Figure 11.1:

1. From the generic operational flow diagram previously developed, various ways of implementing each system function were identified. These different ways would constitute alternative system elements to be considered. A starting point for such elements was provided by the elements contained in the current system.
2. Alternative ways of implementing each function were then combined to form the various alternative systems, each of which is a feasible way of implementing the generic operational flow diagram.

11.4.3 Systems Evaluation

A performance versus cost comparison was made in evaluating the demonstration alternatives. The system selection criterion used was to choose the system alternative that could meet the requirements of the demonstration at minimum cost and minimum disruption of the operational personnel. For this reason the Navy message system was selected. This system offered the added benefit that if the demonstration was successful, it could be converted to an operational system faster than any of the other alternatives.

APPENDIX 11.A ANALYSIS OF SHIP-TO-MEMPHIS TRANSMISSION SYSTEMS USING NAVY VOICE CIRCUIT

System Descriptions

A second communications system concept considered was to transmit the data signal from the ship to the CMI computer using a Navy voice circuit in the same way that the CONUS CMI now operates. This system would operate as follows. The set of test messages to be transmitted would be stacked on the OpScan 17. When a GapSat voice circuit would become available, the OpScan 17 would be patched into the circuit and operated, transmitting this signal to the fleet center with subsequent transmission of the message over telephone lines to the CMI computer at Memphis. The success of this concept depends on the availability of a GapSat voice circuit. Although the concept may be feasible from a technical viewpoint, it may not be practical. In any case three different system designs were considered using this concept, and these are described here for completeness.

System 1. Complete Store and Forward System

Figure 11.A.1 is a flow diagram showing the flow of digital data from the ship to the CMI computer. Again there are three major functions involved in transmitting CMI messages to shore over this path:

- F_1 is the function that stores and converts the CMI messages into electronic form for transmission to the CMI computer center.
- F_2 is the standard Navy voice circuit, which transmits the message from the ship to the fleet center.
- F_3 is the voice circuit that goes from the fleet center to the CMI computer center.

Each of these functions is now described.

F_1—CMI Message Input to GapSat. The objective of function F_1 is to convert the information on the CMI messages into an electronic signal and send this as a modulating signal to the GapSat terminal. This system uses the same preliminary procedure described under the Navy message system:

- The LS takes his CMI tests to the location where the OpScan 17 and teletype are located.
- Using the specially prepared test messages, the LS tests the availability of the OpScan 17, reading the results on the teletype.

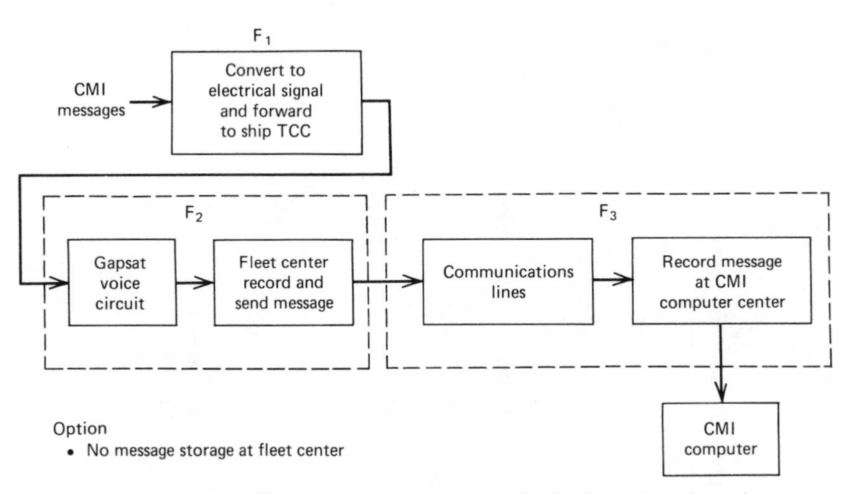

Figure 11.A.1. Ship-to-shore transmission via GapSat voice channel.

Table 11.A.1 *Resources Required for Navy Voice Circuit System*

Equipment	Cost ($1000)	O&M Personnel	Man-Hours/Week
1. SST	Navy provides service	Navy provides service	—
a. Investment			
b. Installation			
c. Spares			
d. Teardown and restore			
2. OpScan 17 (2)			
a. Investment	2 @ 9.3 = 18.6	Learning supervisor operates	Full time (7 mo)
b. Spares	6.5	Navy ET maintains	2 hr/wk (7 mo) = 61 hours
Total 2	25.1		61 man-hours + learning supervisor (7 mo)
3. Terminet 1200 (2)	4.2/ckt × 2 = 8.4	Navy ET maintains programmers'; Navy provides ($40K)	1 hr/wk (7 mo) = 30 hours
a. Investment	2.0		
b. Spares	10.4		
Total 3			30 man-hours
4. Ship TCC		RM request fleet center record message	3 @ 2 min/day = ½ hr/wk (7 mo) = 15 hours
Total 3			15 man-hours

5. NAVCAMS			
Magnetic tape recorder		F.C. RM prepares recorder	3 @ 10 min/day = 2½ hr/wk (7 mo)
a. Investment	To be provided by NETISA[a]	F.C. RM records message	3 @ 1 min/day = ¼ hr/wk (7 mo)
b. Spares		F.C. RM obtains AUTODIN	3 @ 10 min/day = 2½ hr/wk (7 mo)
Total 5			160 man-hours
6. CMI computer center			
Software changes	40.0		
Total 6	40.0		
Grand total	77.5 + cost of modified magnetic tape recorder[b]		236 man-hours + learning supervisor (7 mo)

[a]Commercial version cost = $9K + 1.8K spares.
[b]If the hybrid system is used, the cost of the terminet is eliminated, but the cost of software changes to the CMI computer may increase to $70K, increasing the total cost to $95.1K. Total cost must also include cost of return path using Navy message system. Shipboard cost = (2) + (3) + (4) + (5).
[c]Administrative terminal function.

The real set of CMI tests are then stored on the OpScan 17 and the LS requests a voice circuit for the short time required (less than ½ minute) from the TCC.

When the voice circuit is shortly to become available, the LS is notified. He connects the OpScan 17 to a telephone through either an acoustic coupler or a direct wire, as determined by the ship. When the radioman informs the LS that the voice circuit is now his, the LS speaks into the telephone and gives the identification of this message, including ship, date, time, CMI message number, to the CMI computer at Memphis; he then starts the OpScan 17 readout. The LS listens to the tone signals to make certain the message is entering the voice circuit satisfactorily and completely. When the message has been completed, he either states this over the circuit or hangs up.

F_2—Transmit Message to Fleet Center. The message is thus transmitted over GapSat to the appropriate fleet center (Honolulu for the East Pacific, Guam for the West Pacific, Norfolk for the Atlantic, and Naples for the Indian Ocean), where it is recorded on magnetic tape, following a net control request for this service from the ship's technical control center to the CAMS technical control over the orderwire.

F_3—Transmit Message from Fleet Center to CMI Computer Center. The next step is to relay the recorded signal to the CMI computer center, which can be done in the following ways.

In *Alternative F_{31}* a dedicated telephone line (costing $500 per month) connecting the Norfolk fleet center to the CMI DataNet at Memphis would allow forwarding of the recorded signal by a radioman in the fleet center (in the same way that the CW operator or sub broadcast operator handles these types of Navy messages). The signal would be recorded on magnetic tape at Memphis in exactly the same way that Memphis handles courses with Pensacola, using the same equipment (dedicated line to magnetic tape recorder IBM 2968 into their DataNet), and then would go into the CMI computer. The only disadvantage of this approach is the manpower required for the fleet center to operate the magnetic tape equipment.

Alternative F_{32} would use a standard commercial long-distance call from the Norfolk Fleet Center to the magnetic tape recorder at Memphis. This would be less expensive than the dedicated line for the short time use involved.

Alternative F_{33} would use conventional AUTOVON for the telephone line. This might involve adjusting the levels of the AUTOVON line. The AUTOVON system is not always properly compensated; hence levels are not uniform and adjustments may be needed. At each station through which the voice circuit passes, a certain noise floor and distortion level exist. The difference between these two is the dynamic range. The nominal voice level is set between these two extremes so that good voice quality is obtained at all times. Variable attenuators are used to make the adjustment at each location. The main cause of level variation is the traffic load on the system. The operator at each station adjusts his incoming and outgoing level to nominal values, altered only by the need to compensate for the channel loading condition. Too high a level at a given point will allow crosstalk to enter the channel. The total time required to make this level adjustment is from 20 to 40 minutes.

In *Alternative* F_{34}, if the fleet center is at Honolulu, F_3 could either be the AUTOVON or the ARPA network (Aloha Net) from the University of Hawaii to CONUS. Currently this network goes to Montgomery, Alabama. Thus we would also need phone lines from the Honolulu Fleet Center to the Aloha Net, and from Montgomery to Memphis, using either of the options described previously.

Return Path—CMI Center to Ship. The return message path was also analyzed, and it was found that the identical approach could also be used for this return path. That is, as soon as each of the CMI message replies becomes available (approximately 6 seconds after receipt of each original message) it is immediately transmitted back to the fleet center over the return path, where it is recorded on a separate track of the magnetic tape recorder. When the voice channel next becomes available, the recorded reply signal is transmitted from the fleet center to the ship, where it is patched into a terminet located at the learning center, and the CMI narrative reply messages are converted into page copy.

System 2. Total Path Continuity System

A second system using a voice circuit sets up the entire voice path (of F_1, F_2, and F_3) and then transmits the OpScan 17 signal directly to the Memphis computer. This eliminates the need for the tape recorder. In this system the LS again contacts the ship's TCC requesting a voice link to Memphis. About 30 minutes before this voice link will be available to the LS, the ship's technical control contacts the fleet center technical control over the orderwire and requests a voice path to the Memphis CMI computer. Fleet center technical control dials up such a path using any of the options delineated previously (AUTOVON, ARPANET, leased lines, etc.) and sees that the necessary line level adjustments are made for satisfactory transmission of the signal.

When this path is satisfactorily connected, technical control notifies the ship's TCC (over the orderwire) that the path is available, and when the satellite voice channel next becomes available the radioman notifies the LS to transmit the message, as previously described. Following transmission of this voice link, the rest of the operations occur, as previously described.

System 3. Reply via Navy Message System

Both preceding voice circuit alternatives require a terminet on board the ship to receive the reply messages. One way of avoiding this is to use a hybrid system, one that uses the voice circuit from ship to shore but uses the Navy message system to produce the narrative reply messages. This approach requires that the CMI computer have the ASCII to Baudot translator, as described previously.

Evaluation of Navy Voice Circuit Systems

In the case of the voice circuit approach for transmitting the CMI message, the alternatives differ in the way of implementing F_2, the Navy/AUTOVON (or other voice circuit mode), since F_1 and F_3 are the same for both systems. These two alternatives for F_2 are

- F_{21}, store signal at fleet center and forward later.
- F_{22}, connect GapSat to rest of voice circuit path, adjust line, and allow transmission to occur simultaneously, avoiding forwarding.

Of the two, the preferred system design alternative recommended for the demonstration is F_{21}, which offers the distinct operational advantage of being able to transmit the message at any time the GapSat voice circuit is available for approximately one minute, making it unnecessary to coordinate GapSat availability with the availability of the rest of the path. However, F_{21} does require a magnetic tape recorder at the fleet center to record the signals[15]; and it does impose the store and forward work requirements on the fleet center personnel. This involves setting up the recorder, recording the signal, disconnecting the recording, and maintaining the recorder. All other fleet center efforts have to be done for both systems (e.g., arranging for the rest of the path and adjusting the line levels, if necessary).

The resources required for this system are listed in Table 11.A.1.

[15] It is assumed that the one multitrack tape recorder can be modified at minimal cost to play back the CMI messages on one track and record the replies on another simultaneously. If not, a second tape recorder will be required at the fleet center.

APPENDIX 11.B DATA REQUIREMENTS OF THE DEMONSTRATION SYSTEM

Average Data Transmission Requirement for Each Type of Message

The following data represent current operating experience at the CMI computer center for all courses on line at the time of this analysis:

CMI Messages Sent (from Student to CMI Computer)
There were an average of 18,000 student inputs to the CMI computer per day. Since there were 3000 students enrolled and each student studied 6 hours per day, an average of one CMI message was sent for each hour of training. Based on the total data handled per unit time, the average CMI message sent for all CMI courses currently on line contains a total of 81 characters (including student and lesson identification).

CMI Message Reply (from CMI Computer to Student)
There is one reply for each message for each hour of student training. The average CMI message reply contains a total of 1600 characters, although for one of the CMI courses a reply can be as large as 12,000 characters.

Administrative Message Reply (from CMI Computer to Learning Supervisor)
There are a number of administrative messages currently sent to the learning supervisor in response to his queries. The final designation of which messages will be available to the learning supervisor during the demonstration and the specifications of the format of such messages are made during the next phase.

The most important information for the learning supervisor to have is the names of all students not performing up to expectations. Thus "management-by-exception" reports can be provided, as described in Chapter 8 by combining the following factors:

- Prediction of the times by which each module should be successfully completed. These times are based on the results of the student's battery of recruit testing and his training schedule.
- Amount of time that will be allowed to pass beyond the predicted time before a student is classified as deficient and his name sent to the learning supervisor.

Most of the rest of the administrative messages to the learning super-

Table 11.B.1 *NTCS Format of CMI Message Sent from Ship to CMI Computer, Memphis: Delivery in Tape-to-Card (Data Pattern Format) to NETISA Detachment, Memphis*

RTCUDAZZ RULYSAA1234 2081300 MTMS-UUUU-RUCIFMA.
ZNR UUUUU
R 061300Z OCT 76
FM USS JOHN F KENNEDY
TO RUCLFMA/NETISA DET NAVAL AIR STATION MEMPHIS TN
BT
UNCLAS //N01500//
COMISAT CMI STUDENT INSTRUCTION DATA
1. T E X T (40 lines maximum)
BT
#1234
RTCUDAZZ RULYSAA1234 2081300 MTMS-UUUU NNNN

visor pertain to other matters and will probably not be needed during the demonstration.

CMI Message Batching and Formatting

Having determined the length of each of the two types of CMI messages, the next step was to estimate the total volume of data to be sent each day during the demonstration. This estimate was needed to indicate how much of a communications load we would place on the ship's communications system.

Contact was made with the command to determine the standard Navy message form that each of these two types of CMI messages must follow if the Navy telecommunications system was to be used.

Table 11.B.1 illustrates the format[16] of a student message from the ship (U.S.S. *Kennedy*, for example) to the CMI computer at Memphis, as provided by COMNAVTELCOM. Each message sent consists of a 208-character header (or address),[17] a message text, which cannot exceed 40

[16]Since the message shown in Table 11.B.1 is manually generated, no carriage return symbol is shown, as is the case for the computer-generated message shown later.

[17]This is the message heading, which contains special symbols such as the date and time of message, and information telling the AUTODIN system how to route the message automatically to the addressee.

lines, each line containing about 62 characters, and a 59-character trailer. Since the average CMI message sent contains 81 characters, including identification of the student, course, and lesson, batching of student messages is required for efficient data transmission. In fact, the most efficient data transmission would be obtained by using a special end-of-message character printed on the OpScan sheets, running all CMI messages together into one large tape, and letting the NAVMACS[18] A-Plus computer on the ship divide the total message into separate segments of 40 lines each. Using an end-of-message character, the average CMI message sent would contain 82 characters. Based on 62 characters per line and a maximum block of 40 lines of text, the number of CMI messages sent N_s that could be batched in a single message is on the average:

$$N_s = \frac{62 \times 40}{81 + 1} = 30.2 \text{ CMI messages sent per Navy message sent}$$

Thus the total data requirements of a Navy message containing 30 CMI messages sent in the text can be calculated as follows:

Header (see Table 11.B.1)	208 characters
Text (30 × 82)	2460 characters
Trailer (see Table 11.B.1)	59 characters
Total	2727 characters

Based on an assumption of 30 students engaged in an average of 2 hours per day of training, the daily data requirements of the CMI messages sent would be two of the preceding messages, or 5454 characters per day (sending 30 CMI messages in each batch). The communications efficiency (CE) of this process may be calculated as follows:

$$\text{CE} = \frac{\text{text characters}}{\text{total characters}} = \frac{2460}{2727} \times 100\% = 90.2\%$$[19]

However, if there were only 20 CMI messages sent in the batch (and hence three messages sent per day), each message would contain

Header	208 characters
Text (20 × 82)	1640 characters
Trailer	59 characters
Total	1907 characters

[18]Navy Modular Automated Communications System.

[19]This assumes no retransmissions or service messages because of transmission errors.

Thus the total data requirements of the three messages would be 5721 characters per day. The communications efficiency of this process would be (1640/1907) × 100% = 86.0%.

Finally, if the CMI messages were sent in batches of 10, each message would contain

Header	208 characters
Text (10 × 82)	820 characters
Trailer	59 characters
Total	1087 characters

The total data requirements of the six messages would be 6522 characters per day, with a communications efficiency of

$$\frac{820}{1087} \times 100\% = 75.4\%$$

The data requirements of the reply messages can be calculated in the same way. The number of CMI message replies N_r that could be batched depends on their length. If the length of the CMI message reply is 1600 characters of text and one end-of-message character,

$$N_r = \frac{62 \text{ characters} \times 40 \text{ lines}}{(1600 + 1) \text{ characters}}$$
$$= 1.55 \text{ CMI message replies per Navy message reply}$$

Assuming that only one CMI message reply of 1601 characters of text was sent, the total number of characters sent in each Navy message reply would be

Header (see Table 11.B.2)	237 characters
Text (1601)	1601 characters
Trailer (see Table 11.B.2)	54 characters
Total	1892 characters

The communications efficiency of this process is

$$\frac{1601}{1892} \times 100\% = 84.6\%$$

Based on an assumption of 60 of these CMI message replies being sent each day, the daily data requirements of these 60 Navy message replies would be (60)(1892) = 113,520 characters per day. Thus the total daily data requirements for the demonstration would be 118,974 characters for all messages sent and replied to.

Table 11.B.2 *NTCS Format of Reply Message from CMI Computer to Ship: Originated in Card-to-Tape by NETISA Detachment, Memphis*

RCTUDAZZ RULYSAA1234 2801300 0050-UUUU-RUISJFK <[a]
ZNR UUUUU <
R 061300Z OCT 76 <
FM NETISA DET NAVAL AIR STATION MEMPHIS TN <
TO USS JOHN F KENNEDY <
BT <
UNCLAS //N01500// <
PASS TO COMISAT LEARNING SUPERVISOR <
COMISAT CMI STUDENT INSTRUCTION DATA <
1. T E X T 40 LINES BLOCK <
BT <
RCTUDAZZ RULYSAA1234 2801300 0050-UUUU NNNN <

[a]The symbol < indicates the carriage return.

For greater communications efficiency it should be possible to transmit a batch of CMI message replies in which the text of any CMI message reply extending beyond the fortieth line would continue on the next Navy message reply. In this case all but the last Navy reply message would have the following characteristics:

Header	237 characters
Text (62 × 40)	2480 characters
Trailer	54 characters
Total	2771 characters

$$\text{Communications efficiency} = \frac{2480}{2771} \times 100\% = 89.5\%$$

In this case the total number of characters in the Navy message replies sent each day is approximately:[20]

$$\frac{(60)(1601)}{0.895} = 107{,}330 \text{ characters}$$

which would result in a total data transmission requirement over the satellite for the demonstration ranging between 112,784 and 113,852

[20]This ignores the lower efficiency of the final message in each batch of reply messages transmitted each day. The actual total is 107,409 characters for the 39 Navy message replies.

characters per day. This corresponds to between 94 and 95 average 1200-character Navy messages to be transmitted each day.

Periodically, the learning supervisor (LS) will request some administrative reports such as the complete progress of a student. Such reports are currently printed on the administrative terminal for the LS. Our review of the current system shows that

- Such reports are very lengthy.
- "Management-by-exception" reporting would provide the LS with information on those students needing his attention with much less data transmission.

Since the data transmission using the satellite must be minimized, we plan the following actions

- During the demonstration, all administrative reports will be transmitted to the LS by mail, unless we find during the design phase that the size of these messages is small. The effect of this delay on student management will be measured.
- The operational system will be designed to provide "management-by-exception" reporting.

Hence the requirement for LS replies via satellite during the demonstration will be either zero or negligible.

Calculation of Response Time Required

The third important characteristic determining the response time required is the time between when the student's test is submitted for scoring and when his schedule calls for his continuing with the next lesson. Several scenarios described below illustrate this relationship.

Figure 11.B.1*a* illustrates the time sequence of events in a typical day of a student training in a CONUS schoolhouse. He studies the training material and once each hour he takes a test. The test is then inserted in the OpScan 17 terminal for scoring and feedback (indicating that he advances to the next module or studies additional material from the current module). Since this process is repeated over the 6- or 8-hour day, any delay in obtaining the feedback message is time completely lost to the student (since he normally has nothing else to do while waiting for the feedback).

Figure 11.B.1*b* illustrates the day's activities of a student training for 1 hour per day on a ship. Assuming that he takes the test toward the end of

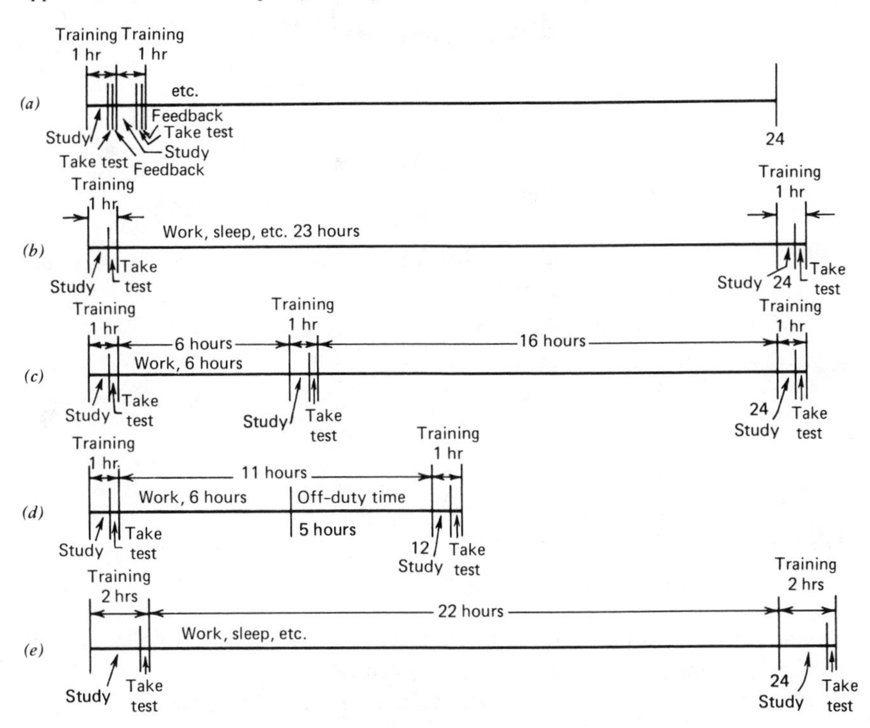

Figure 11.B.1. Time sequence of training and work activities. (*a*) Schoolhouse training. (*b*) Training on ship 1 hour/day. (*c*) Training on ship 2 hours/day. (*d*) Training on ship 2 hours/day. (*e*) Training on ship 2 hours/day.

the training hour, he must have the test results returned to him within 23 hours. Thus the maximum total response time is 23 hours, including the time lost due to batching, waiting for a number of tests to be collected prior to transmission.

Figure 11.B.1*c* illustrates a scenario in which 2 hours of instruction per day are allowed, and programmed as the first hour and last hour of an 8-hour work shift. Thus the intervening time of 6 hours in one case and 16 hours in the other is available for obtaining feedback on the test.

The demonstration will be used to determine if the 6-hour response time is achievable. If not, other possibilities are available:

- Figure 11.B.1*d* illustrates that by delaying the start of the second hour of training by 5 hours, the response time required is increased to 11 hours.

- Figure 11.B.1*e* is another scenario in which modules are divided into 2 hours of length for one test so that the test results are not required for 22 hours.

If the 6-hour scenario of Figure 11.B.1*c* is imposed, there are two possibilities for meeting the response time:

- Construct the course so that the student takes 2-hour modules, hence converting the scenario to Figure 11.B.1*e*.
- Construct the course in two unrelated tracks. The student takes a 1-hour module from one track, then a 1-hour module from the second track. Alternating modules effectively changes the scenario to Figure 11.B.1*a* for each track, thus permitting a time response of 23 hours for each track.

APPENDIX 11.C DATA REQUIREMENTS FOR THE OPERATIONAL SYSTEM

This appendix extends the analysis described previously for the demonstration system to the data requirements of the operational system. These requirements are a function of the following characteristics:

- The number of student training hours.
- The average number of characters in CMI message sent (same as demonstration: 82).
- The number of CMI messages sent in each batch (same as demonstration: 10, 20, or 30).
- The average number of characters in a CMI message reply. NETISA says this could be reduced to 600 to 800 characters in two ways.
- First, an abridged form of reply could be developed and used as a separate track for remote students. Second, additional compression could be used, resulting in a 62-character line.
- The number of CMI message replies transmitted in each batch (same as demonstration: 10, 20, or 30).
- Batching of the texts of the CMI message replies to use all or most of the 40 lines (2480 characters) available, thus providing highest communications efficiency.

To obtain an order of magnitude of the amount of data to be transmitted in an operational system, a "worst case" analysis[21] was performed based on the following values of these characteristics:

- Ten CMI messages sent would be transmitted in one batch, giving 820 characters of text, and results from 10 student training hours.
- Ten CMI message replies would be transmitted in one batch.
- Each CMI message reply would be assumed to have 801 characters, including the end of message symbol. Thus the total batch of the 10 CMI message replies would contain 8010 characters.
- Each Navy message reply would contain the maximum of 2480 characters of text.

Based on these assumptions, the Navy message sent (10 CMI messages) would contain 1087 characters and would be transmitted at an efficiency of 75.4%.

The data requirements of the Navy message replies are calculated as follows:

- The total number of characters of text is (10)(801) = 8010.
- The maximum number of characters of text in a Navy message reply is (62)(40) = 2480.
- The total number of Navy message replies is 8010/2480 = 3.23, or four messages. (This is equivalent to three messages of 2480 characters of text and a fourth message of 570 characters of text.)
- The total number of characters of header and trailer in each reply message is 291.
- Thus the total number of characters in the four Navy message replies is 8010 + 4(291) = 9174.
- The communications efficiency of the reply is 8010/9174 = 87.3%.
- Thus the total data transmitted for 10 student training hours is 1087 + 9174 = 10,261 characters.
- This equates to 1026 characters per student training hour. A planner may wish to use this planning factor: Each training hour requires a round trip transmission of 1000 characters. Note that this is less than one average Navy message of 1200 characters.

[21]"Worst case" in terms of the maximum amount of data to be transmitted. Similar analyses could be made for other sets of characteristics.

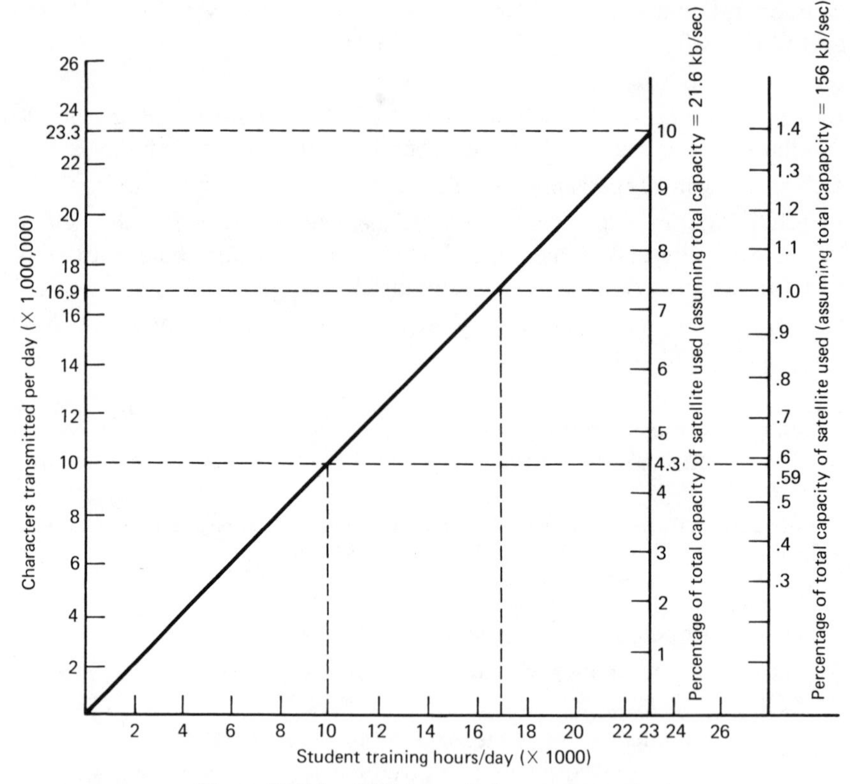

Figure 11.C.1. CMI transmission requirements.

Data Requirements as a Function of Satellite Capacity

Another way of expressing the data requirements is as a percentage of the entire transmission capacity of a specific satellite. This was done for the Fleet Satellite Communications System (FleetSatCom) using available information about its total transmission capacity.

The FleetSatCom system will have nine channels, each 25 KHz wide, suitable for message traffic. Originally, each of the nine channels was to carry a 2.4 Kilobits (kb)/sec data rate. However, a time division multiple access (TDMA) system has been designed[22] that can provide a total of about 65 simultaneous 2.4 kb/sec circuits on the nine 25 KHz channels,

[22]J. D. Bridwell and I. Richer, "A Preliminary Design of a TDMA System for FleetSat," MIT Lincoln Laboratory, Group 67, Technical Note 1975-5, March 12, 1975, NTIS Document AD-A-007-823.

for an average data rate of 17.33 kb/sec per channel for a "representative mixture of ships, aircraft, submarines and shore stations."

Present Navy plans are to use Channel 1 for fleet broadcast (15 channels of shore-to-ship TDM), with the other channels computer controlled and interactive. Two CUDIXS channels will be provided. Each will serve 10 major ships with duplex transmission, plus 50 small ships with ship-to-shore transmission.

A bit rate figure for the entire satellite has been derived but is not felt to be realistic "because of functional assignments—the number of people in the net and the traffic." Hence although the actual FleetSatCom channel capacity is not yet established, it will be somewhere between 2.4 and 17.33 kb/sec for each of the nine 25-KHz channels. Based on this, the total FleetSatCom capacity is between 21.6 and 156 kb/sec.

Thus in one 24-hour period, the total capacity of the satellite, assuming AUTODIN II requiring eight bits per character, is between

$$\frac{(21.6\ \text{kb/sec})(24)(3600)}{8\ \text{bits/character}} = 2.3328 \times 10^8\ \text{characters/day}$$

and

$$\frac{(156\ \text{kb/sec})(24)(3600)}{8\ \text{bits/character}} = 1.6848 \times 10^9\ \text{characters/day}$$

Thus each 1000 student training hours per day in the satellite area requires between

$$\frac{10^6\ \text{characters}}{2.3328 \times 10^8\ \text{characters}} = 0.0043 = 0.43\%\ \text{of total capacity}$$

and

$$\frac{10^6\ \text{characters}}{1.6848 \times 10^9\ \text{characters}} = 0.00059$$

$$= 0.59\%\ \text{of the total of the FleetSatCom satellite}$$

Conversely, using 1% of the satellite would provide between 2300 and 16,900 student training hours per day and using 10% of the satellite would provide between 23,000 and 169,000 student training hours per day. Both this and the previous functional relationship are shown in Figure 11.C.1, including the two limits of satellite capacity.

12

Evaluating the Operational System

This chapter describes the approach developed for performing the economic analysis of a proposed operational, self-paced, computer-managed instruction (CMI) via satellite system and its alternatives. The objective of this analysis is to determine whether satellite-delivered CMI to shipboard or remote land sites is as economical as CMI in the learning center environment. Here the term *economic analysis* is synonymous with systems analysis, cost-benefit analysis, or systems evaluation. The output of the analysis is structured information that aids decision makers in evaluating the various training system options available to them on the basis of their training objectives and costs, risks, and uncertainties. The specific objective of the analysis is to compare a proposed operational system, designed to provide CMI instruction at remote sites but linked to the CMI computer at Memphis, Tennessee, via a Navy communications system, with other Navy training systems that can provide such CMI instruction. Such a comparison would be used to determine the preferred system for different future time periods and student load conditions in developing a master implementation plan.

Although this systems evaluation follows many of the steps described in the production planning case of Part II, it involves additional complexities not present in that case. The economic analysis is based on the following assumptions:

- Only training system alternatives utilizing CMI need to be evaluated. Training systems using instructor-managed instruction (IMI) or traditional classroom instruction are not considered. This is a reasonable assumption since the Navy has already evaluated CMI training against these other two types and has found CMI training to be superior.
- The performance and cost data used in the evaluation are based on the current CMI system as given in the training centers, the COMISAT

demonstration, and the U.S.S. *Gridley* minicomputer[1] CMI demonstrations, as extrapolated to fully operational systems.

This chapter presents both the approach to be followed and the information and data to be collected in implementing the analysis.

12.1 ANALYTICAL APPROACH

The economic analysis is performed using the same principles for the evaluation of alternatives as were developed in Chapter 4.

12.1.1 Identify Alternative System Concepts

The first step is to identify the alternative ways of delivering Navy CMI training, for which performance and cost data will be available at the time of the evaluation phase. As shown in Table 12.1 there are several distinct CMI training system concepts.

ALTERNATIVE 1

All students are trained at training centers. This is the current system using the CMI computer at Memphis with land lines to various Navy training centers in the United States. Each of the remaining alternatives consists of some students being trained at the training centers and the rest being trained at remote sites in the following ways.

ALTERNATIVE 2

The CMI computer at Memphis is connected by suitable means of communication to the remote sites (the current COMISAT project).

ALTERNATIVE 3

A minicomputer at each remote site is used to perform the same functions as the CMI computer; thus this system requires neither the Memphis computer nor the same lines of communication as Alternative 2. This system would be designed after the shipboard command management and

[1]This is a competing CMI alternative now being demonstrated on the U.S.S. *Gridley*. It is assumed that the performance and cost characteristics of this system, as obtained from both demonstrations, will be made available for our economic analysis.

Table 12.1 *Alternative CMI Training Systems Being Considered*

All students trained at training centers (Alternative 1)
Some students trained at training centers. Others trained by:
- CMI Computer plus lines of communication to remote sites (Alternative 2)
- Minicomputer at each remote site (U.S.S. *Gridley*) (Alternative 3)
- CMI computer plus lines of communication to some remote sites, minicomputer at other remote sites (Alternative 4)
- Alternative 2 plus microprocessor at each remote site (Alternative 5)
- Instructor-managed instruction at remote sites (Alternative 6)
- CMI Minicomputer at each remote site (Alternative 7)

readiness system as demonstrated on the U.S.S. *Gridley*. This system performs a number of administrative functions in addition to the training function. Hence we shall have to pay special attention to relating total costs to the various functions this system provides.

ALTERNATIVE 4

A mix of some students trained at those sites that will have the minicomputer anyway because of the nontraining applications it provides (Alternative 3) with the remaining students using the CMI computer (Alternative 2). Although this alternative is really a mix of Alternatives 2 and 3, it may merit consideration if a minicomputer is to be procured for some sites anyway, since the incremental cost for the training system may be small.

ALTERNATIVE 5

A fifth alternative uses the same approach as Alternative 2 but attempts to keep satellite use time required to a minimum. Two implementation possibilities exist within this alternative. The first is to redesign the narrative replies so that they convey the same basic information but require fewer characters than currently estimated (600 to 800 characters). The second approach is to reprogram the Memphis CMI computer to send student reply information to the remote site in coded form rather than narrative text. The information is then fed into a microprocessor at the remote site, which translates the reply into the normal narrative text format of the current CMI system and prints it on a separate terminal. It is assumed that the first possibility will be included as part of Alternative 2 as new CMI courses are designed. We could evaluate the second possibility (the use of a microprocessor as a reply translator) if requested to do so.

Two other alternatives are also possible at remote sites:

ALTERNATIVE 6

Instructor-managed Instructon (IMI), in which the instructor (or chemically treated test paper) is used to grade the tests and remedial instruction loops, is available as part of the audiovisual training package. This option may be superior if only a few students on the ship are participating in training.

ALTERNATIVE 7

A minicomputer devoted only to CMI training can be used.

12.1.2 Identify Alternative System Concepts to Be Analyzed

Only Alternatives 1 to 4 are considered in the evaluation, since performance and cost data for the other alternatives will not be available by the evaluation phase of this program.

12.1.3 Define the Training System Objectives

The next step in the analysis is to define the operational objectives all system alternatives will be designed to meet. In the case of the COMISAT program we can define these objectives in the following way. At some future time when an operational CMI system is required

- There will exist a set of CMI courses available for instruction (i.e., all training materials will have been developed, programmed, coded, and validated in the CMI format).
- These courses will be designed to fill the following training needs:

  A-school courses[2] given to students immediately upon completion of their recruit training.

  A-school courses given to selected students already assigned to a ship who have never had such training but could profit from it. Some of

[2]A-school is the Navy designation for "initial skill training," referred to as advanced individual training and technical training by the Army and Air Force, respectively.

these students would normally be assigned to training centers as "fleet returnables." Because of manning limitations, the remainder of these students will not have the opportunity to attend A-schools.

Other courses of a more general nature, such as General Damage Control, normally given at the ship rather than at training centers.

CMI courses that would provide the same educational content as correspondence courses now given to students at sites.

- The requirement of the CMI training system can be defined as providing a given number of graduates from each of these courses for each operational year of the CMI training system. This requirement is defined as the system demand function.

12.1.4 Identify Differences Among Alternative Systems

When these various types of courses are delivered to the available students using the four different training system alternatives defined previously, certain differences in performance and total cost of delivering the training are expected to exist. An identification of these differences and an explanation of how they can be measured in the analysis are given below and illustrated in Figure 12.1.

Factor 1: Student Training Time in CMI Course

The man-hours required for students complete a course satisfactorily may differ among training systems. These times will be converted into equivalent student billet costs by determining over how many weeks the training extended and what proportion of this time was spent in training, as compared with other duties. Student billet costs contribute to the hierarchy of costs as shown in Figure 12.1.

Factor 2: Standard Workweek

The standard workweek while stationed at sea is greater than while stationed at a training center. Thus student billet costs per hour are less at sea than in the United States.

Factor 3: Other Student Costs

Other student costs such as transportation and living expenses while traveling, which also occur while the student is in training, may also differ among systems.

Factor 4: Student Attrition

Certain students may not graduate from the course. The time spent by these students while in training and all other resources these students

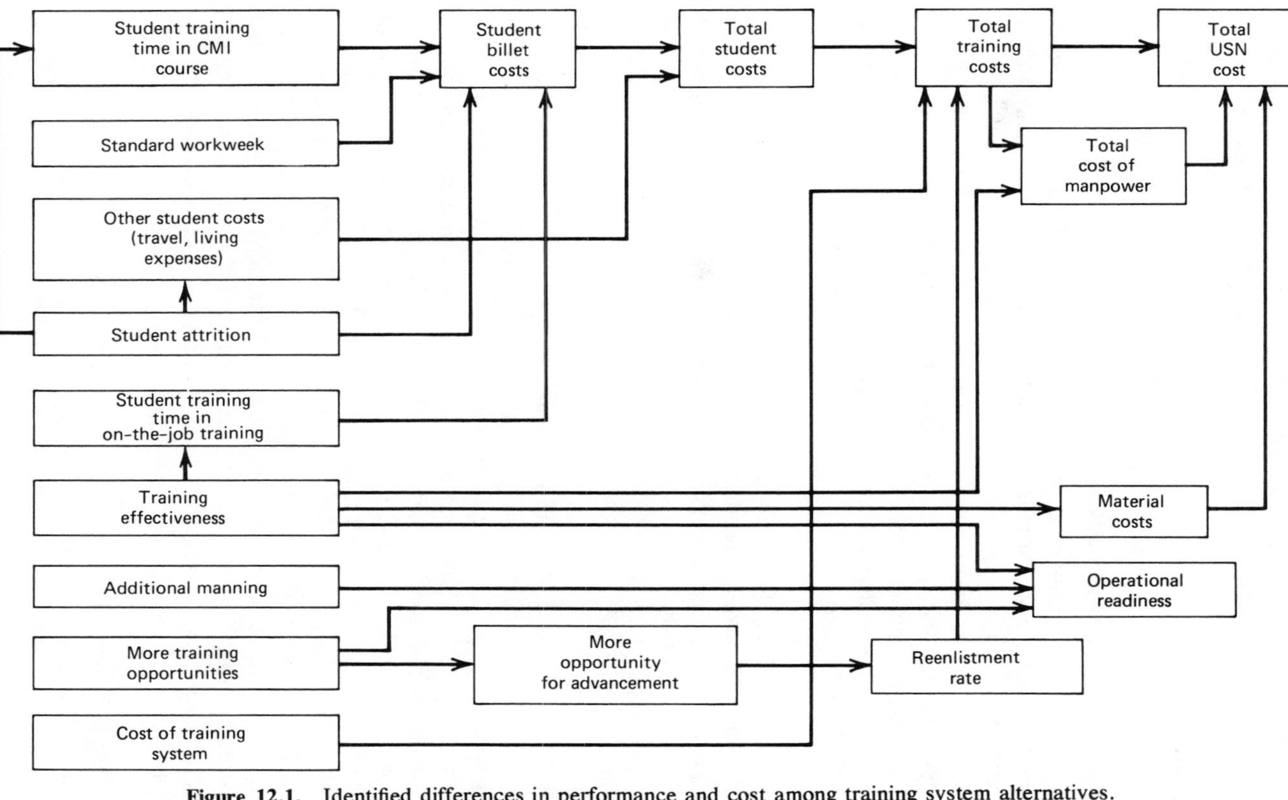

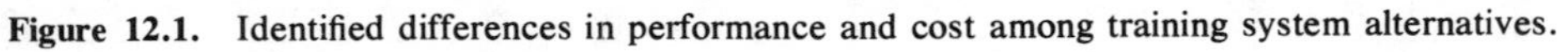
Figure 12.1. Identified differences in performance and cost among training system alternatives.

require during this time may differ among systems. All of these resources will be converted into equivalent student costs and added to the costs of the graduating students. It is important that this measurement only include those students who do not graduate because of academic reasons. Current attrition measurements include other reasons, such as health and misconduct, which are not relevant to this analysis.

Factor 5: Student Time in On-the-Job Training

Some of the alternatives may provide synergistic effects on other concurrent functions. For example, conducting CMI training concurrently with on-the-job training (OJT) and work assignments may result in the OJT being completed more rapidly, particularly if the CMI course was designed around the specific set of equipment in place at the job site rather than around a general set of equipment at the training center. Such time reduction would also reduce the equivalent student billet costs attributed to OJT in the same way as previously described for a CMI course.

Factor 6: Training System Costs

The costs of developing, installing, operating, maintaining, and supporting each training system alternative may differ. These costs (not including the student costs previously described) are defined as training system costs. Such costs are obviously a function of the amount of student training provided.

Factor 7: Additional Manning at Remote Sites

At the same time that training occurs on the ship, the student is also able to spend part of his time doing productive work under direction of his supervisor. This difference among system alternatives is measured by using standard cost accounting techniques. The proportion of the standard workweek used for the CMI course and OJT is charged against productive time, except for one adjustment factor. We also determine the "average trainee unproductive time," defined as that time when there was not productive work that could be assigned to the student because of his lack of training. Unproductive time also includes taking longer than a "standard time" to complete a task. The productive man-hours provided by a student are converted to the equivalent of additional men "on board."

Factor 8: Training Effectiveness

One training system alternative may result in more effective training than another, as expressed in differences in test scores. However, to analyze the value of such differences, research would have to be performed relating test scores and later production on the job in terms of being able to do productive work faster or better or to complete OJT faster. These benefits could result in less manpower being required as well as less waste of materials. It may not be possible to quantify this difference among systems during the demonstration phase.

Factor 9: More Training Opportunities
Providing more training opportunities to those who would normally not be able to take such courses aids in two ways. First, it provides the individual further opportunity for advancement, thus increasing the reenlistment rate and reducing replacement training costs. In addition, it increases the operational readiness of the ship. It may not be possible to quantify this difference among systems during the demonstration phase.

12.1.5 Method Used to Compare and Analyze Differences

Two conclusions become apparent from consideration of these nine differences previously discussed and structured in Figure 12.1. First, some of these factors may provide advantages beyond decreasing the total cost of providing a given number of graduates from a set of Navy courses. Hence to include these factors there is a need to expand the system demand function beyond the training system objectives originally considered. Second, some of these factors may result in lowering total Navy costs in providing a given level of operational readiness (the ability of the Navy to perform its missions); yet some of these factors would tend to increase operational readiness, such as those that decrease total training time or decrease the time required for an individual to do a work task satisfactorily. Since we must relate operational readiness and total cost, one way of doing this is to pivot on a constant level of operational readiness (say, the current capability). We might consider that any savings in personnel time would equate to less personnel required in the Navy, resulting in decreased manpower cost. One way to include this is to convert any manpower savings into dollar savings by calculating the cost per hour for the personnel at the grades being trained. Specifically, compare the total system lifetime costs of obtaining the required CMI course graduates using the CMI satellite system with all other ways of doing the same job. In addition, determine the economic value of the additional benefits described and the additional costs which any of these CMI training system alternatives provide that other system alternatives do not provide.

As indicated previously, it is assumed that all the data required to make such calculations are obtained as performance and cost data based on an extrapolation of actual CMI system operations of

- The current training center CMI system.
- The satellite CMI system demonstration.
- The minicomputer system.

We shall perform the analysis by

- Designing each of these system alternatives so that each produces the same required number of course graduates.
- Analyzing the nine differences identified among systems and calculating for each system design alternative:

  The total student costs attributed to training.

  All training system costs.

  All other benefits obtained (as described previously) that differentiate one system alternative from another.

12.2 CONSTRUCTING THE SCENARIOS

Figure 12.2 illustrates the time sequence of activities involved in training, travel, and work under three hypothesized scenarios. The importance of these scenarios is that they help identify the differences among system alternatives by showing the different activities that may occur or differences in time to accomplish an activity.

The variable T represents the time in weeks for an activity and M represents the man-hours devoted to it. The first scenario represents the current CMI training at training centers, with two travel periods and expenses (to A-school and to the ship). The second scenario is for those courses in which the student could travel directly to the ship and then begin the course, requiring only one travel period. The third scenario, requiring two travel expenses, is for those courses that must be begun at the training center, followed by transfer to the ship where the course is completed, along with the other duties depicted in the second scenario. In addition, total student costs are affected by

- How many man-hours of duty time (as opposed to free time) were spent by the successful graduate in completing the course?

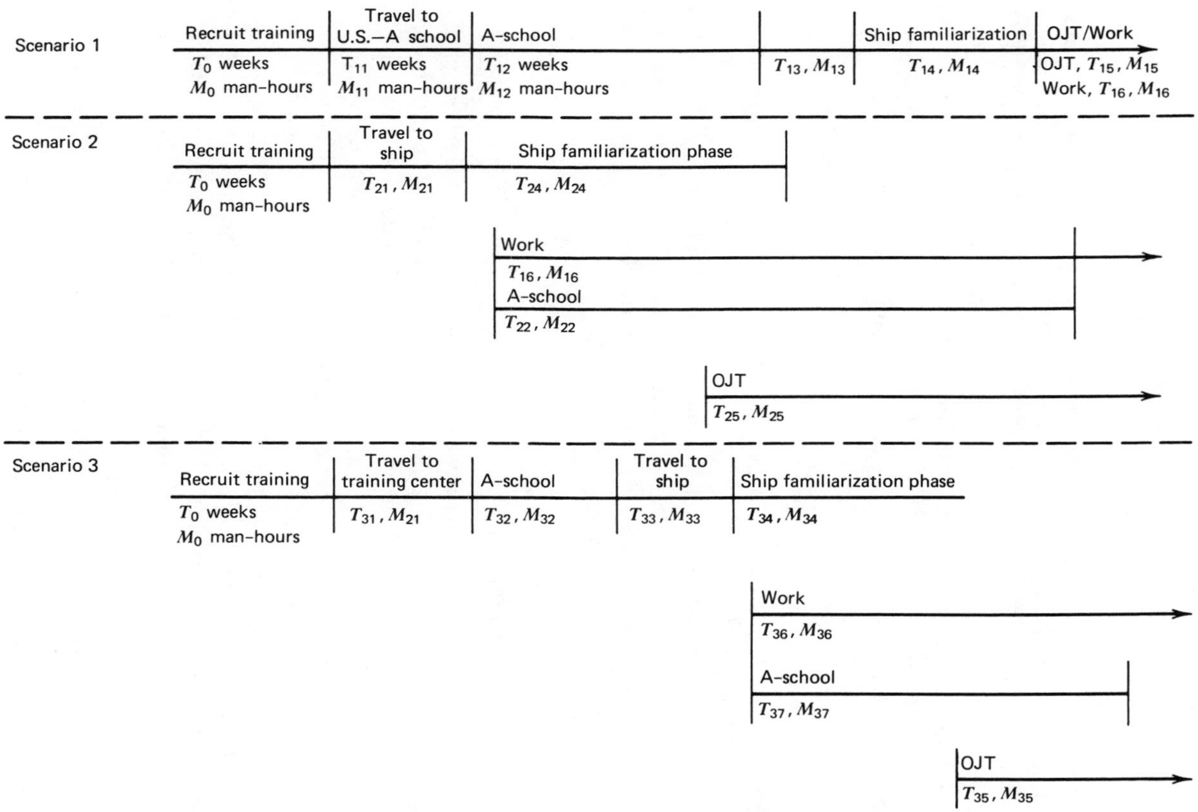

Scenario 1
Recruit training | Travel to U.S.–A school | A-school | | Ship familiarization | OJT/Work
T_0 weeks, M_0 man-hours | T_{11} weeks, M_{11} man-hours | T_{12} weeks, M_{12} man-hours | T_{13}, M_{13} | T_{14}, M_{14} | OJT, T_{15}, M_{15}; Work, T_{16}, M_{16}

Scenario 2
Recruit training | Travel to ship | Ship familiarization phase
T_0 weeks, M_0 man-hours | T_{21}, M_{21} | T_{24}, M_{24}
Work
T_{16}, M_{16}
A-school
T_{22}, M_{22}
OJT
T_{25}, M_{25}

Scenario 3
Recruit training | Travel to training center | A-school | Travel to ship | Ship familiarization phase
T_0 weeks, M_0 man-hours | T_{31}, M_{21} | T_{32}, M_{32} | T_{33}, M_{33} | T_{34}, M_{34}
Work
T_{36}, M_{36}
A-school
T_{37}, M_{37}
OJT
T_{35}, M_{35}

Figure 12.2. Training and work activities.

- What is the standard work time per week expected of the student?
- What is the attrition rate?
- How many student man-hours were spent before attrition?

By comparing the man-hours required to complete the various training and other activities for the training center CMI system with that required for the remote site systems, we can determine time (and cost) differences between systems. The analysis assumes that there is a requirement for A graduates of a given CMI course for a particular operational year (the system demand function). However, because of attrition, it is assumed that B additional students must enter the course during this year.

12.3 CALCULATING STUDENT COSTS

Student costs are defined as all expenditures spent directly on the student because he has attended the course. They include billet costs (salary, benefits, and living expenses per week) and travel costs. These costs are for both graduates and nongraduates of the course.

12.3.1 Recruit Training Cost

Let T_0 be the average time (in weeks) for recruit training. Although this time may be the same for all students who enter A-school, some students who fail to graduate A-school are discharged and hence have no value to the military. Therefore the total cost of recruit training for these discharged students must be considered as a part of the training costs. Thus the man-hours (M_0) devoted to training are the same as T_0, since this is a full-time activity. Some of the B students who do not graduate are discharged because their enlistment contract was based on their assignment to this type of work, for which it turned out they were not qualified. Thus the recruit training cost attributable to later A-school attrition (CRT) is

$$\mathrm{CRT} = D(T_0\,\mathrm{BC} + \mathrm{CUR})$$

where D = number of students discharged
BC = weekly student cost (salary, benefits, and living expenses)
CUR = unit cost of recruiting one enlistee

12.3.2 Student Billet Cost During Travel

Let T_{11} be the average travel time from recruit training center to A-school,[3] T_{13} be the average travel time from A-school to the ship, and T_{21} be the average travel time from the recruit training center to the ship for Alternatives 1 and 2. All times are in weeks. The total billet costs during travel (BCT) for Alternative 1 are

$$\text{BCT} = [(A_1 + B_1)T_{11} + (A_1)(T_{13})]\text{BC}$$

and for Alternative 2 are[4]

$$\text{BCT}_2 = [(A_2)(T_{21})]\text{BC}$$

where A and B are the number of course graduates and nongraduates, respectively, of system alternatives 1 and 2, and the difference in travel times is reflected as shown.

Obviously the third scenario requires the same travel time as the first.

12.3.3 Transportation Costs

The remaining cost of transportation (CT) depends on the distances involved:

$$\text{CT}_1 = (A_1 + B_1)\text{TC}_{11} + A_1\,\text{TC}_{13}$$
$$\text{CT}_2 = A_2\,\text{TC}_{21}$$

where TC_{11}, TC_{13}, and TC_{21} are the average transportation costs (including per diem living expenses while en route) from the recruit training center to CONUS A-school, A-school to ship, and recruit training center to ship, respectively.

12.3.4 Student Cost During Training

Figure 12.3 illustrates how an individual would spend his productive time under scenarios 1 and 3.[5]

Let PSC be the proportional student billet costs (salary, benefits, and living expenses) attributed to CMI training for the A graduates:

[3] Leave time is not included in these calculations, since it accrues to the student and is eventually used by him.

[4] This assumes that all nongraduating students from system 2 (B_2) remain on the ship and are productive in some other rating whereas the travel costs for the nongraduating students from system 1 (B_1) are a complete loss.

[5] The ship familiarization phase has been omitted in this illustration.

$$\text{PSC} = \frac{A(\text{MHS})(P)(\text{BC})}{\text{MHW}}$$

where PSC = proportional student billet costs attributed to training
A = number of graduates
MHS = average man-hours required to complete the course successfully
P = proportion of work time devoted to course training
BC = total billet cost (weekly salary, benefits, and living expenses)
MHW = average man-hours per week of training performed

Following are examples of how this calculation is to be made:

Example 1: Training Center Training
Assume that at the training center a course requires an average of 240 hours (MHS) for each of 400 graduating students (A) to complete. Each student goes to

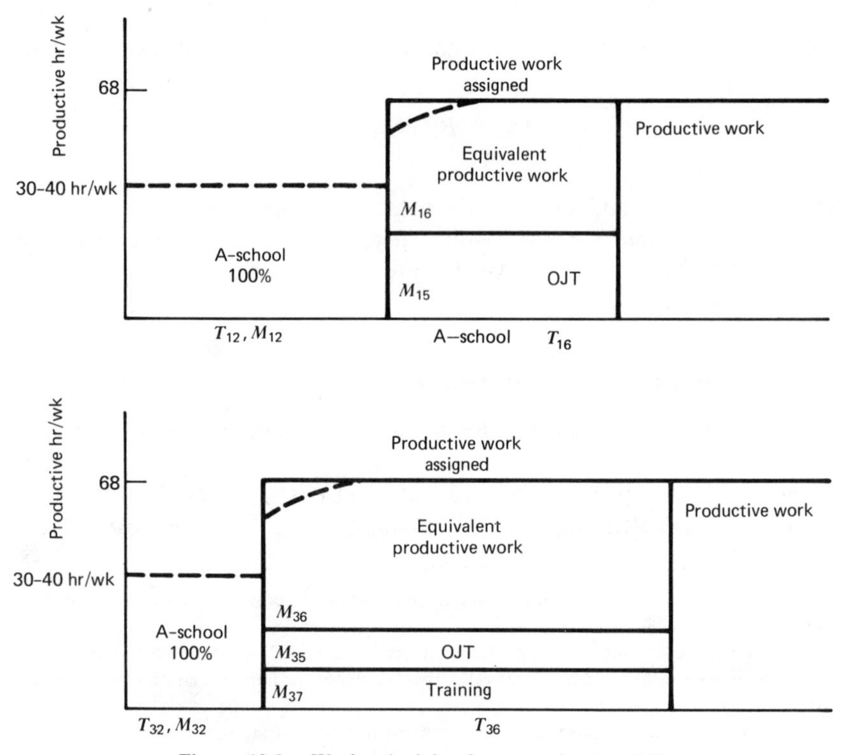

Figure 12.3. Work schedules for scenarios 1 and 3.

school an average of 30 hours per week (MHW), and this constitutes a full-time work assignment ($P = 1$). Total salary, benefits, and living expenses are \$300 per week (BC). Thus

$$\text{PSC} = \frac{(400)(240)(1)(\$300)}{30} = \$960{,}000$$

for 400 successful graduates.

Example 2: Shipboard Training
Assume that the 400 graduates had been on board ships, had taken the courses for 10 hours per week on duty time, and that the total course required an average of 300 man-hours. The proportional student costs for the 400 graduates attributed to training would be found as follows.

The average number of weeks required to complete the courses is 300/10 = 30 weeks. However, only P proportion of these 30 weeks' total salary is chargeable to training. Assume the total standard workweek for a watch stander is 74 hours, of which 6 hours are devoted to military training and service diversions. Hence these 6 hours can be considered as overhead to both the work and the training. Thus the proportion of total available work time chargeable to training is

$$P = \frac{10}{68} = 0.147$$

Hence

$$\text{PSC} = \frac{(400)(300)(0.147)(\$300)}{10} = \$529{,}200$$

for the 400 graduates.

Example 3: Shipboard Training
In Example 2, if the 400 graduates spent 10 hours per week taking the courses but only 5 hours per week on duty time (the other 5 hours per week on "free time"), P is now equal to 0.0735 and,

$$\text{PSC} = \frac{(400)(300)(0.0735)(\$300)}{10} = \$264{,}000$$

However, salary, benefits, and living expenses for nongraduates must also be paid for the B nongraduates while they are attending the course. Therefore let

$$\text{PSN} = \frac{(B)(\text{MHN})(P)(\text{WS})}{\text{MHW}}$$

where PSN = proportional student costs of B nongraduates
MHN = average man-hours of training performed before attrition

and P, WS, and MHW are defined as before. Thus TSC, the total student costs for the A graduates, is TSC = PSC + PSN.

It should be explicitly noted that in this analysis we are including the cost for all students who do not graduate from the course. However, we are not assigning any benefits for those who do not graduate. This is a reasonable assumption, since attrition in general takes place early in the training period. This method of treating the costs of all nongraduates is based on the assumption that all nongraduates complete their enlistment at some other assignment. However, as indicated previously, if the service policy is to discharge nongraduates following their attrition from school, the cost of recruit training (including recruiting and transportation costs) must also be included in the cost analysis.[6]

12.3.5 Synergistic Effect of A-School, OJT, and Work

A hypothesis we make (and requiring statistical validation during the demonstration) is that following an integrated program combining theory (A-school CMI), OJT, and work assignments will enable a person to achieve a level of proficiency in less total training man-hours than taking theory first followed by both OJT and work assignments. Thus, in addition to measuring the man-hours to complete A-school satisfactorily, we should also measure the total man-hours to complete OJT (up to a given level of proficiency).

The following measurements will be conducted during the demonstration:

- Measure the man-hours required to complete A-school (M_{12} versus $M_{32} + M_{36}$), as described previously.[7]
- Measure the man-hours required to reach various defined levels of competence through OJT and work done in parallel (M_{15} and M_{35}). OJT continues informally indefinitely, but for the purposes of this analysis, we define OJT in the same terms as a particular course. That is, the student keeps advancing over increasing levels of competence and is thus able to perform increasing sets of tasks. With this in mind, we would like to measure the time taken to reach certain levels of competence through OJT and work experiences. Obviously, M_{15} and M_{35} will themselves vary as the competency level parameter changes.
- Determine the equivalent man-hours of unproductive time lost during

[6]The Air Force has such a policy.

[7]This analysis is made by comparing the first and third scenarios of Figure 12.2.

the man-hours assigned for productive work (M_{16} versus M_{36}). By *productive work* we mean all work tasks done that are part of normal day-to-day duties (even if they are accomplished within the context of OJT). On the other hand, pure OJT activity may include typical work elements, but they are only done as practice. For example, a student may practice tuning a transmitter (off-line, main power or high voltage off). This time is counted as part of OJT training. On the other hand, when he advances to the point that his supervisor asks him to tune the transmitter for actual operation under a petty officer's supervision, the student's time is counted as productive, and the supervisor's time is counted as OJT training supervision. By *equivalent unproductive man-hours* we mean that amount of the man's time when he was available for productive work (M_{16} and M_{36}) but could not perform productively, either because of a lack of productive work at his skill level or because his skill level prevented him from doing the work at some reasonable time standard. Obviously, this is determined by the following factors:

The supervisor assigning work to the student in accordance with his current capabilities as he progresses through course training and OJT.

The capability mix of the entire division. If the proportion of new students is too high, the available lower-level work may be insufficient to occupy this group fully, and the equivalent productive work load will be as shown by the dotted lines of Figure 12.3. This problem decreases as time goes on and the capabilities of the students increase.

To maximize the productive work obtained from these semitrained students, some restructuring of the division's work duties may be required. Figure 12.4 shows a "frequency distribution" of the total work load in a division such as the communications division. This figure represents a listing of all work tasks being done in the division, ranked from the least complex to the most complex (the way OJT would probably be taught), and the amount of man-hours per year required by each task. We also show all of the tasks that a fully qualified individual (one who completes the level of training and OJT previously defined) can do. Because a member of the division assigned to OJT progresses from left to right in his qualifications as he advances in his OJT program, it should be possible, by relating each trainee's work assignments to his progress through OJT, to reduce the Equivalent Unproductive Time to nearly zero. With the fleet now manned at some factor less than 100%, any additional manning (say up to the 100% level) transferred from A-school to the ship for training as well as productive work, would increase the amount of

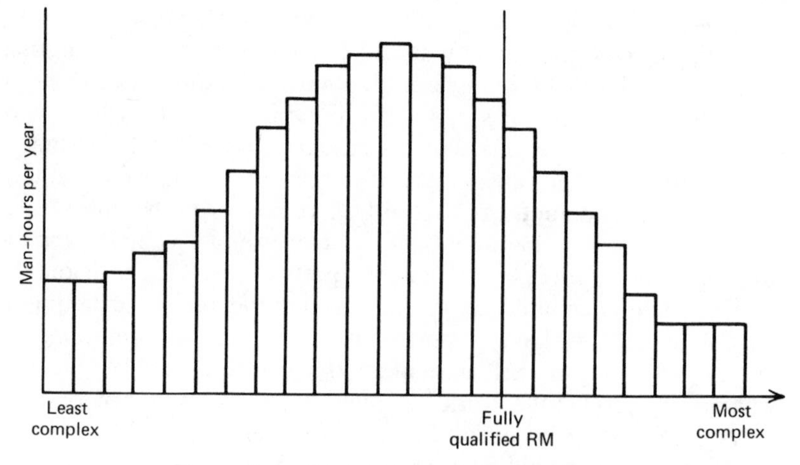

Figure 12.4. Work task distribution in division.

productive work capability based on the following assumptions, which will be validated during the demonstration:

- Work assignments could be adjusted within the division to utilize the students as they progress through OJT.
- Supervision of the students during OJT and their productive work will not be excessive.

Thus the final analysis of each operational system would contain data of the man-hours required to

- Complete A-school.
- Complete a given level of OJT.
- Accomplish productive work provided by the students.
- Supervise OJT.

12.4 CALCULATING TRAINING SYSTEM COSTS

This section of the report will describe how to estimate all the other costs required to develop, acquire, install, operate, maintain and support each system alternative during its entire system life.

The systems analyst, working with a designated contact point at the CMI computer center, a typical training center, and the CMI minicomputer system project will design each system alternative for varying system demands and determine the resources required for each alternative design. To aid in gathering and storing the required data, the data for each system design should be accumulated in a matrix format similar to that shown in Table 12.2. Basically, column 1 lists the names of all resources to be used over the entire system life, with columns 3, 4, and 5 listing the amount of each resource expended over time (column 3—preoperations, column 4—each year of operations,[8] column 5—postoperations, giving the residual value of any investment resources remaining). Hence the amount of each resource required over time can be inserted in the appropriate entry of the matrix.[9] The following steps will be followed in accumulating the data:

1. Classify all entities that comprise the system alternative.

Column 1 of the table lists the various entities and other resource categories comprising the system in a three-level hierarchical form. The first level consists of the three main system functions, as listed:

- EDP system.
- Communications system.
- Instructional system.

Under each system function are listed the three key system subfunctions requiring resources:

- Operations.
- Maintenance.
- Support.

The third level consists of the four major categories of entities needed to implement each subfunction:

- Equipment and computer software. Include any interface equipment required.
- Materials and supplies.
- Personnel.
- Facilities and utilities to house and support these entities.

[8] If the resources expended each year are not uniform, a separate entry will be made for each year of operations.
[9] Note that some entries will be zero, since that resource is not required at that time.

Table 12.2 *System Design and Cost Data Base*

SYSTEM DESIGN ALTERNATIVE NO. ______________
SYSTEM DESIGNATION ______________

(1)	(2)	(3) Resources required			(4)			(5)			(6)	(7)
System Entities	Numbers of Equipment Required	Preoperations			Operations, Maintenance, and support (Per Year)			Postoperations (Residual Value)			Billet Title	Additional Billets Required
		Time	Manhours	Dollars	Time	Manhours	Dollars	Time	Manhours	Dollars		
1.0 EDP system												
1.1 EDP system operations												
1.1.1 Equipment												
1.1.2 Materials and supplies												
1.1.3 Personnel												
1.1.4 Facilities												
1.2 EDP system maintenance												
1.2.1 Equipment												
1.2.2 Materials and supplies												
1.2.3 Personnel												
1.2.4 Facilities												
1.3 EDP system support												
1.3.1 Equipment												
1.3.2 Materials and Supplies												
1.3.3 Personnel												
1.3.4 Facilities												

2.0 Communications system												
2.1 Communications system operations												
2.1.1 Equipment												
2.1.2 Materials and supplies												
2.1.3 Personnel												
2.1.4 Facilities												
2.2 Communications system maintenance												
2.2.1 Equipment												
2.2.2 Materials and supplies												
2.2.3 Personnel												
2.2.4 Facilities												
2.3 Communications system support												
2.3.1 Equipment												
2.3.2 Materials and Supplies												
2.3.3 Personnel												
2.3.4 Facilities												
3.0 Instructional system												
3.1 Instructional system operations												
3.1.1 Equipment												
3.1.2 Materials and supplies												
3.1.3 Personnel												
3.1.4 Facilities												
3.2 Instructional system maintenance												
3.2.1 Equipment												
3.2.2 Materials and supplies												
3.2.3 Personnel												
3.2.4 Facilities												
3.3 Instructional system support												
3.3.1 Equipment												
3.3.2 Materials and supplies												
3.3.3 Personnel												
3.3.4 Facilities												

Note that a hierarchical numbering system is to be used for each entity within the hierarchy. Any required entity or resource that cannot be included in this classification system should be included in a category labeled "other" or some more appropriate label. The main objective is to make certain that all entities are identified for later quantification. The systems analyst will work with the systems contact point in completing this table.

2. Determine how many units of each type of prime mission equipment listed in column 1 are required as a function of system capacity, and list in column 2 along the appropriate row for system operations. Exclude spares.

The total capacity or size of the system is a function of the following key characteristics:

- The number of CONUS and remote sites to be operated (S).
- The number of students per year trained at each site (ST).
- The average amount of student learning hours per day at each site (LH).

Since the key objective of the analysis is to provide the decision makers with information concerning the total costs, the basic resource data collected must also be collected as a function of these characteristics. These data will then be manipulated to show total costs for different system capacities.

Several examples of how column 2 is to be completed for the different system designs follow:

- *Satellite terminal (if not supplied).* One may be required per remote site.[10]
- *Test sensing equipment (OpScan 17).* One may be required per remote site.
- *Student terminal.* A minimum of one terminal may be required at each remote site when the minicomputer system is used, or if the Navy communications system is not used. However, if the number of student instruction hours exceeds a certain level, additional student terminals may be required. The need for such additional terminals for

[10]Some system alternatives may link remote sites together by some form of communications and hence only require one satellite terminal for a cluster of sites.

Table 12.3 *Calculating Terminal Usage*

(1)	(2)	(3)	(4)	(5)	(6)
Course Title	Average Student Instruction Time per Course (Hours)	Average Student and Instructor Terminal Time per Course (Minutes)	Number Students Simultaneously Enrolled in Course	Amount of Student Instruction per Course (Hours per Day)	Total Terminal Time per Day (Minutes per Day)
Avionics	500	400	10	2	(10)(2/500)(400) = 16 minutes
Aviation machinist mate	1000	600	20	5	(20)(5/1000)(600) = 60 minutes
Total time					76 minutes

obtaining student replies in "real time," as opposed to batch processing, may be calculated as follows:

Although a terminal is theoretically available for student use 1440 minutes per 24-hour day, even with close system control one terminal cannot be utilized this efficiently. Current practice in the schoolhouse or the demonstration must determine the maximum acceptable usage obtainable from a terminal before a second terminal is required. This upper limit will be a compromise among such factors as the number of students using the terminal, average hours of student instruction per day, the student "control system" used to minimize waiting, and the maximum acceptable student waiting time. Suppose that for a given type of student control it is found that one terminal can be used for a maximum of only 500 minutes per day to avoid excessive student delays. We must next compare the average estimated student usage with this threshold level to see if it will be exceeded. This calculation, shown in Table 12.3, is made in the following fashion:

List in column 1 all courses that will be taken simultaneously.[11]

List in columns 2 and 3 the average student instruction time required for each course listed.

List in columns 4 and 5 the number of students taking instruction in each course and the amount of instruction time taken each day.

[11]Since the objective of this calculation is to determine the maximum terminal usage required, if the total student work load is apt to change over the next year, several calculations may be required to find this maximum.

Calculate in column 6 the total student terminal time required per day for each course listed, as follows:

$$TT = (S)\left(\frac{ST}{CT}\right)(TTC)$$

where TT = total terminal time required per day for the course in minutes (average value)

S = number of students taking course

ST = student instruction time each day in hours.

CT = total instruction time required for the entire course in hours (average value).

TTC = total terminal time required during the course in minutes (average value).

Compare the total terminal time required per day (TT) with the maximum acceptable terminal time. If TT exceeds the maximum, additional terminals are required.[12]

Printer requirements may be calculated using the same method as described for the student terminal.

- *Communications requirements.* Communications channel time requirements, as a function of student instruction hours, should be indicated.
- *Maintenance, test, and repair equipment.* The maintenance equipment required must be related to the prime mission equipment to be repaired. Its degree of sophistication helps determine the mean time to repair and should be based on the trade-offs among the following three items to minimize total maintenance subsystem costs and keep equipment availability within acceptable bounds:

  The maintenance equipment cost.

  Man-hours per repair returned.

  Maintenance personnel salaries.

3. Determine how many units of each type of equipment listed in column 2 are required as spares and list in column 2 along the appropriate row for system maintenance.

The number of spares required will be a function of:

[12]During the demonstration this maximum value must be determined for one and more terminals. Theoretically the value for n terminals will be greater than n times the value for one terminal.

- Operational usage.
- Frequency of failure.
- Mean time to repair.
- Navy policy.

4. Determine if any of the required equipment will normally be located at the remote site and, if so, if it can be used by the training system.

Whether such equipment can be used for training purposes will normally be determined by the amount of time required. Obviously, specialized equipment such as the OpScan 17 will have to be purchased. However, if a standard Navy teletype is required for a short time per day, the use of a spare available at the site may be possible.

The use of common-purpose equipment such as AUTODIN presents special problems of costing. The amount of loading (number of characters per day transmitted) the training system produces on the communications system will determine whether any additional costs will have to be borne by the AUTODIN system. If so, such costs must be charged to the training system.

The use of service-provided assets such as the Navy FleetSatCom system presents similar problems. Again, the key is to determine how much loading the training system imposes. Several ways in which such costing has been handled in the past are listed below:

- *Option 1.* The communications load is small and the capacity is available so that no other user is deprived. Hence, no costs are allocated to the training system.
- *Option 2.* Some users are deprived of the communications channel and must use alternative means of communication. The costs of these alternative means are added to the training system costs.
- *Option 3.* An entire satellite system devoted to training is straighforward to cost since all the costs are borne by the training system.
- *Option 4.* An expanded FleetSatCom system with spare satellites made available for training until a malfunction of the tactical satellite occurs requires the use of spare satellites. This is a more complex option to cost, since, by definition, the spares are needed anyway, yet they can serve the additional function of training. If such use of the spare satellite shortens its life, this loss could be charged to the training system. If not, it is a free resource.

The appropriate option to use in this project will be determined by our particular scenario and will be confirmed after discussions with Department of Defense personnel.

5. Determine all resources required prior to the beginning of operations for such things as equipment or software development, acquisition, installation, and training. List each in column 3 of Table 12.2 on the appropriate row.

Having determined which resources are required, the cost analyst will determine the cost of these resources as well as the time required to complete the function. Each resource requirement should be related to the number of system components involved.

Three measures of resources are listed under each column—time, man-hours, and dollars. "Time" represents the total time in months required for completion of the activity such as research, development, test, and evaluation (RDT&E), for procuring prime mission equipment, or for installation. Time is also used to determine "man-hours," the O&M manpower resources required. Lastly, "dollars" are used to represent the out-of-pocket costs of each activity. Obviously, man-hours can be converted to the equivalent dollar cost, by knowing the rate of pay.

If high uncertainty exists, the system designers should also provide some indication of the range of uncertainty in each estimate by providing three numbers: the expected value, the optimistic value, and the pessimistic value.

Although Table 12.2 is an overall data collection form, all supplementary data explaining how the final data are arrived at should be attached as backup information. Examples of such data are:

- *RDT&E cost.* A fixed value of cost, independent of student usage.
- *Investment cost.* The cost of each piece of equipment as a function of volume (the average unit cost of equipment may decrease as the volume increases).
- *Installation cost.* A fixed value for each type of installation required.
- *Facilities required.* The size and location of the facilities required. This is a function of the type of remote location, the number of student training hours per day, and the equipment and materials involved.
- *Training of personnel.* The cost of training personnel to operate, maintain, and support the training system. This is a function of the number of personnel required to be trained.

6. In the same fashion as described in Step 5, determine all resources

required during the operational phase. List these on an annual basis in column 4. Again, certain rows will be empty. All resources should be listed. If annual costs differ from year to year, use a separate costing sheet for each year.

The systems designers will indicate the appropriate resources required for each of these activities, expressed as some function of system capacity or size. Five examples are described below:

Learning Supervisor Work Load. The primary functions of the learning supervisor should be listed as backup information. They will probably include custody and distribution of instructional materials, counseling students, and quality control checks of the system. The remote site may prefer to assign the administrative duties to the training officer and to delegate the function of providing specialized technical assistance to senior petty officers to whom the students are assigned for work duties. In such cases these resource requirements must be included in the cost analysis so that any additional billets required may be added to the ship manning document. The average instructor-student contact work load per week for any given course will have to be estimated as some fraction of the number of student training hours on that course. Data obtained during the demonstration and from the current training center can be extrapolated. Obviously, this total work load is the sum of contact hours over all courses being taken during the week.

Communications Operator Work Load. If communications personnel are required to any major extent over and above their normal duties, their training system work load must also be included. For example, the communications operator primarily performs quality control (QC) checks, maintenance work (see next section), and message handling. The QC weekly work load can be estimated as the product of the average estimated unit time required for each QC check and the average number of QC checks made per week. The message-handling work load can be estimated as the product of the average time required for each message (both originating and reply) and the average number of messages handled per week. In all three cases the incremental man-hours required attributed to the training work load will be estimated.

Maintenance Work Load. The man-hours required per year for each piece of equipment are determined by the preventive maintenance (PM) schedule and by unit times required. The corrective maintenance (CM) man-hours required are the product of the frequency of failure and the mean time to repair (which in turn depends on the maintenance test and repair equipment specified).

Support Work Load. This work load is determined by the various types of support services required. It is the product of the unit time required for each service and the frequency of the service.

Supplies and Other Materials. These will be as required by system performance characteristics.

7. Determine the net resources available at the end of the assumed 8 years of operations. List these in column 5 of Table 12.2 in the rows of the entities they are associated with. Again, some rows will be empty.

These net resources should include both expenditures for disassembly of equipment as well as the residual value (if any) of the equipment.

8. Determine the billet titles that will do the personnel work loads listed in columns 3 through 5 and list this information in column 6.

Part of the system design process is to cluster related work elements, thus defining all positions required. It is particularly important to note all new positions required. If any work will be done by a position already assigned as part of the current system, no cost for this work is attributed to the system.

9. Determine the additional number of billets required and list this information in column 7.

By summing the number of man-hours per week required for each activity done by the same billet title, we shall find the total work load for each billet for a given size system. Dividing this by the standard work time available in the standard workweek, we can find the number of billets required for each billet title[13] and the additional number of billets required for each system design.

12.5 DETERMINING TOTAL SYSTEM LIFE COSTS

The training system cost data base of Table 12.2 and the student-associated costs previously described must now be combined to calculate the total cost stream over the entire system life for each system alternative being analyzed. In making such a calculation, the following considerations must be taken into account:

[13]Watch standers are treated on the basis of the standard number of men required for each watch position.

- Total system costs are a function of system capacity.
- System costs have two elements: man-hours per year, which will have to be converted to billets, and total nonmanpower dollars per year. By factoring in billet costs, the total system costs can be expressed as equivalent dollars per year.
- Total system costs are a function of the assumed system life.
- Total system costs of each alternative must be compared.

The total costs for each system alternative will be compared with one another on the same basis. Each of these elements will now be described.

12.5.1 Total System Costs Are a Function of System Capacity

As indicated previously, the total costs of each system will be a function of three basic parameters:

- Number of remote instructional sites activated.
- Number of students being trained per year at each site.
- Amount of student instruction per year at each site.

To permit the decision makers to examine as many options as they desire, total costs shall be calculated as the sum of four main costs, as shown in Figure 12.5.

- Cost 1 is the total cost associated with the central CMI system in Memphis, and consists of two parts. Cost 1.1 comprises the fixed costs of the central CMI system. These costs must be borne irrespective of the number of remote sites, the amount of courses actually used at these sites, the number of students, and the rate at which they take instruction. One such example is the cost of developing and coding a CMI course. Cost 1.2 comprises the additional costs of the central CMI system that are incurred because of the three basic parameters (number of remote sites activated, students, and instruction rate).
- Cost 2 is the total cost of each remote site. It consists of two main parts. Cost 2.1 comprises the fixed costs associated with the procurement and installation of all equipment for a standard minimum size learning center at a particular site,[14] including

  Satellite terminals at each remote site (if not already provided by some other communications need at the site).

[14]Obviously, these costs may be different for a carrier as compared to a destroyer, since this minimum size may vary.

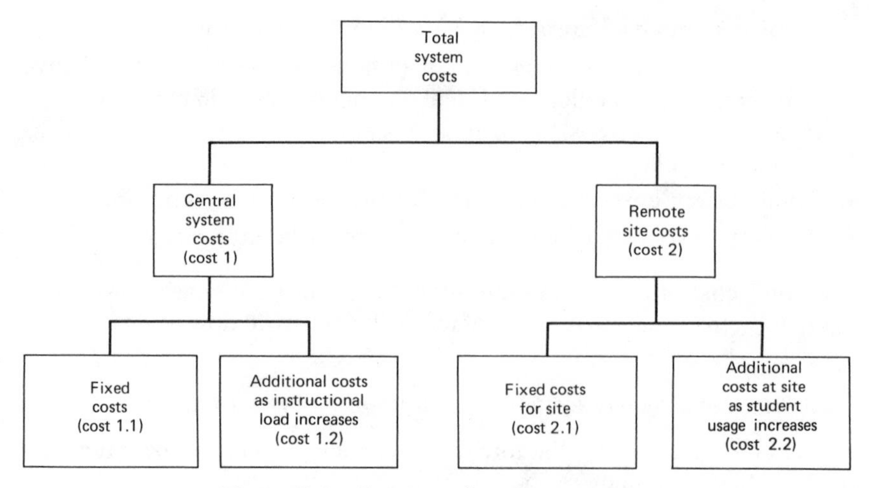

Figure 12.5. Structure of total system costs.

Standard-sized, outfitted learning center (appropriate to the particular type of remote site).

Cost 2.2 consists of all other costs associated with number of students and instruction rates, including all the additional costs of an enlarged learning center for more students, additional terminals required, all operations, maintenance, support, and personnel costs, since they depend on the amount of student usage.

Thus the total cost of all remote sites equals the sum of the Cost 2 elements for each of the remote sites involved.

12.5.2 System Costs Have Two Elements

With the four main costs of Figure 12.5 in mind, the cost stream of any given option may now be calculated as follows:

- Start with the annual costs associated with Cost 2.1 (for each type of remote site). Determine the total dollar costs to include the cost of the total additional billets required,[15] as obtained from Table 12.2.
- Next, calculate Cost 2.2, the total annual costs as a function of different amounts of students and usage (as measured in student-hours of training per year), as also obtained from Table 12.2.

[15]Manpower resources required can be separated from dollar costs if desired.

- Next, calculate the total annual costs associated with Cost 1.1, including course development.[16]
- Next, calculate the total annual costs associated with Cost 1.2, as a function of the number of remote sites and student usage.

12.5.3 Total System Costs Are a Function of the Assumed System Life

Most cost analyses of this type assume a time horizon for operations and maintenance of at least 5 years. TAEG (the Chief of Naval Education and Training's training analysis and evaluation group) has recommended a cost analysis on the basis of 8 years of operations. Generally, the planning horizon is long enough that any residual value of the equipment can be ignored since dismantling and shipping and the discount factor reduce this value generally to a negligible amount. If for a particular system alternative this assumption is not true, the system designer should make certain that he provides an estimate of the residual equipment value in column 5 of Table 12.2 so that it can be included in the cost analysis.

12.5.4 Total System Costs Must Be Compared

The total cost streams associated with each system alternative, designed to provide the same number of student graduates each year, will be presented and compared with one another. The total system life cost streams associated with each system design will also be used to make the following additional cost calculations:

- *Present value of total system cost.* This cost is the discounted value of the total system life costs for each alternative configured to provide the same number of student graduates each year. This cost will be calculated by using the cost stream calculated previously (with manpower costs converted to equivalent dollars) and an appropriate value for the discount rate. Ten percent is generally used for this type of system, as indicated by the Office of the Secretary of Defense.
- *Present value of remote site cost.* Another cost calculation that should be important to a decision maker is the incremental cost of a remote site as a function of student usage (present value of Cost 2). This can be obtained as a total cost stream (as a function of student usage) and then converted to a present value using the discount rate. The results

[16]These, and other nondifferentiating costs, could be excluded if a more simplified cost analysis is acceptable.

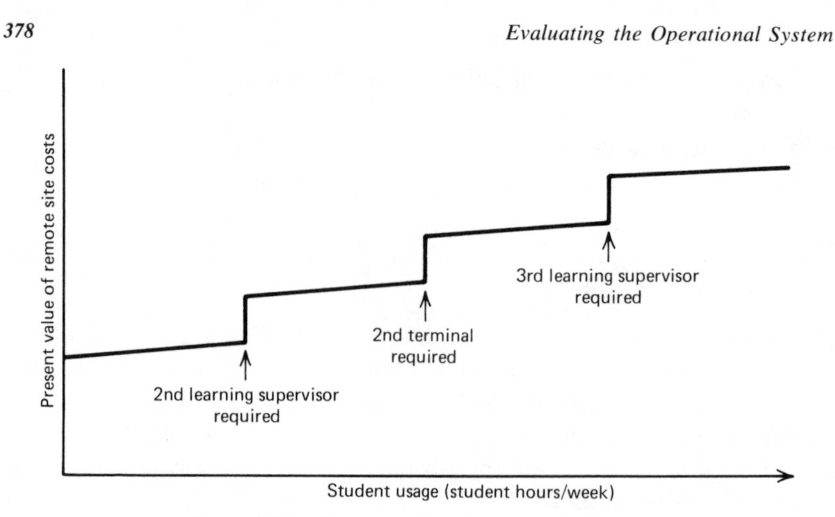

Figure 12.6. Present value of remote site costs.

expected for the latter type of calculation are illustrated in Figure 12.6. The locations of the break points shown are particularly important, since they show design limits where additional entities are required.

- *Average cost per hour of student training*. This may be calculated by dividing the present value of total costs by the total number of student hours of instruction (or its present value) given to successful graduates only, taking into account system attrition. In general, the cost per hour diminishes as student usage increases.

12.6 CONCLUDING REMARKS

At the beginning of this chapter we identified nine differences among the system alternatives. Six of these differences were treated as factors that were then translated into cost streams for a given (8-year) student training load, each cost stream contributing to the total system cost stream. The six factors are:

- Factor 1, the student training time in the course, and Factor 2, the number of productive hours in the standard workweek. These are combined in determining the total student billet costs while the student is enrolled in the course.
- Factor 3, other student costs (transportation and living expenses while traveling).

- Factor 4, student attrition costs.
- Factor 5, student time in OJT.
- Factor 6, training system costs.

Because of the resource constraints of the demonstration it may not be possible to quantify the benefits of the remaining factors—7, 8, and 9—and these benefits may have to be expressed in nonmonetary terms. Evaluation of these benefits is now discussed.

12.6.1 Other Benefits Available

The preceding analysis takes into account the readily measurable differences among the training system alternatives affecting the total cost of developing, installing, operating, maintaining, and supporting each system, as well as the total student costs to graduate the same number of students, taking into account differences in attrition and training time required to complete the different courses as well as different standard workweeks in force at different training sites. As mentioned previously, there are several other differences between training centers and remote site training that may yield additional benefits for remote site training. These include:

- Additional manning at remote sites (Factor 7).
- Differences in training effectiveness (Factor 8).
- More training opportunities (Factor 9).

These differences among systems may be evaluated in the following ways:

12.6.1.1 Additional Manning at Remote Sites

If the current system trains these students at schoolhouses, and job site training will bring some of these students to the job site where they are available for work, this benefit may be measured as the additional productive work obtainable per year (nontraining man-hours in peacetime; all man-hours in a contingency when training will cease). These times can be converted into the average additional manning available to the unit.

Although this additional manning does have a monetary worth (the current equivalent cost of manpower), and some analysts might evaluate this benefit by subtracting the value of the additional manpower from the total cost stream, it is not really an out-of-pocket saving that such a subtraction would represent. Hence the additional manning will be kept as a separate measure.

12.6.1.2 Training Effectiveness

As indicated previously, it may not be possible to express the difference among training systems quantitatively without extensive job-related performance evaluations.

12.6.1.3 More Training Opportunities

At the very least, we shall identify the number of personnel at remote bases for whom programmed CMI courses could offer training opportunities for which the only alternative—a return to a training center—is not really available because of manning limitations. The impact of this training on operational readiness or reenlistment rate could be estimated through quantitative survey techniques.

12.6.2 Sensitivity Analysis

Various analyses will be performed showing the differences in total costs among the alternatives as various factors or assumptions vary. Although the total student usage per year and number of remote sites are the key parameters identified thus far, the demonstration will identify others.

12.6.3 Developing the Master Plan

Based on the analysis described, a decision maker will be able to select the preferred system given the demand function of

- Types of courses to be offered.
- Number of students to be trained (over time).
- Remote site where students may be located.

The time dimension will be inserted into the decision-making process, and the transition plan will be developed.

Each of the alternative operational systems (except the current system) would have to be phased into the total Navy environment, requiring not only resources but time for testing, development, procurement, and installation. Hence the results of the evaluation will indicate:

- The preferred system.
- The time phasing for implementing the preferred system at the desired remote sites, in accordance with the implementation strategies decided on (part of the system design).

Since, in general, there may be considerable delay between now and the time the preferred system achieves full operational capability, we shall also explore transitional alternatives, such as:

- Operating the current system until full operational capability is achieved by the preferred system.
- Going from the current system to one or more intermediate systems and then to the preferred system.

The costs and benefits of each transition plan will be calculated by using the economic model described.

12.7 SUMMARY OF PROBLEM-SOLVING PROCESS FOR COMISAT ECONOMIC ANALYSIS PROBLEM

The generic problem described in Chapter 12 involved the development of an economic or cost-effectiveness analysis model for evaluating the final operational system based on the performance and cost data to be gathered during the demonstration phase. Hence many of the principles of systems evaluation in Case 1 apply. The following steps introduced in Case 1 were followed:

- Identify all alternatives available.
- Design each alternative so that it meets the objectives selected.
- Determine the total system life cost of meeting the objective.
- Determine the economic value of any of the differences among the systems not reflected in the objective selected, and select that system whose net cost is lowest.

In this case the primary objective of the system was to obtain a given number of course graduates over a given period of time. It was relatively straightforward to calculate all the costs attributed to course training for all alternatives considered. However, as indicated in Figure 12.2, there were several characteristics that on-site training could provide that training center training could not:

- Possibly shorter time required to complete OJT.
- Higher training effectiveness.
- Additional manning at the job site at the time of a contingency.
- More training opportunities available leading to a higher rate of reenlistment.

These were not able to be quantified and were only described qualitatively.

The exact system demand function could not be exactly specified in terms of:

- The number of students to be trained.
- The number of student training hours involved.
- The number of training sites to be used.

Thus the total analysis was formulated using parameters for these variables.

13

Implementing the Economic Analysis Model

The problem in this chapter deals with the same situation as that presented in Chapters 11 and 12. While the economic analysis approach developed in Chapter 12 was useful as a starting point in conducting such an evaluation, the objective of Chapter 13 is to illustrate the analytical simplifications that can be made when time and analytical resources are extremely limited. We show how a "best efforts" type of analysis can be made when all the desired data are not available.

13.1 PROBLEM AS GIVEN

You are now completing Phase II of the COMISAT program and for the past 4 months have been contributing to the detailed design of the demonstration phase. Your major role has been to do a more detailed design of the information-handling system, linking the remote site selected for the demonstration with the CMI computer in Memphis. Several important events have taken place during this phase that are affecting the future course of this program. First, the operational Navy fleet commanders have been reluctant to provide a ship for the COMISAT demonstration, largely because of fear of having personnel on board ships who have not completed their initial skill training. However, the project was able to find a remote shore station interested in the potential of the program and willing to participate in the demonstration. The station commander views the program from a different perspective than the operational scenario initially conceived. Every year he sends a number of his personnel to U.S. training centers to take skill progression training, the next level of training, which generally occurs from 1 to 3 years after initial skill training. Although this training does increase the capability of personnel, if a station commander sends an individual to school on TDY (temporary duty), not only does the station lose his services during this school time but his travel and living costs come out of the station budget for TDY. Since these funds are generally limited, most stations cannot afford to send many of their personnel to school. Instead they wait until the end of the person's current tour of duty (2 to 3 years), when he is assigned to school under permanent change of station (PCS) status. Travel and living funds then come out of the Navy's Bureau of Personnel budget.

As you are completing your portion of the Phase II report describing what needs to be done to implement the information-handling system linking the demonstration site and the CMI computer in Memphis, and determining the associated costs and the time schedule involved, you are presented with another problem. Because of a change in emphasis in the COMISAT program from the higher-student-volume initial skill training to the lower-volume specialized skill training, there is now some doubt that COMISAT will provide the cost savings originally contemplated. You are asked to provide an economic rationale for the COMISAT program, based on whatever data you can readily put together. You indicate that, although you did have an economic model available for comparing COMISAT with other alternatives (as described in Chapter 12), you do not have any of the data required to implement this model, since the data were to be collected during the subsequent demonstration and evaluation phases. The project manager realizes this is true but he replies that you can make some reasonable assumptions regarding these data. Your analytical report can strongly emphasize that the results provided are preliminary and will be refined as new data become available during the demonstration. At least, you can provide some order-of-magnitude results regarding possible savings due to COMISAT, as compared with the conventional CMI system. He further indicates that the project officer insists on some preliminary economic analysis and that the future of the project may very well depend on the results of your analysis.

Describe how you would approach this problem, using the analysis described in Chapter 12 as a guide.

13.2 PROBLEM DEFINITION AND ANALYTICAL APPROACH

The major advantage the analyst had in solving this problem is that he had achieved a good understanding of the evaluation problem during the previous phase. This was evidenced by the "ideal" evaluation approach (described in Chapter 12) having been developed and approved by the client during phase 1. Thus the current approach was to review this evaluation approach and determine how much of the data required could be collected and analyzed during the short time remaining. These data consisted of performance and cost characteristics of alternatives, as well as all other data regarding the environment (including the system demand function).

13.2.1 Statement of the Problem

Based on the additional information found since the economic analysis approach of Chapter 12 was completed, it was possible to redefine the current COMISAT economic analysis problem in the following way:

Most of the Navy's courses are presented using traditional instruction (TI). However, there are courses in a self-paced instruction format, some of which are already programmed onto the CMI computer at Memphis and available for CMI instruction at locations such as Memphis, Great Lakes, San Diego, and Orlando. In the future the Navy intends to expand the CMI capability by (1) developing and programming additional courses into CMI, and (2) enlarging the CMI computer to accommodate the increased course and student loads. It is possible that the COMISAT project could play a significant role in these future plans, thus affecting the need for an additional training center and the operations of the existing training centers.

13.2.2 Identification of Alternatives

It was obvious that certain alternatives identified in Chapter 12 could not be evaluated because of the unavailability of performance or cost data. These included the decentralized system being tested on the U.S.S. *Gridley* and any other system requiring the design of a minicomputer. In addition, discussions with the client indicated that we should include the option of the "do nothing" case (the current system). Hence three alternatives were developed:

Alternative 1 is to continue the current and programmed training system of giving all courses at the five training centers, some at CMI classrooms outfitted with CMI terminals for those CMI courses currently developed, most in TI classrooms for the remaining non-CMI courses.

Alternative 2 is to expand the CMI concept by developing more courses into CMI format and presenting these at the training centers. All other courses in the total curriculum would be given in TI or self-paced format, the current method.

Naval experience indicates that utilization of CMI at a training center yields:

- A reduction in course time.
- A reduction of instructional and support personnel.
- An increase in student end-of-course achievement levels.
- A reduction in student attrition, as described in Chapter 12.

Thus the major task in designing Alternative 2 is to determine which additional courses should be converted into CMI format, considering the following criteria:

- The course can be converted into CMI format.
- The course will have the approval of operational commanders to be given to their personnel at remote sites, and time will be given them to take the course.
- The total cost of delivering courses selected for CMI (in terms of the number of graduates required) is less than that of the training method now being used.
- The CMI benefits obtainable as previously described.

Alternative 3 is identical to Alternative 2 in that selected courses are converted to CMI. Now, however, all courses are given at remote sites equipped with COMISAT terminals. Those students who are to take the CMI courses but who are not stationed at COMISAT-equipped remote sites would take the CMI courses at training centers that would be provided with additional CMI terminals to accommodate them. All other students would take the TI courses at a training center as in Alternative 1.

The major task in designing Alternative 3 is to determine not only the courses to be converted into CMI format but also which remote sites should be outfitted with COMISAT terminals. Obviously, lower-population sites may not provide sufficient benefits to offset the costs of outfitting and operating a COMISAT site.

Thus the basic question of what cost savings COMISAT can offer as compared with the current CMI program can be answered by comparing the relative costs of the three alternatives postulated, since all three will be designed to accomplish the same end objective: providing the required number of graduates from all courses.

This cost analysis will be done as follows using the same type of systems evaluation approach as was previously used for Case 1 (Chapters 4–7) and Case 4 (Chapter 12).

- First, having identified the major alternatives to be considered, the key cost elements that relate to the total system life cost of these alternatives will be identified.
- Second, these different values of cost will be calculated. Since we are really interested in the differences in total cost among the three alternatives, we can calculate the incremental cost between two alternatives.

13.2.3 Cost Elements to Be Considered in Analysis

The next step in the analysis was to determine what cost elements we felt were the most important ones to be included in the analyses and for which data might be obtainable within the time available. The starting point for this is a listing of the cost elements one would like to include in an "ideal analysis" (where all data desired are available). Chapter 12 described such a list. Next the analyst should concentrate on the key cost elements, particularly those whose numerical values would differ most among the alternatives being considered. The total cost model should be built in a modular fashion, so that the analyst can use whatever data might be gathered within the time available, adding new cost data at a later date as more data become available.

Some 10 cost elements were considered in the analysis (Table 13.1). In changing from Alternative 1 (base case) to Alternative 2 (CMI system), the following costs will increase as follows:

- The cost to develop and code current non-CMI courses into CMI format ($C_{12} - 0 = C_{12}$) is a major investment cost if CMI course programs are to be obtained.
- CMI program maintenance ($C_{22} - 0 = C_{22}$), the average annual cost required to make any necessary changes in each course program.
- The cost of expanding the size of the CMI computer to accept the additional courses and students ($C_{32} - 0 = C_{32}$).
- The cost of additional CMI terminals for the additional students taking CMI courses at the training centers ($C_{42} - 0 = C_{42}$).
- The cost of maintenance for these additional CMI terminals ($C_{53} - 0 = C_{53}$).

On the other hand, the following costs will be reduced in changing from Alternative 1 to Alternative 2; hence these can be considered as CMI "cost savings":

Table 13.1 *Cost Elements for Viable Alternatives*

Cost	ALTERNATIVE 1 TI Courses at Training Centers	ALTERNATIVE 2 CMI Courses at Training Centers	ALTERNATIVE 3 CMI Courses at Some Remote Sites
1. C_1: Develop and code non-CMI courses into CMI format	$C_{11} = 0$	C_{12}	$C_{13} = C_{12}$
2. C_2: CMI program maintenance	$C_{21} = 0$	C_{22}	$C_{23} = C_{22}$
3. C_3: Expand CMI computer	$C_{31} = 0$	C_{32}	$C_{33} = C_{32}$
4. C_4: CMI terminal investment	$C_{41} = 0$	C_{42} (training center terminals)	C_{42} (remote site terminals and some training center terminals)
5. C_5: CMI terminal maintenance	$C_{51} = 0$	C_{52} (terminal maintenance at training centers)	C_{53} (terminal maintenance at remote sites and training centers)
6. C_6: Student time required (equivalent cost)	C_{61}	C_{62}	C_{63}
7. C_7: Living expenses at training center	C_{71}	C_{72}	$C_{73} = 0$
8. C_8: Travel expenses to training center	C_{81}	$C_{82} = C_{81}$	$C_{83} = 0$
9. C_9: Cost of instructor and support staff	C_{91}	C_{92}	C_{93}
10. C_{10}: Training equipment and materials	C_{101}	C_{102}	C_{103}
Total			

- Since it is expected that a CMI course can be completed in less time than its predecessor TI course, the equivalent cost of the students' time required will be reduced under Alternative 2, the cost for CMI, by $(C_{62} - C_{61})$.
- If the students' time is reduced under CMI, his living costs at the training center $(C_{72} - C_{71})$ will also be reduced.

All other costs for the CMI system were assumed to be the same as for the TI system. These include:

- Travel expenses to the training center and return $(C_{82} = C_{81})$.
- The cost of instructors and support staff $(C_{92} = C_{91})$; this is a conservative assumption, since there is probably some savings for CMI here, but no data describing these savings could be assembled.
- The cost of training equipment and materials $(C_{102} = C_{101})$.

Hence these cost elements are omitted from further consideration in this analysis.

13.3 COMPARING THE COSTS OF THE EXPANDED CMI SYSTEM WITH THE CURRENT SYSTEM

The first evaluation was a comparison of an expanded CMI system with the current system (Alternative 1 vs. Alternative 2).

13.3.1 Determination of Courses Economically Feasible for CMI at the Training Centers

The various costs associated with each system were a function of the various characteristics of the training courses to be given in future years (the systems demand function). Hence the next step in the analysis was to determine the student work load for the planning horizon under consideration. Ideally, this would involve listing all courses taught during the past year, including their enrollment, course length, and attrition rate, and other information that could be used to estimate which of these courses was likely to be given in future years and expected enrollment characteristics. Since there were several thousand courses taught in the preceding year, such data were not readily available to the analyst. Contact with the Navy Education and Training office in Washington revealed that a list of the total number of Navy students trained in the past 2 years and estimated to be trained in each of the future years was available from the most recent annual military manpower training report. The necessary data on annual overall training loads are summarized in Table 13.2.

13.3.1.1 The Candidate Training Segment

Six distinct training segments were identified by these data. There were three categories of courses offered: initial skill training, skill progression training, and functional training. Initial skill training consisted of Class A courses offered immediately after recruit training. Skill progression training was made up of more advanced Class C type courses, offered to sailors who have already had some work experience and are preparing

Table 13.2 *Navy FY 78 Training Loads*

Category	PCS	TDY	Total
Initial skill training			
1. Entrants	157,833	6,269	164,102
2. Graduates	156,073	6,214	162,287
3. Average course length (days)	42	42	42
4. Loads average on board (AOB)	19,152	570	19,722
5. Percentage of Loads[a]	57%	2%	59%
Skill progression training			
1. Entrants	30,042	32,867	62,909
2. Graduates	29,336	32,330	61,666
3. Average course length (days)	51	51	51
4. Loads average on board (AOB)	7,274	2,526	9,800
5. Percentage of loads[a]	22%	8%	30%
Functional training			
1. Entrants	22,226	329,340	351,566
2. Graduates	21,611	320,619	342,230
3. Average course length (days)	4	4	4
4. Loads average on board (AOB)	1,224	2,863	4,097
5. Percentage of loads[a]	4%	9%	13%
Grand total			
1. Entrants			578,577
2. Graduates			566,183
3. Average course length (days)			—
4. Loads average on board (AOB)			33,619
5. Percentage of loads[a]			100[b]

[a]Percentage of loads = percentage of total CNET population enrolled.
[b]A 2% error due to roundoff is found if the individual numbers are added.

themselves for an advanced rating. Finally, functional training consisted of short-duration training in shipboard duties given at the ports of fleet concentration such as San Diego.

There were also two types of students: those on permanent change of station (PCS) and those on temporary duty (TDY). Those on permanent change of station included recruits going from recruit training to a training center and then to fleet or shore stations, and sailors going from fleet or shore stations to a training center and then to new duty stations. Temporary duty occurred when the sailor left his station for a short period of time, such as to go to a training center, and then returned to his permanent station.

The most likely candidates for COMISAT instruction were those taking skill progression training. The students in these groups were productively employed before their training and could continue to carry out their duties, although at a reduced level, if they were to receive training on board ship or at a land base. The group taking initial skill training on TDY was not included in the analysis even though it might actually benefit from COMISAT, because it is relatively small and data on initial skill courses were not readily available for use in this analysis. Initial skill students on PCS were not likely candidates for CMI because their lack of skills made them of little use while training at the site, and operational commanders had been found to oppose the idea of placing untrained men at their sites or on board their ships. Finally, functional training segments were not considered in the study, since this instruction is primarily port-side team training and is essentially already being delivered to the remote site. Hence in this analysis the only students considered as candidates for COMISAT are those in the skill progression groups.

13.3.1.2 Analysis of Highest-Volume Skill Progression Courses

Having identified the candidate training segment, the next step was to obtain data on course enrollments. The only obtainable data on individual enrollments for skill progression courses were the title, number of graduates, and course length for each of the 10 highest-volume skill progression courses. Average attendance and student man-days for the skill progression courses had to be estimated.[1] The derived enrollment information for these 10 courses is given in Table 13.3.

An analysis of the net savings obtained from each of these 10 courses showed that each of the courses does make a positive contribution to the total net savings. However, the contribution does decrease as the volume of students decreases. The next question then was to determine how

[1]See Appendix 13.A for details of this analysis.

Table 13.3 *Ten Specialized Skill Courses Producing Most Graduates*

Title	Number of Grads	Average Attend-ance	Length (Calendar Days)	Student Man-Days
1. Instructor Basic	3110	3141	24	75,384
2. Career Information Counselor	1323	1336	26	34,736
3. Nuclear Propulsion Plant Operator Mechanical[a]	1174	—	182	—
4. Sonar Electronics Intermediate	960	970	117	113,490
5. Air Conditioning and Refrigeration	611	617	53	32,701
6. Nuclear Propulsion Plant Operator Electrical[a]	608	—	182	—
7. International Morse Code	586	592	82	48,544
8. Nuclear Propulsion Plant Operator Reactor Control[a]	512	—	182	—
9. Marine Gas Turbine Basic	401	405	40	16,200
10. Surface Explosive Ordnance Disposal Refresher	400	404	28	11,312
Total (10 courses)	9685			
Total (7 courses)	7391			332,367

[a]Not suitable for CMI format.

many subsequent courses (beyond the tenth) might also provide additional net savings. Since the only course data we had available pertained to these 10 highest-volume courses, it was decided to estimate the enrollment data for additional lower-volume courses by fitting a curve through the known populations for the 10 highest-volume courses and extrapolating.[2] The length of all courses beyond the tenth was assumed to be 51 days, the average course length.

With all data on enrollment and course length complete, the courses eligible for conversion to CMI were identified by calculating and comparing costs and savings to determine those that have a net savings.[3] The following two sections describe the procedures and assumptions.

Incremental Costs of Changing to CMI System at the Training Centers. The costs associated with the transfer of traditional courses to CMI

[2]See Appendix 13.A for details of this analysis.

[3]See Appendix 13.B for further details of this analysis.

for training center use include development and coding (C_{13}), course maintenance (C_{23}), computer leasing (C_{33}), terminal purchase (C_{43}), and terminal maintenance (C_{53}). Before a course can be made available on CMI, it must be developed into the proper CMI format and coded for the CMI computer. Assuming the materials are less than 10% audiovisual, the development cost was estimated to be \$2930 per hour of instruction, using past CNET experience. The cost of coding is between \$200 and \$300 per instruction hour, and a mean coding cost of \$250 per hour was used in the calculations. The estimated total development and coding cost per hour (C_{13}) was \$3180. However, since CNET is planning to redevelop courses for some 70 to 80 ratings in the next 4 to 7 years, those courses would cost \$1130 per hour even if they were not developed for CMI. Thus C_{11} is not equal to zero as originally assumed, but is really \$1130 per hour. This results in an incremental cost per hour of training for developing a CMI course ($C_{13} - C_{11}$) of

$$\$3180 - \$1130 = \$2050$$

It was assumed that by expending 5% of the incremental development and coding costs (\$2050 per hour of training) each year, the CMI program could be adequately maintained for 12 years. Thus $C_{23} = 0.05C_{13}$. For an average course of 255 hours, annual course maintenance is

$$(255)(\$2050)(0.05) = \$26,137$$

The computer leasing costs have two aspects: (1) expansion of the computer main frame and (2) expansion of the number of peripherals, particularly the disk packs. The current main frame accommodates 6000 "average on board" (AOB) students.[4] The planned expansion to a Model 60 computer will provide for a maximum capacity of 16,000 AOB students. It is assumed that this expansion would accommodate the additional skill progression courses. If the assumption is not true, the computer could be expanded to a Model 80, at additional cost. With the current mix of courses on the computer, one disk pack would be required for each additional 1000 AOB students, assuming an average of one student response per hour. This would cost \$800 per month. However, if the volume of students per course decreased, as it would if many additional courses were added, the requirement might be more than one disk pack per 1000 students. In this analysis it is assumed that an additional 1.5 disk packs would be required for each 1000 students. Note that because

[4]Universities would call this student work load "full-time equivalent" students. This unit is obtained by dividing the total student training hours provided in one year by 52 times the number of training hours per week given a full-time student.

the final results have low sensitivity to the required number of disk packs, the relatively arbitrary nature of the figure used is of little consequence. Thus the annual cost of disk packs for each 1000 AOB students is approximately

$$(\$800)(12)(1.5) = \$14,400$$

Since the computer main frame will be expanded whether or not it is used for CMI, no computer expansion costs are incurred by development of courses into CMI. Therefore the only computer cost involved is the leasing cost of the disk packs.

The purchase of terminals is less expensive than leasing, based on the cost data in Table 13.4. The purchase price for the terminal cluster is $14,250 and the annual maintenance is $1764. The present value of the maintenance over an assumed 8 years life is $9873, assuming a 10% discount rate.

One cluster is usually required for each 60 full-time students, and it is assumed that clusters could be shared. Therefore a proportional cluster cost is assigned to each course on the basis of the AOB student load. A course with 30 students would be charged half the cost of one cluster under those circumstances.

Incremental "Savings" by Going to the CMI System at the Training Center. The savings achievable (or costs avoided) by implementing CMI

Table 13.4 *Naval Training Center CMI Cluster Equipment and Associated Costs*

Item	Purchase Option		Lease Option (per month)
	Invest-ment	Mainte-nance	
One OpScan 17 (basic terminal)	$ 8,998		$261
Maintenance		$ 69	69
One OpScan Automatic Feed	652		27
Maintenance		5	5
One Terminet 1200	4,200		132
Maintenance		68	68
One GDC 202-9D Modem	400		13
Maintenance		5	5
Totals	$14,250 +	$147	$580

courses at a training center accrue for the most part from reduced student pay and living costs due to decreased training time ($C_{61} - C_{63}$ plus $C_{71} - C_{73}$). There are other expected sources of savings, such as reduced instructor time, but those were not considered significant enough to justify the additional effort required to include them in the analysis. Hence the savings identified here are conservative estimates.

A previous analysis of eight courses that underwent instructional system development (ISD) showed an average saving of 20% of student man-hours of attendance time, as compared with the time required using traditional instruction. This planning factor was used throughout the analysis, and the sensitivity of the results to the parameter was tested as described later. Using the 20% student time saving factor, the man-hour savings for each course were converted into equivalent student salary savings.[5]

While attending a training center course a student is provided food at the same cost to the Navy as at his operational site. However, he or she is also given $2.50 per day in cash to cover miscellaneous, out-of-pocket living expenses. Because of the reduced training time, 20% of these living expenses would be saved for each course.

The total costs and savings for each course were computed in this manner and discounted at a 10% rate over the assumed 8-year life. The present value of each course's costs and savings are listed in columns 13 and 17 of Table 13.5. Since the development and coding investment produces a course with a lifetime of 12 years, the discounted residual value of development and coding was treated as a credit to CMI and listed in column 18. Therefore the resulting present values consistently represent an 8-year program. Column 19 lists the present value of net savings for each course (column 17 plus 18 minus 13).

As column 19 indicates, this technique identified the top 17 courses as feasible for CMI development (the eighteenth course resulted in a net loss). However, 3 of the first 10 courses involve nuclear training, and it is assumed that these would not be operationally practical for CMI. This same planning factor was applied to courses 11 to 17. Thus, on the average, only 70% of these courses were considered to have potential for CMI development and only 70% of these net savings were credited to CMI. This analysis shows that using Alternative 2 with these courses on CMI would result in a present value of net savings of $6,701,712.

The analysis of the sensitivity of the results to the value of this parameter is discussed later under uncertainties.

[5]See Appendix 13.B for further details of this analysis.

Table 13.5 *Analysis of High-Volume Skill Progression Courses*

(1) Course Number	(2) Grad-uates	(3) Annual Average Students	(4) Course Time (Hours)	(5) Average Students on Board (AOB)	(6) Investment Cost (Course Development and Coding)	(7) Investment Cost (Training Center Terminals)	(8) Terminal Maintenance Cost/Year	(9) Program Maintenance Cost/Year	(10) Computer Leasing Cost/Year
1	3110	3141	120	206.5	$ 246,000	$49,044	$6071	$123,000	$2974
2	1323	1336	130	95.18	226,500	22,605	2798	13,325	1371
3	—	—	—	—	—	—	—	—	—
4	960	970	585	310.8	1,199,250	73,815	9138	59,962	4476
5	611	617	265	89.60	543,250	16,530	2370	27,162	1290
6	—	—	—	—	—	—	—	—	—
7	586	592	410	138.0	840,500	32,775	4057	42,025	1915
8	—	—	—	—	—	—	—	—	—
9	401	405	200	44.38	410,000	10,540	1305	20,500	639
10	400	404	140	30.99	287,000	7,360	911	14,350	446
11	447	415	255	63.08	522,750	14,981	1855	26,137	908
12	427	431	255	60.26	522,750	14,312	1772	26,137	868
13	409	413	255	57.71	522,750	13,706	1697	26,137	831
14	393	397	255	55.46	522,750	13,172	1631	26,137	799
15	379	383	255	53.49	522,750	12,704	1573	26,137	770
16	366	370	255	51.65	522,750	12,267	1518	26,137	744
17	355	359	255	55.1	522,750	11,899	1473	26,137	721
18	343	346	255	48.4	522,750	11,495	1423	26,137	697
Total of top 7 courses	7391			910.45					
Total next 7 courses	2776			391.75					
Next 7 courses 70% of total	1943			274.22					
Grand total (top 7 plus 70% of next 7)	9334			1184.67					

Table 13.5 *(Continued)*

(1) Course Number	(11) Total Annual Cost	(12) Present Value of Annual Costs	(13) Present Value of Total Costs	(14) Value of Student Time Saved/yr	(15) Total Living Cost Savings/ Year	(16) Total Annual Savings	(17) Present Value of Total Savings	(18) Residual Value of Development and Coding	(19) Present Value of Net Savings
1	$21,345	$119,468	$ 414,512	$416,491	$37,694	$454,185	$2,542,073	38,253	2,165,814
2	17,494	97,916	387,021	191,041	17,370	209,311	1,171,514	41,441	825,934
3	—	—	—	—	—	—	—	—	—
4	73,576	411,805	1,684,870	626,745	56,721	683,466	3,825,359	18,548	2,159,037
5	30,823	172,516	732,296	180,697	10,352	197,049	1,102,883	84,475	455,062
6	—	—	—	—	—	—	—	—	—
7	47,997	268,639	1,141,914	268,130	24,265	292,395	1,636,530	130,698	625,314
8	—	—	—	—	—	—	—	—	—
9	22,444	125,619	546,159	89,503	8,099	97,603	546,284	63,755	63,880
10	15,707	87,912	382,272	62,496	5,656	68,152	381,447	44,628	43,803
11	28,901	161,739	699,490	127,207	11,512	138,719	776,410	81,288	158,208
12	28,777	161,065	698,127	121,516	10,997	132,513	741,675	81,288	124,836
13	28,666	160,444	696,900	116,393	10,532	126,925	710,399	81,288	94,787
14	28,567	159,881	695,811	111,840	10,121	121,961	682,616	81,288	68,093
15	28,481	159,408	694,862	107,856	9,762	117,618	658,308	81,288	44,734
16	28,399	158,349	693,966	104,156	9,426	1,136,582	635,718	81,288	23,040
17	28,331	158,567	693,218	101,026	9,143	110,169	616,616	81,288	4,686
18	28,257	158,154	692,399	97,611	8,833	106,441	595,767	81,288	−15,344
Total of top 7 courses			5,289,044				11,206,090	421,798	6,338,844
Total next 7 courses			4,872,374				4,821,742	569,016	518,384
Next 7 courses 70% of total			3,410,662				3,375,219	398,311	362,868
Grand total (top 7 plus 70% of next 7)			8,699,706				14,581,309	820,109	6,701,712

13.4 COMPARING COSTS OF COMISAT SYSTEM WITH THE EXPANDED CMI SYSTEM (ALTERNATIVE 3 VS. ALTERNATIVE 2)

13.4.1 Determination of Feasible COMISAT Sites

Although implementation of CMI at the training centers would result in net savings, COMISAT provides an opportunity for additional savings by delivering the courses to operational land bases and ships, thus eliminating students' traveling and living expenses. Certain costs and savings occur to the same degree through both COMISAT and CMI at the training center. Such elements do not aid in differentiating between the two alternatives and were not explicitly calculated in this analysis. Hence the cost elements discussed below consist entirely of those differentiating costs and savings.

13.4.1.1 Costs to COMISAT

The costs to COMISAT consist of the purchase or leasing price of the terminals, the cost of annual maintenance, and the cost of software changes to the CMI computer, required to interface with AUTODIN. The required equipment and its costs can be found in Table 13.6. Each termi-

Table 13.6 *COMISAT Terminal Equipment and Cost*

	Purchase Option		
Item	Investment	Maintenance (per Month)	Lease Option (per Month)
One OpScan 17 (Basic Terminal)	$ 8,998		$369
Maintenance of one		$69	69
One OpScan Automatic Feed	652		27
Maintenance		5	5
One Device 273[a]	1,620		
			74
One LRC Character[a]	180		
Maintenance of total equipment		6	6
Total	$11,450 +	$80	$550
Spares[b]	$ 6,529		

[a]Required for compatibility with AUTODIN.
[b]Based on list of spares recommended by OpScan. These spares are required for all ship sites and land sites where a terminal is maintained by Navy personnel.

nal was assumed to have a lifetime of approximately 8 years. Since leasing over this period is much more expensive than purchase, only the purchase option is considered in this analysis. Quantity discounts are available as found in Table 13.7. The maintenance costs in Table 13.6 are for sites in the continental United States (CONUS) 50 miles or less from an OpScan service location. Other CONUS sites may be maintained either by OpScan or by specially trained Naval electronic technicians (ET's). Overseas and ship sites must be serviced by ET's.

The cost of service by OpScan personnel for sites greater than 50 miles from the service center consists of

- Travel cost of 18 cents per mile.
- Labor cost of $30 per hour, including travel time.
- An approximate annual parts cost of $600 obtained from OpScan's price for a service warranty.

Labor cost was calculated assuming a mean time to repair of 2.5 hours per breakdown.

Service by an ET involves the following costs:

- An approximate labor cost of $75 per breakdown, assuming the same total OpScan labor cost per breakdown as for OpScan maintenance personnel.
- An approximate annual parts cost of $400 per year based on the price of a service warranty with a 30% return on investment removed.
- A spare parts inventory that can be purchased at an undiscounted base price of $6529.

Table 13.7 *OpScan Discounts from GSA Schedule*

Units	Percent Discount
1st	0
2nd–30th	5
31st–60th	10
61st–90th	15
91st–120th	20
121st–150th	25
151st–180th	30
181st–210th	35
211th on up	40

The OpScan maintenance cost on a time and materials basis is less than the cost of the Navy maintaining the equipment itself for CONUS sites less than 90 miles from an OpScan service center.[6] Five categories of sites, each with different maintenance costs, are derived in the analysis. These categories and their 8-year discounted maintenance costs, assuming three breakdowns per year, can be found in Table 13.8.

13.4.1.2 COMISAT Savings

The COMISAT savings considered in this analysis consist of student travel costs to and from the training centers, student living costs while training at the training centers, and training center terminals that would no longer be required. Travel cost is assessed differently for PCS students then for TDY students. The travel cost for a student attending training on TDY includes round-trip travel from site to training center and meals required during travel time. One-way travel expenses from site to training center, plus meals during travel time, are required for a student attending training on PCS. Also, when a student on PCS travels to attend a course of under 140 days duration, he is permitted transfer of household effects at a cost of $291. This amount increases if the time is greater than 140 days.

Practically all travel cost is a function of the distance between the original station and the training center location. Since the exact mileage distribution is difficult to obtain, the Navy Bureau of Personnel uses an average distance of 1500 miles for planning purposes. This factor is used here. Mileage data were combined with the current schedule of airline tariffs, obtained from the Civil Aeronautics Board. The one-way cost of a 1500-mile trip, including tax, was found to be $137. The total travel expense that could be saved if the entire student population was trained at remote sites is $3,374,054 per year.

If COMISAT is employed, there is also a saving of $2.50 per calendar day of training for incremental living expenses that would not have to be spent at a training center. Thus the incremental living cost saved for the entire student population is $862,920 per year.

The total gross travel and living savings per year (TGS) for training at remote sites, as compared with CMI training at the training center, are

$$\text{TGS} = \$4,236,974$$

or a present value over 8 years of

$$\text{TGS}_{\text{pv}} = \$23,714,343$$

The number of training center terminals needed would be reduced if

[6]See Appendix 13.C for details of this analysis.

Table 13.8 *Site Categories and Present Value Maintenance Costs for the Low-Maintenance Case*

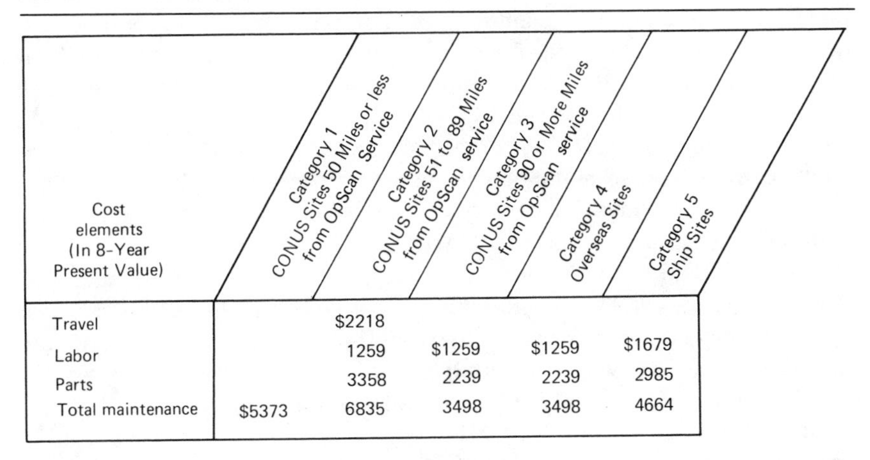

Cost elements (In 8-Year Present Value)	Category 1 CONUS Sites 50 Miles or less from OpScan Service	Category 2 CONUS Sites 51 to 89 Miles from OpScan service	Category 3 CONUS Sites 90 or More Miles from OpScan service	Category 4 Overseas Sites	Category 5 Ship Sites
Travel		$2218			
Labor		1259	$1259	$1259	$1679
Parts		3358	2239	2239	2985
Total maintenance	$5373	6835	3498	3498	4664

some students were trained at their work site. One training center terminal is required for every 60 AOB students. Hence there is a saving of one training center terminal for each 60 AOB students reached through COMISAT. A purchase price of $14,250 and a maintenance cost of $147 per month would be saved by eliminating a training center terminal, as was found in Table 13.4. As indicated earlier, the purchase option is the least expensive and is used to estimate the savings.

The COMISAT costs and savings for each category were utilized to determine a break-even site population.[7] The distribution of students over Navy sites could not be obtained; hence it was assumed to be identical to the enlisted man distribution. A comparison of the break-even population for a category and each site population in that category determined which sites could be cost effectively developed for COMISAT. Sites with populations greater than the break-even value qualify. This procedure resulted in the identification of 306 feasible sites.

13.4.2 Determination of Courses Justified by COMISAT Savings

The next step was to derive an estimate of the number of feasible courses, assuming the number of sites remains fixed at the level determined previously. To this end the costs and savings due to COMISAT are reexamined.

[7]See Appendix 13.B for details of this analysis.

Since only the feasible courses from the first 17 courses can be economically developed for CMI independently of COMISAT, the development and coding, course maintenance, computer leasing costs, and student time and living savings for these courses are not attributed to COMISAT. However, the development of any courses beyond the seventeenth would not be feasible without COMISAT, for these courses are justified only by savings achieved at remote sites. For this reason the development, coding, course maintenance, computer leasing costs, and student time and living savings due to the use of CMI for the eighteenth and higher courses must be attributed to COMISAT.

One analytical concern should be noted for considering costs and savings associated with the eighteenth to *N*th courses. As before, all cost elements are considered over an 8-year period. It is also important to note that the annualized values not only represent costs and savings for the first 8 years but also apply to a program continuing 8 years.

Once a course is set up for CMI at remote sites, it will also be used in this form at the training center to train students whose base is not outfitted with a terminal. Hence COMISAT must be charged for the additional training center terminals required to teach the course. The total cost to COMISAT is the sum of the development and coding, course maintenance, computer leasing, training center terminal, and COMISAT terminal costs. The savings credited to COMISAT include the travel and living costs for all courses, student time and incremental living savings for courses justified only by remote site savings, and training center terminal savings for the first seventeen courses.

The net COMISAT cost expression that includes all the preceding costs and savings is

Net COMISAT cost = Development and coding
+ course maintenance
+ computer leasing
+ training center terminals
+ COMISAT terminals
− travel and living
− student time
− incremental living
} for courses justified by COMISAT savings

+ COMISAT terminal
− travel and living
− training center terminals
} for courses justified by training center savings

This expression can be rewritten so that the only known factor is the number of courses that should be developed. Maximizing the net cost expression yields the optimum number of courses for CMI development.[8]

13.4.3 Determination of Additional COMISAT Sites

With the number of courses fixed at this new level, the feasible sites are again identified using a break-even population as previously described. The iteration between courses and sites was continued until the net savings changed by less than 5% from one cycle to the next. The results obtained by this procedure are discussed below.

13.5 KEY FINDINGS AND RECOMMENDATIONS

13.5.1 Reduction in Navy Training Costs

The analysis showed that full implementation of the COMISAT concept could result in present value net savings of \$18 to \$35 million as compared with the CMI system. These savings are based on an 8-year program and a 10% discount rate and depend on the set of conditions assumed, as shown in Figure 13.1. The top half of Table 13.9 shows the incremental investment and savings-to-investment (or benefits to investment) ratio associated with the savings for a representative case analyzed.

The major source of COMISAT savings used in this analysis is the reduction in travel and living expenses achieved by having the students remain at their operational land site or ship. The analysis shows that on the order of 30% or more of skill progression students who ordinarily travel to a training center could be served by COMISAT. This means that more than 20,000 students a year, representing more than one-tenth of the total number of Navy students trained per year would be affected. Table 13.9 lists the number of sites and trainees served for the representative case.

The uncertainties accounting for the variation in the preceding results have not been resolved, as discussed in the next section. However, even the lower values of savings are large enough to warrant serious consideration by the Navy of implementing COMISAT and are certainly large enough to justify the proposed demonstration.

These conclusions are also important for the other services. If equivalent savings can be achieved for each of the other two services, which it is

[8]See Appendix 13.B for details of this analysis.

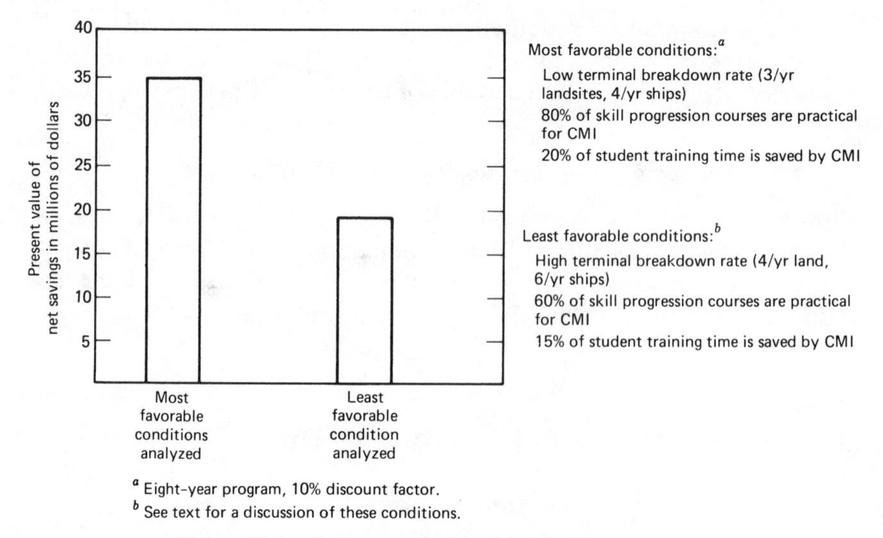

[a] Eight-year program, 10% discount factor.
[b] See text for a discussion of these conditions.

Figure 13.1. Summary of potential COMISAT savings.

reasonable to expect, then total Defense Department savings could have a present value of between \$54 and \$105 million for an 8-year program.

13.5.2 Uncertainties

The variation in the savings quoted above is due to uncertainties in three areas:

- It is not known what percentage of Navy skill progression courses are suitable for conversion to CMI. Values of 60 to 80% were examined in this analysis. A value of 70% was picked as nominal based on discussions with Navy training personnel and a review of courses. A 10% variation was used to analyze the sensitivity of the results to the nominal assumptions.
- The frequency of terminal breakdowns under operational conditions on land and at sea is not known with certainty. Manufacturer estimates were increased by one-third on land (from three to four per year) and one-half at sea (from four to six per year) to assess the sensitivity of the results.
- The amount of student time saved by having CMI instead of conventional instruction is estimated between 10 and 20%. Both cases were considered here.

Table 13.9 *Characteristics of a Representative COMISAT Program*

Financial	
Present value net savings (eight year program, 10% discount)	$29,777,000
Annual Savings	$10,858,000
Incremental investment	$31,582,000
Savings-to-investment ratio	1.94
Operational	
Number of CMI courses	69
Number of sites	
Ships	411
CONUS land bases	80
Overseas land bases	47
Number of students	20,881
Proportion of CMI students served at sites	96.7%

The analysis shows that the resultant change in savings attributable to (1) the change in annual terminal breakdown rates from three to four on land and four to six on ships, (2) a 10% change in the proportion of courses convertible to CMI, and (3) a 5% change in student time savings with CMI would each result in a change of about five million dollars in present value savings. That corresponds to an annual change in savings of about $1 million.

Table 13.10 lists the results for the two cases that were developed in the greatest detail. The cases show the effect of different remote site terminal breakdown rates on the savings achieved, on the number of courses that can be delivered, and on the number of ships and land sites served. There is a decrease of $5 million in present value savings for increases in breakdown rates of one-third on land and one-half at sea. The greatest impact is at sea where 83 fewer ships could be served with the higher breakdown rates. The number of courses that can be delivered decreases from 69 to 67 because of the higher maintenance costs. In both cases, more than 90% of the students who normally travel to a training center to attend the feasible CMI courses would receive the training at their base or ship.

These two cases assume that 70% of Navy skill progression courses are operationally suitable for conversion to CMI and that CMI requires 20% less time than conventional instruction. The effect on savings of varying the latter two assumptions is shown in Table 13.11. A 10% change in the

Table 13.10 *Summary of Results*

Item	Low Maintenance[a]	High Maintenance[b]
Number of CMI courses	69	67
Number of sites	538	434
Ships	411	328
CONUS	80	66
Overseas	47	40
Proportion of CMI students served at the sites	96.7%	92%
Annual net savings	$10,858,000	$10,190,000
Present value of savings (8-year program)	$29,777,000	$26,384,000
Incremental investment	$31,582,000	$32,096,000
Savings-to-investment ratio	1.94	1.82

[a]The low-maintenance case assumes that three breakdowns occur each year on land and four breakdowns occur each year at sea.
[b]The high-maintenance case assumes four breakdowns per year on land and six at sea. This case also assumes that because of higher breakdown rates spare terminals will be required at all sites serviced by Navy personnel.

number of courses suitable for CMI results in a change in the present value of the savings of about five million dollars.

The effect of a change in student time savings because of the use of CMI is greater when there is a higher proportion of courses suitable for conversion. When 60% of the skill progression courses are suitable for CMI, a 10% change in student time savings (from 10 to 20%) changes the present value savings by $7.398 million (from $17.627 to $25.025 million). However, when 80% of the courses are convertible, the same change in student time savings changes the present value of savings by $9.864 million (from $25.913 to $35.777 million).

The lowest estimate of present value savings is $14 million, assuming high maintenance, 60% course convertibility to CMI, and 10% student time savings with CMI.

The savings results are sensitive to other assumptions, but not to the same extent as those just discussed. For example, the results in Tables 13.9 to 13.11 assume that land sites within 30 miles of a central location can be served by one terminal at that location. Courier service would carry inputs to the terminal and return messages to the base. If that is not feasible, present value savings would be reduced by $1 million.

Table 13.11 *Sensitivity of Present Value Savings to Time Savings and CMI Convertibility Assumptions: Low-Maintenance Case*

Percentage of Skill Progression Courses Convertible to CMI	Percentage of Student Time Saved		
	10%	15%	20%
60	$17,627,000	$21,326,000	$25,025,000[a]
70	$21,770,000	$26,085,000	$30,401,000
80	$25,913,000	$30,845,000	$35,777,000

[a]The values in this table are estimates based on a closed form expression for net COMISAT cost. Hence this number differs slightly from the present value savings figure in Table 13.10.

13.5.3 Communications Requirements

Another concern is that communications lines be available. As the number of CMI students increases, the capacity required will increase the Navy communications required, particularly from the CMI computer to the remote sites. To assess this, the capacity required was estimated[9] based on the characteristics of the CMI messages. The total communications load at the average site is estimated to be 0.28 characters per second, and at the CMI computer, 150 characters per second. This would appear to be a reasonable load for the AUTODIN II system.

13.5.4 Recommendations

Since the economic analysis shows large potential savings, it is important to test a major assumption, that the COMISAT concept is workable at a remote site. It is necessary to show that supporting personnel assignments can be carried out as assumed for the analysis, that students can take the time on site for training, and that, under those conditions, the training is effective. The proposed demonstration would help confirm that the concept is practical and desirable.

Additional investigation of the economic benefits of COMISAT should be conducted:

- The Navy courses should be surveyed in detail to determine suitability for CMI, enrollments, and lengths.
- A better approximation of equipment breakdown rates should be ob-

[9]See Appendix 13.D for details of this analysis.

tained. In the event that these rates are higher than those in the high maintenance case explored here, the manufacturer should be consulted in an attempt to improve maintainability characteristics.

- An investigation should be made to determine whether the average travel and living expenses used in this analysis adequately reflect the range of costs and the usual payment procedures.
- The availability, cost, and feasibility of courier service between nearby sites should be investigated to obtain an improved estimate of the extent to which clustering of sites is practical and profitable.
- The available capacity of AUTODIN at each site should be assessed to ensure that the additional demand due to COMISAT can be adequately met. In particular, the AUTODIN II capacity to the CMI computer will have to be large enough to accommodate the new load.[10]

13.6 SUMMARY OF KEY PRINCIPLES

The first step is to identify the system alternatives to be considered in the evaluation and to have an "ideal" evaluation model that would be used if all data were available. For this model we used the economic analysis model developed in Chapter 12.

The next step is to make a preliminary survey of what data relating to this model could be made available within the time limits. This means that either the remaining data required by the "ideal" model must be obtained by extrapolation from available data, including the "expert opinion," or the factor must be ignored in the actual model used.

The remaining steps follow the steps of the systems evaluation developed in Case 1:

- Design all systems to accomplish the same objective (providing the required number of course graduates forecasted). Here the two key design variables were how many courses could be economically justified and how many terminals would be required at the training centers and at remote sites.
- Calculate the costs of each system. In this case we actually calculated the difference in total costs when changing from one system to another (either from the current system to a more intensive CMI system or from this CMI to a COMISAT system).

[10]The net cost of these new lines was assumed to be the same as the additional communications cost for the CMI system enlarged to include these skill progression courses.

The concept of a "learning curve" was introduced. Although this concept or model was originally developed as a method of estimating reduced production costs as production quantities are increased, it was used in this case since the data followed this model.

Further illustration of the use of sensitivity analysis was provided. Such analyses are important, when the analyst is confronted with data uncertainties, to see what impact such uncertainties have on the final decision.

Note that the overall analysis was conducted in a modular fashion; thus if additional alternatives were presented or additional or more accurate data were provided, the analysts would only have to change those parts that were affected by the new data.

APPENDIX 13.A ESTIMATION OF COURSE ENROLLMENT DATA

As mentioned in Chapter 13, the data on individual enrollments for skill progression courses were limited to the title, length, and number of graduates of the 10 courses with the highest volume. The lack of data necessitated generation of populations for courses and approximation of such enrollment quantities as average attendance and student man-days. This appendix provides further details about the procedures used to arrive at these estimates.

Generating Course Populations

The basic assumption underlying the generation of estimates of course volumes is that the derived populations follow the same function as the population data of the 10 highest-volume courses. With this assumption, the required populations can be obtained by fitting a curve through the populations for the 10 highest-volume courses and extrapolating this curve to obtain the needed populations. These steps are now described in greater detail.

Fitting the Curve

As Figure 13.A.1 illustrates, the course populations seem to follow an exponential curve. A learning curve was used to approximate these populations because such a curve is exponential in shape, is relatively simple to derive and manipulate, and provides a close approximation to the data. This appendix contains a discussion of the learning curve and its properties.

A process follows a learning curve if the accumulated average of the

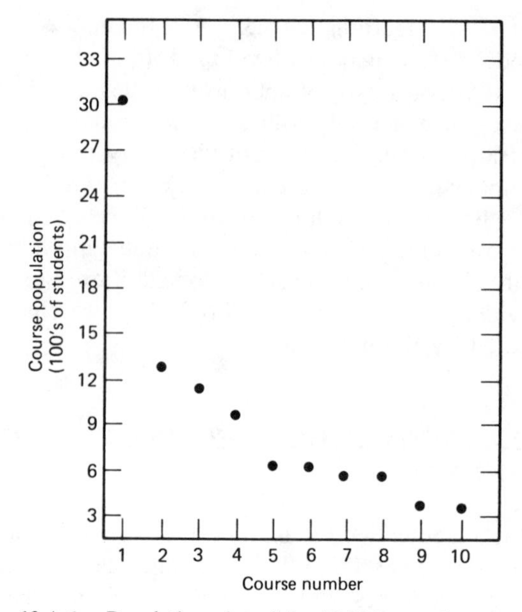

Figure 13.A.1. Population plot of the 10 highest-volume courses.

dependent variable decreases to a constant percentage of the previous accumulated average value whenever the independent variable is doubled. For example, if the populations of the high-volume courses follow a learning curve, then by doubling the number of courses, the accumulated average student population is decreased to a constant percent of the previous accumulated average population. The constant percentage mentioned above is called the learning rate. A learning curve with a 70% learning rate is plotted in Figure 13.A.2.

The primary use of the learning curve is to relate the average man-hours required to do a job as a function of the number of times that similar jobs are done. Thus, applying Figure 13.A.2 to this manpower problem, it can be seen that with each doubling of work done, the average man-hours required decreases to 70% of the original amount required (before doubling).

The reader should realize that there is no reason why the course population should follow a learning curve. All we are doing is (1) attempting to fit this type of function to the limited course population data available and (2) assuming that the next set of data also follows this same decreasing function.

To determine whether the course population data follows a learning

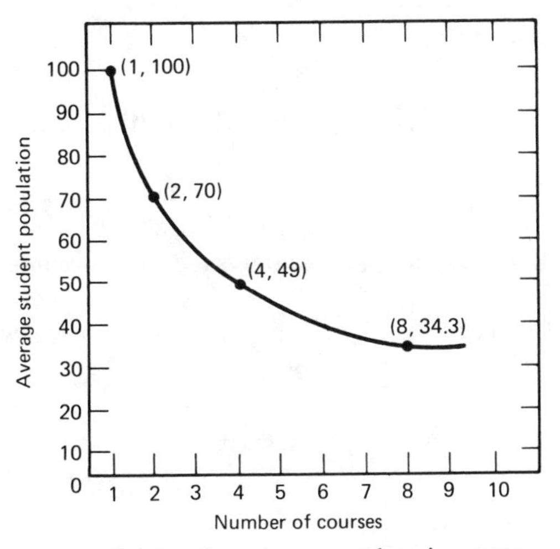

Figure 13.A.2. Seventy percent learning curve.

curve the accumulated average populations for each course were determined and used to examine the learning rate. Table 13.A.1 contains these accumulated average populations, which were found by dividing the cumulative population by the course number. The learning rate can now be estimated by performing the calculations given below, where A_i is the accumulated average population of the ith course.

$$\frac{A_2}{A_1} = 71\%$$

$$\frac{A_4}{A_2} = 74\%$$

$$\frac{A_6}{A_3} = 69\%$$

$$\frac{A_8}{A_4} = 68\%$$

$$\frac{A_{10}}{A_5} = 67\%$$

It was noted that all these ratios are fairly close to 70%. In fact, their average is 69.8%. Hence a 70% learning curve seemed to provide a reasonable approximation to the course population data and was used in obtaining subsequent course populations.

The next step in arriving at these course populations was to determine the algebraic form of the particular learning curve to be used. The algebraic form of a general learning curve is

$$Y_N = \frac{a}{N^b}$$

where, for purposes of this analysis, Y_N is the accumulated average number of students over N courses; a is the number of students in course 1; N is the number of courses; and b is an exponent associated with the learning rate. In this equation Y_N and N are variables, a is a constant whose value is known to be 3110, and b is a constant whose value must be derived.

To derive b the learning curve formula was first applied to the second and fourth courses, which yields

$$Y_2 = \frac{3110}{(2)^b}$$

and

$$Y_4 = \frac{3110}{(4)^b}$$

respectively. Dividing the first of these equations by the second results in

$$\frac{Y_2}{Y_4} = \frac{3110/(N)^b}{3110/(2N)^b} = (2)^b = \frac{1}{0.70}$$

because of the relationship between the accumulated average popula-

Table 13.A.1 *Actual Cumulative and Course Populations*

(1) Highest-Volume Course Number	(2) Sums	(3) Accumulated Average
1	3110	3110
2	4433	2216.5
3	5607	1869
4	6567	1641.75
5	7178	1435.6
6	7786	1297.67
7	8372	1196
8	8884	1110.5
9	9285	1031.67
10	9685	968.5

tions. Applying the logarithmic function to both sides of the equation, the value of b was found to be 0.516. Hence the specific form of the learning curve applied in this analysis is

$$Y_N = \frac{3110}{N^{0.516}}$$

Extrapolating the Curve

The accumulated average population for the eleventh and subsequent courses can be determined by substituting the course number for N in the formula previously derived. These results are shown in column 2 of Table 13.A.2 for courses 10 to 18. However, the course population, not the accumulated average course population, is required. Course population can be obtained from the accumulated average population by employing the following steps:

- For course 10 and all subsequent courses, multiply the accumulated average by the course number N to find the cumulative number of students in each of the N courses. The results of these calculations are shown in column 3 of Table 13.A.2.
- Subtract the cumulative population for the $N - 1$st course from the cumulative population for the Nth course to find the number of students taking the Nth course. The resulting course enrollments can be found in column 4 of Table 13.A.2.

Table 13.A.2 *Approximate Cumulative and Course Populations*

(1) N Course Number	(2) Accumulative Average	(3) Fitted Cumulative	(4) Course Population
10	947.9	9,479	
11	902.4	9,926	447
12	862.8	10,353	427
13	827.9	10,762	409
14	796.8	11,155	393
15	768.9	11,534	379
16	743.8	11,900	366
17	720.9	12,255	355
18	699.9	12,598	343

Approximation of Enrollment Quantities

Two enrollment quantities were approximated in this analysis: average attendance and student man-days.

Average Attendance

The average number of people enrolled in a course at any given time was determined from the number of graduates, using data from Table 13.2. The average number of students is

$$\frac{\text{Entrants} + \text{graduates}}{2} = \frac{62{,}909 + 61{,}666}{2} = 62{,}287$$

Average attendance can be obtained from the number of graduates by

$$\frac{\text{Total average attendance}}{\text{Total number of graduates}} \text{ (number of graduates in the course)}$$

$$= \frac{62{,}287}{61{,}666} \text{ course graduates} = 1.01 \text{ course graduates}$$

Student Man-days

The number of student man-days required by a course was determined by multiplying the average attendance by the course length. With the exception of courses 1 to 10, all courses were assumed to require 51 days, the average course length for skill progression courses.

APPENDIX 13.B DETAILS OF THE CALCULATION OF COSTS, SAVINGS, BREAK-EVEN POPULATIONS

The components of costs and savings were described in general terms in Chapter 13. This appendix presents the details of the techniques used to calculate each cost element so that the reader may verify the analytical results.

Costs and Savings Incurred through Development of CMI Courses for Training Center Use

Course Development, Coding, and Maintenance

As discussed in Chapter 13, developing a course for CMI instruction and coding it for use on the computer requires an incremental expenditure of $2050 per hour of training. The number of training hours in a course was determined by first multiplying the number of calendar days by 5/7, to obtain the number of working days required by the course. Multiplying the result by 7, the number of training hours in a day, yielded the hours of training required by a course. The development and coding cost were then found by

$$\text{Development and coding cost per course} = (\$2050)(\text{calendar days})(5/7)(7) = (\$10{,}250)(\text{calendar days}) \quad (1)$$

The annual course maintenance cost was estimated by taking 5% of this incremental expense of development and coding.

Computer Expansion Cost

The components of computer expansion costs were also discussed in Chapter 13 and an annual cost of $14,400 for each 1000 AOB students was determined. AOB can be found by dividing the average annual attendance by the number of successive courses in a year, which is 365 divided by the course length in calendar days. Therefore

$$\text{AOB} = (\text{Average annual attendance})\left(\frac{\text{course length in days}}{365}\right)$$

and

Annual computer leasing cost

$$= \left(\frac{\$14{,}400}{1000}\right)(\text{average annual attendance})\left(\frac{\text{course length in days}}{365}\right)$$

$$= \$0.0395\ (\text{average annual attendance})(\text{course length in days})$$

Terminal Costs

For each 60 AOB students taking CMI at the training center an additional terminal must be purchased and maintained. Assuming courses can share terminals, the number of training center terminals required to service a course can be found by dividing the number of AOB students in the course by 60.

The investment and maintenance costs of a training center terminal, from Table 13.4 in Chapter 13, are $14,250 and $147 per month, respectively. The 8-year present value of total cost of one training center terminal is

$$\$14{,}250 + 5.597(\$147)\ 12 = \$24{,}123$$

Therefore the present value of a training center terminal cost for any course can be found from

$$\text{Present value} = \frac{\text{course AOB}}{60}(\$24{,}123) \tag{3}$$

Savings in Student Time and Living Costs

The saving in student man-hours of training with CMI compared to conventional training is the product of the number of AOB students, the percentage of time saved by CMI, and the annual rate of pay for each student. The annual salary, including basic pay, quarters, incentive and special pay, retirement pay, and miscellaneous expenses, was assumed to be $10,090, the salary of a student with an E-4 rating. As indicated in Chapter 13, a 20% decrease in training time was assumed. Thus the annual saving in student salary is

$$(\text{AOB})(0.20)(\$10{,}090) = \$2018 \quad (\text{AOB}) \tag{4}$$

Incremental living savings were derived in a similar manner by multiplying the average annual number of students, average number of days saved, and incremental living expense saved. The average number of days saved was calculated as the product of the number of calendar days in a course and the 20% time savings factor. The incremental living cost saved is $2.50 per day. Thus

$$\begin{aligned}\text{Annual living cost savings}\\ &= (\text{average number of students})(0.20)(\text{calendar days})(\$2.50)\\ &= \$0.50\ (\text{average number of students})(\text{calendar days}) \end{aligned} \tag{5}$$

COMISAT Costs and Savings for Training Center Justified Courses

Terminal Cost

To teach the first 17 courses through COMISAT, the only costs are for terminal purchase and maintenance. For the low-maintenance case, the undiscounted terminal investment cost is the number of sites multiplied by the unit terminal price of $11,450. For the high-maintenance case with assignment of spare terminals, the same calculation is made but the number of terminals to be purchased is increased by the spares.

Spare parts investment cost was determined in a similar manner as terminal cost, using a spare parts unit cost of $6529. The discount for a particular number of terminals or spare parts sets was found by taking the sum of the discounts in Table 13.7, weighted by the number of units in each discount group. For example, if 80 units are needed the total discount is

$$0(1) + 5\%(29) + 10\%(30) + 15\%(20) = 745\%$$

of the unit terminal cost. The group and cumulative discounts can be found in Table 13.B.1.

Maintenance Costs

Maintenance costs depend on the site category and the maintenance assumptions of each case. However, with the exception of category 1 (CONUS sites 50 miles or less from a service location) unit maintenance rates for the high-maintenance case can be obtained from those of the low-maintenance case by multiplying by 4/3 for land sites and 6/4 for ship sites. Hence this discussion is concerned only with the low-maintenance case.

Category 1 sites would be serviced by OpScan at a cost of $80 per month or $960 per year. Category 2 sites (CONUS sites between 50 and 90 miles from service) would also be maintained by OpScan but on a time-and-materials basis involving costs of travel, labor, and parts. Travel charges are 18 cents per mile plus $30 per hour of travel for each failure. If an average travel speed of 45 miles per hour is assumed,

Average total travel cost per year = (number of annual breakdowns) ($0.18) (miles to the site) + (number of annual breakdowns) (1/45) ($30) (miles to the site)

Since this cost varies with the distance to the site, a mean travel cost of $2218 for all sites in this category was calculated. Labor cost is $30 per hour and the mean time to repair is 2½ hours; thus

Average annual labor cost = ($30) (2½) (number of annual breakdowns) = $75 (number of annual breakdowns)

The only information available concerning the cost of parts was

Table 13.B.1 *Group and Cumulative Quantity Discounts*

Units	Percent Discount	Number in Group	Group Discount (%)	Cumulative Discount (%)
1st	0	1	0	0
2nd–30th	5	29	145	145
31st–60th	10	30	300	445
61st–90th	15	30	450	895
91st–120th	20	30	600	1495
121st–150th	25	30	750	2245
151st–180th	30	30	900	3145
181st–210th	35	30	1050	4195
211th–up	40	n	$40n$	$4195 + 40n$

OpScan's price of $600 for a service warranty for the second year. This was taken as the annual parts cost. Naval personnel would service all other remote (category 3, 4, and 5) sites, thereby incurring costs of labor and parts. The total labor cost per repair for service by site personnel was assumed to be the same as OpScan's total labor cost.[11] Therefore, labor cost is, as before, $75 (number of annual breakdowns). The average annual cost of replacing used parts from the inventory, based on a failure rate of three per year, was estimated by assuming that OpScan's $600 per year warranty price was derived on the basis of the price for parts used and their expected return on investment for a perpetual inventory. It was further assumed that they require a 30% return on their parts investment of $6529 and that this inventory is used by the OpScan repairmen in servicing 10 sites. The desired return on investment is $1960 or approximately $200 per year per site. Thus the average cost of parts to the OpScan company is $400 per year for each terminal when three annual failures occur. The same parts cost was assumed for service by Navy personnel.

COMISAT Savings

The savings achievable by utilizing COMISAT to teach the first 17 courses consist of three components: student travel, living expense at the training center, and the cost of terminals no longer required at the training center. The average cost of a one-way trip to a training center is $137. In addition, a student on PCS is allowed $291 to cover transfer of household effects. Thus the total travel costs saved per student for PCS are $141.50 + $291 = $432.50 and for TDY are 2($141.50) = $283.

Taking into account the relative frequency of PCS and TDY,[12]

Average travel cost

$$= \left(\frac{30{,}042}{62{,}909}\right)(\$432.50) + \left(\frac{32{,}867}{62{,}909}\right)(\$283) = \$354.39 \text{ per entrant}$$

or

$$= \left(\frac{62{,}909}{61{,}666}\right)(\$354.39) = 1.02(\$354.39) = \$361.48 \text{ per graduate}$$

$$\text{Total travel savings} = (\$361.48)\ (\text{number of graduates}) \qquad (6)$$

The living savings are the product of the number of student-days and $2.50, the daily living allowance at the training center. There is a savings

[11]This is a conservative assumption, permitting the ET to take a longer mean time to repair than the Opscan repairman since the ET's hourly cost is less.

[12]The data in the proportions come from Table 13.2.

of one training center terminal for every 60 AOB students reached through COMISAT. Each training center terminal requires a $14,250 investment and a present value cost of $9873 to maintain over 8 years. Hence for training center terminals

$$\text{Savings} = \frac{\text{AOB}}{60}(\$14{,}250 + \$9873) = \$402.05\ \text{AOB} \tag{7}$$

Calculation of Break-Even Populations

To determine which sites should be outfitted with COMISAT terminals it was necessary to calculate the population for which the cost of delivering training to the site is equal to the savings achieved, because all site populations greater than this would result in savings. This section describes the derivation of the break-even population for each category of site. The formula used to derive the break-even population is

$$P_{\text{be}} = \text{TC}_{\text{pv}}E/\text{TGS}_{\text{pv}}$$

where P_{be} = break-even population
TC_{pv} = present value of equipment
E = enlisted personnel population
TGS_{pv} = present value of travel and living savings

The present value of total gross travel and living savings for those courses economically justified for use at the training centers was found to be $23,714,343 when discounted at 10% over 8 years. The total enlisted personnel population (E) was given as 370,346. Thus the only quantity needed to determine the break-even population (P_{be}) was the present value of the total cost of a site (TC_{pv}) discounted over 8 years. TC_{pv} is the sum of the terminal and spare parts cost without quantity discounts, and the 8-year present value of maintenance. Two factors, total number of terminals purchased and site category, influence the value of TC_{pv}. For this reason a different break-even population was calculated for each category within each case.

The only exception to the calculation of break-even population by site category occurs in category 2, where the dependence of maintenance cost of mileage makes calculation of one TC_{pv} for the entire category impossible. As a result, decisions about the inclusion of Type 2 sites were made by comparing the total cost of each particular site to the savings it could contribute.

Since quantity discounts are available for terminals and spare parts, the investment cost, and hence the total cost, depends on the numbers of terminal and parts sets required. For this reason the calculations were made iteratively using the following steps:

- Estimate the discount intervals in which the number of terminals and the number of parts sets will fall.
- Using these estimates determine the correct quantity discount factor from Table 13.7.
- Use the discount factors to derive a quantity discounted investment cost of a terminal and a spare parts set.
- Determine TC_{pv} by adding the investment costs to the maintenance cost for the category under consideration.
- Use the equation given above to derive P_{be}.
- Compare the site populations to P_{be} to identify sites whose population is larger than P_{be}. These are the sites feasible for COMISAT development. If the number of terminals and spare parts are within the estimated intervals, the correct break-even populations have been calculated. Otherwise, revise the interval estimates as indicated and repeat the process.

This procedure resulted in the identification of 306 feasible sites requiring an expenditure of \$5.348 million in present value over 8 years.

Costs and Savings in the Iterative Approach

Costs

For courses whose development is justified only by COMISAT, the cost of development, coding, maintenance, leasing, site terminals and the cost of additional training center terminals required by students who cannot be reached by COMISAT must all be charged to COMISAT. With the exception of training center costs, each of these is calculated as described earlier in this appendix. Terminal costs for implementing 70% of the eighteenth and subsequent courses at the training center are derived similar to the savings discussed earlier under COMISAT costs. However, the number of students who must take the new courses at the training center (AOB_s) is used in the formula. Thus the training center terminal cost charged to COMISAT is

$$\text{Cost} = \left(\frac{AOB_s}{60}\right)(\$14{,}250 + \$9873)$$

where AOB_s is the number of students who cannot be reached at their own sites by the eighteenth and subsequent courses.

The gross COMISAT cost for teaching all economically and operationally feasible courses at Navy sites can be found by evaluating the following expression:

> Eight-year present value of COMISAT gross cost = (development and coding for courses justified only by COMISAT [1]) + (computer leasing [2]) + (course maintenance for courses justified only by COMISAT) + (COMISAT terminals) + (training center terminals [3])

Applying the appropriate formula for each cost yields

$$\text{Eight-year present value of COMISAT gross cost} =$$

$$0.7(N - 17)\,(2050)\,(51)\,(5/7)\,(7) - 0.7(N - 17)\,(\$81{,}288) +$$

$$0.7(N - 17)\,(5.597)\,(0.05)\,(522{,}750) + \left(\frac{14{,}400}{1{,}000}\right)(g_{\Delta N})\,(1.01)\left(\frac{51}{365}\right) +$$

$$\text{COMISAT terminal cost} + \frac{g_{\Delta N}(1.01)\left(\frac{51}{365}\right)(\text{PC})}{60}\,(\$14{,}250 + \$9873)$$

where N is the total number of economically feasible courses of which 70% can be converted to CMI, $g_{\Delta N}$ is the number of graduates in the 70% of courses 18 through N convertible to CMI, and PC is the percent of the population that cannot be reached onsite through COMISAT. The preceding expression can be simplified to:

$$\text{Eight-year present value of COMISAT gross costs} =$$
$$411{,}426N + 11.4g_{\Delta N} + 57(g_{\Delta N})(\text{PC}) + \text{COMISAT terminal cost}$$

Savings

The savings COMISAT achieves through development of additional courses consist of travel, living, time, and incremental living savings. Each of these is calculated by using formulas discussed previously in this appendix.

The gross COMISAT savings for teaching 70% of the N highest volume courses at remote sites is

> Eight-year present value of gross COMISAT savings = (travel cost for students of all courses as indicated in equation [6]) + (living costs for students of all courses) + (time costs for students of courses 18 to N as indicated in equation [4]) + (living costs for students of courses 18 to N as indicated in equation [5]) + (training center

terminal costs for students who are taught courses 1 to 17 through COMISAT as indicated in equation [3])

Substituting the proper formulas into this expression yields

$$\text{Eight-year present value of gross COMISAT savings} =$$
$$(5.597)\,(1 - \text{PC})\,(\$361.48)\,(g) + 5.597\,(1 - \text{PC})\,(\$2.50)\,(51)\,(g)\,(0.80) +$$
$$(5.597)\, g_{\Delta N}\,(1.01)\left(\frac{51}{365}\right)(0.20)\,(\$10{,}090) +$$
$$g_{\Delta N}(1.01)\,(0.20)\,(51)\,(\$2.50)\,(5.597) +$$
$$\text{training center terminal costs}$$

where g is the total number of graduates in the N courses convertible to CMI.

This expression simplifies to

$$\text{Eight-year present value of gross COMISAT savings} = 2594g - 2594g\,(\text{PC}) + 1741g_{\Delta N} + \text{training center terminal cost}$$

Subtracting the gross savings expression from the gross cost expression results in net cost. A simplified expression for net cost in terms of N can be obtained by using the identities

$$\begin{aligned} g_{\Delta N} &= 0.7\,[3110N^{0.484} - 3110(17)^{0.484}] \\ &= 2177N^{0.484} - 8578 \end{aligned}$$

But, by definition,

$$\begin{aligned} g &= 9334 + g_{\Delta N} \\ &= 2177N^{0.484} + 756 \end{aligned}$$

The simplified net cost expression is

$$\text{Eight-year present value of net COMISAT cost} = 411{,}426N - 6{,}107{,}356N^{0.484} + 2{,}342{,}017N^{0.484}\text{PC} + 850{,}425\text{PC} + 7{,}497{,}450 + \text{COMISAT terminal cost} + \text{training center terminal cost}$$

PC, COMISAT terminal cost, and training center cost are known once the number of sites is fixed. Thus net cost is a function of the single variable N.

The expression for net cost is useful in two ways. First, by taking its derivative, setting the result to zero, and solving for N, the maximum number of courses that could be justified for CMI is obtained. The net cost expression can also be rewritten in terms of the student time saving and

Table 13.B.2 *Sensitivity of Present Value Savings to Time Savings and CMI Convertibility Assumptions: High-Maintenance Case*

Percentage of Skill Progression Courses Convertible to CMI (%)	Percentage of Student Time Saved		
	10%	15%	20%
60	$14,466,000	$18,079,000	$21,691,000
70	$18,423,000	$22,637,000	$26,852,000
80	$22,379,000	$27,196,000	$32,012,000

operational constraint parameters. With the expression in this form the sensitivity of total cost to the parameters can be easily investigated. The results of this sensitivity analysis were shown for the low-maintenance case in Table 13.11. Similar results for the high-maintenance case can be found in Table 13.B.2.

APPENDIX 13.C DETERMINATION OF MAXIMUM MILEAGE FOR MAINTENANCE BY OPSCAN PERSONNEL

To determine the total maintenance costs, a decision must be made regarding which CONUS sites should be serviced by OpScan personnel and which by Naval (ET) personnel. The cost of service by an ET is constant within each of the maintenance cases defined in the report, and the cost of OpScan service varies only with the mileage to the site from the service location. Hence all sites beyond a determined distance from service should have site personnel handle repairs; all other CONUS sites should be serviced by OpScan.

The break-even mileage for each case can be determined by performing the following steps. The low-maintenance case data are used as an illustrative example.

1. Add the costs of parts, labor, and inventory investment to arrive at an 8-year present value cost of ET repair. The cost of parts for ET repair is $400 per year. The cost of labor for ET repair is 3(2.5)$30 = $225 per year. Inventory investment is $6,529. The 8-year present value cost of ET repair = $6,529 + 5.597 ($625), or $10,027.
2. Add the costs of parts, labor, and mileage to arrive at an expression for the 8-year present value cost of OpScan service as a function of miles to the site.

Cost of parts for OpScan service = \$600 per year
Cost of labor for OpScan service = 3(2.5)\$30 = \$225 per year
Cost of mileage for OpScan service = (\$0.18)(6)d + (1/45)(6)(\$30)d
= \$1.08d + \$4d
= \$5.08d

where d is the mileage from the service location to the site.

Eight-year present value cost of OpScan service
= 5.597(\$825 + \$5.08d)
= \$4617 + \$28.43d

3. Equate these two costs.

$$\$10{,}027 = \$4617 + \$28.43d$$

4. Solve the resulting equation.

$$d = \frac{\$10{,}027 - \$4617}{\$28.43}$$

$$d = 190 \text{ miles}$$

Similar calculations for the high-maintenance case result in equally high mileages. However, response time for this distance would be several hours. To reduce the terminal downtime to more reasonable limits, a maximum distance of 90 miles, with a 2-hour driving time, was chosen.

APPENDIX 13.D REQUIRED AUTODIN II CAPACITY

Each 10 student-hours of training require a transmission of 1087 characters from the remote site to the CMI computer and 9174 characters from the CMI computer back to the remote site. These transmission requirements assume a batching of 10 messages together to increase communication efficiency. To determine the size of the transmission line required to and from the CMI computer, the larger of the two transmission requirements must be used. Using the characteristics of the representative COMISAT program of Table 13.9, the number of student training hours per year is

(20,881 students per year) (0.967) (255 hours)[13] =
5,148,941 student training hours per year

[13]Assuming an average course length of 255 hours.

Thus the total number of characters per year to be transmitted to the CMI computer center is

$$\frac{(5,148,941)(9174)}{10} = 4,723,638,473 \text{ characters per year}$$
$$= 150 \text{ characters per seconds}$$

This transmission requirement should then be converted into an appropriate-sized transmission link between the nearest AUTODIN switch and the CMI computer, and the cost of such a link should then be calculated. The net cost of the AUTODIN lines required should be calculated as the difference between the total cost of the AUTODIN lines and the cost of the additional dedicated lines required to operate the CMI skill progression courses at the training centers. Neither of these costs is available at this time.

Similar calculations could be made for the additional transmission requirements the COMISAT system imposes on the message center at each remote site. For the 538 sites included in the representative case this amounts to an average additional load of 0.28 character per second.

Part VI

PROBLEM RECOGNITION: IMPROVING A BUSINESS OR OTHER ORGANIZATION

Each of the previous cases was concerned with a specific problem that had already been recognized and assigned to a planner for his analysis and solution. It was shown that the same systems planning approach could be applied to each of these different types of problem contexts. In Part VI these same principles of planning are extended to an even more complex situation. Now we describe how to make a diagnostic audit of an organization, arriving at the preferred method of treatment in much the same way a physician performs a physical examination of a patient. Consider the following problem.

PROBLEM AS GIVEN

You are a staff consultant with Management Systems, Incorporated, and are about to meet with Ted Atkins, the new president and general manager of the Acme Electric Corporation, a medium-sized firm that manufactures electric products. You and Ted worked together as planners a number of years ago, and have kept in contact over the years.

At the meeting Ted welcomes you and says,

As you know, I've recently taken on this new job. Since I'm new to the company, one of the first things I want to see done is an audit of this business. Where are we now? What kinds of problems are we facing? I've looked over what was given to me as the corporate long-range plan, and it was a big disappointment to me. It's full of platitudes and shows no real systematic look at this business. It was put together about a year and a half ago by a small planning staff who have since left the company. They worked on it essentially in isolation from the operating managers. Naturally, the managers ignored it after it was released. I for one do not find its underlying assumptions realistic. So I would also like your help to use the initial audit as a first step in a longer-range planning effort. This business needs to plan for the next 5 years. If you get the operating managers involved in the effort, the new plan will become the basis of our actions for the period ahead.

Third, I'd like you to also help us develop a better planning approach. Each year we should be able to update the past year's business plan, and generate the next year's budget. I sense from my observations of this place, and from the business plan I inherited, that this company has not been planning oriented. I think all the data needed to do sound planning may not currently exist. My objective is to improve the planning process in this company by holding each functional manager responsible for planning in this area, even though he delegates this responsibility to someone else in his organization. I also plan to appoint a director of planning. Hence I would like you to demonstrate some improved procedures we can follow in future planning and determine the kind of data we should be collecting in each of the functional areas to implement this procedure.

The problem context described involves a manager or a planner who enters an organization with which he is unfamiliar and is asked to determine the organization's current condition and to generate an improvement plan. Here the main emphasis is on problem recognition because no specific problems have been identified. We begin with the perception that no organization is perfect; hence improvements can always be found. The approach described here can also be used by the planner who is a full-time member of the organization.

In addition to showing a method of recognizing and analyzing problems in a new organization, Part VI has these other objectives:

- To identify the key generic functions present in every organization and to show how each of these functions may be properly "designed" with respect to one another in a systematic fashion to obtain the preferred design for the entire organization using the same principles of planning previously described.
- To use this case as a means of summarizing and coalescing the principles of systems planning developed thus far and to extend these principles to other types of systems and organizations.

The discussions in Part VI differ from the previous parts in that the objective is not to provide a specific solution to the problem as given, but to use it to discuss the key principles of systems planning used in meeting the objectives listed. Other contexts will also be considered as they relate to the principles under discussion.

Finally, to aid in this total summary extensive use of footnotes will be made, citing references to previous material in this book which reinforce the principle being discussed.

Since Part VI serves as a summarization of the complete systems planning process, it was decided to incorporate its summary of key principles as the nucleus of the concluding Chapter 16, to serve as a summary of the key principles of planning discussed and illustrated throughout the book.

14

Planning for a Business Organization: Reducing Costs

14.1 PLANNING APPROACH TO BE FOLLOWED

Although the context previously presented involves a manufacturing-distribution type of organization, the objective of Part VI is to present a systematic approach that can be used in any type of organization. To implement this objective, we first develop the approach for a manufacturing-distribution organization and then show how it may be applied to other types of organizations. The approach can then be used as a checklist of steps to be done.

The key steps to the approach to be followed are listed in Table 14.1.

14.2 INITIATING THE PLANNING EFFORT

14.2.1 Introducing the Project

One of the first steps in getting the planning effort started is to inform senior management of the project's existence and its goals. It is always helpful to ask the top manager to do this at a staff meeting devoted to this subject. You can then explain the objectives of the effort, summarize the approach to be followed, and ask for their cooperation. Such a meeting will help allay fears.[1]

14.2.2 Organizing the Planning Team

It is important that a planning team containing representatives of the key operational areas involved be organized early in the study. This team

[1]Refer to Chapters 3 and 9 for a further discussion of this process.

Table 14.1 *Approach to Be Used in Analyzing an Organization*

1. Initiate the planning effort
   a. Introduce project to key managers
   b. Organize the planning team
   c. Solicit problems or solutions
2. Determine whether the organization is "under control"
   a. Is there a plan available describing the current operations and management system? (If so, obtain a copy)
   b. Does the plan contain standards that can be used to measure current performance of the organization in producing its current products or services in terms of quality, timeliness, and cost?
   c. How is the organization currently doing in terms of this plan?
   d. What are the "customers' " feelings regarding how well the organization is meeting their needs?
3. Determine environmental changes expected
   a. Are changes expected in the environment that will influence the ability of the organization to accomplish its objectives?
   b. Are there opportunities that will arise in the future that may aid the organization to better satisfy the environment and accomplish its objectives?
4. Develop methods for improving the organization (focus on quality, timeliness, and cost)
   a. Reduce the cost of producing the same product or service
   b. Reduce the time to produce the same product or service
   c. Change the cost or timeliness of producing a different product or service
   d. Increase sales revenue by offering:
      (1) Same quality product at lower cost or time
      (2) A lower-quality product at lower cost or time
      (3) A higher-quality product at different cost or time
5. Develop an improved management control system for evaluation of the performance of the organization
   a. Develop an improved current and long-range plan
   b. Develop an improved management control process for implementing the plan and adjusting to both today's and a changing environment
   c. Determine the data and reports needed by management to properly control their operations
   d. Develop the management information system providing such reports

should consist of selected individuals having knowledge and past experience in key operations. These individuals amplify the resources available to the planner, particularly when data gathering is required. The team representatives serve as focal points in providing the necessary information and data, either through their own efforts or through assignments to others in their organization possessing such data. They also provide a means of validating the information obtained. Lastly, they will no doubt develop alternative ways of improving business performance. Getting the involvement of the operating components who eventually must implement the plan aids in gaining acceptance of the final plan. The planner will focus much of his own effort on providing leadership in organizing and implementing the planning process.

14.2.3 Structuring the Organization

To identify the areas of the organization that should be represented on the planning team, two structures of the organization should be obtained or developed. The first structure is the table of organization, which identifies the organizational functions and the names of those in charge of these functions. The second is an operational flow model showing the key activities involved.

Figure 14.1 is a generic structure that applies to all manufacturing-distribution type organizations. This model is a further extension of the basic input-output model (Figure 2.3) described in Chapter 2. The total manufacturing-distribution system can be viewed as consisting of a set of subsystems, related as follows:

1. The major part of the organization is defined as the "main line system," consisting of those elements required to obtain production parts and materials from vendors (the incoming distribution system), to produce the current products (the primary production system), and to distribute these to customers (through the outgoing distribution system). Distribution may include delivering the product or service directly to the customer, as in the case of a telephone company or electric power utility, or to a point at which the customer picks it up, as in a retail store.
2. The remaining parts of the organization (consisting of maintenance, accounting, etc.) are defined as the support systems. They service the main line system in performing its current functions. New product and development planning is the support activity that develops new products and new methods for producing and distributing future products.

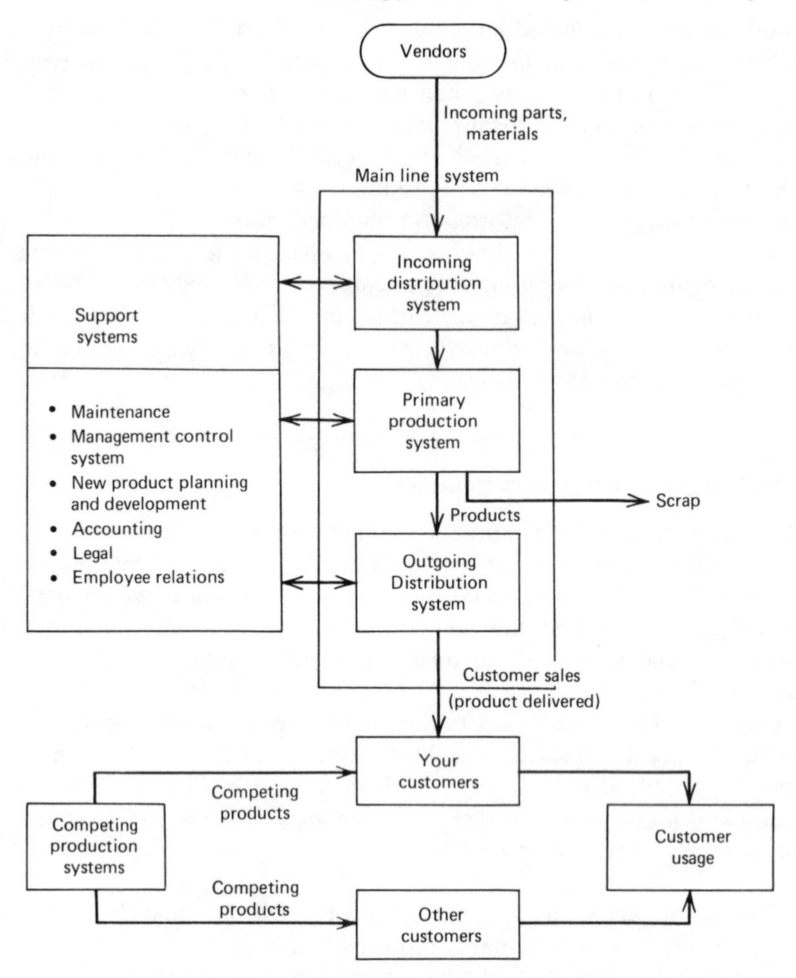

Figure 14.1. Generic structure for all organizations.

This part of the system is defined to include the areas of market research and product planning (to recognize new product needs), advance engineering, research, and development (to develop such products), and product engineering (to design these products to meet production capabilities).

3. The management control system[2] provides a way of measuring the four main characteristics of the products offered by the organization:

[2]See Case 2 (Chapter 8) for further discussion.

   a. Quality of product or service, in terms of final acceptability standards (as set by the organization, based on perceptions of customer acceptability).
   b. Timeliness of delivery to the customer by the outgoing distribution system.
   c. Cost of production and distribution.
   d. Financial conditions of sale to customer, which may include such conditions as selling price, delivery price, leasing arrangements, credit terms.

   The first three characteristics can be used as management control standards for the organization. Characteristics a, b, and d may be defined as the conditions of sale, since they represent the terms and conditions of a sales agreement.
4. An additional output consists of the various types of scrap resulting from the manufacturing-distribution process that must be reused or disposed of by the organization.
5. A sequence of events must take place to operate this system properly:
   a. Marketing must set the financial conditions of sale, based on factors such as the size of the perceived market and the characteristics of competing products.
   b. This results in a forecast of sales volume that aids production control (part of the management control system) in setting the production schedule.
   c. The primary production system manufactures products according to this schedule.
   d. Based on this schedule, incoming parts and materials required to operate and maintain the total organization are procured from vendors through an incoming distribution system, which selects the appropriate supply vendors and determines the timing of orders to the vendors. Some of these parts and materials are used by the primary production system to manufacture the products provided by the organization. The rest are used by all the support systems, such as accounting, maintenance, and so on.

14.2.4 Three Analytical Approaches

There are three different analytical approaches which can be used to recognize and understand the problems of the organization. They can be characterized as follows:

1. Gathering perceptions of problems from organization's personnel
2. Auditing the organization
3. Functional analysis of the organization.

Each method will be described in the following sections. Since each method provides its own contribution to problem recognition, each should be used in the total analysis.

14.3 GATHERING PERCEPTIONS OF ORGANIZATION'S PROBLEMS

14.3.1 Initial Gathering of Perceived Problems and Possible Improvements

Using the approach described in Chapter 3, an initial set of problems being faced by the organization is gathered as perceived by the key management personnel, the planning team, and others in the various operations. Also gathered are other ideas for improvement of the organization, including those proposals and recommendations for changes made in the past and rejected but still felt to be worthwhile by someone. Another source of ideas may be found by asking whether someone thinks that something competitors are doing should be done by this organization.

14.3.2 Relating Problems and Solutions

All problems identified and solutions proposed are then structured by using a hierarchy-of-objectives or means-to-an-end model. This structure can provide the following benefits. First, it provides a "first cut" to problem recognition, as well as an initial focus to the planning efforts. Second, a proposed solution can be related back to some perceived problem, further enlarging the initial problem list. Since the problems collected may only be symptoms of a more complex condition, clustering the set of initially recognized problems and solutions is helpful so that all relevant symptoms may be considered at one time.

14.3.3 Developing Initial Priorities for Planning Effort

The planning team next sets initial priorities for the subsequent planning effort, based on the group's perception of the importance or urgency of each problem cluster identified, the potential improvements and benefits that are possible, and an estimate of the analytical resources required for solution.

14.3.4 Continuing the Planning Effort

With the initial list of problems and possible solutions, the planning effort can now continue, using the planning approach of Table 7.1, culminating in specific proposals and the cost effectiveness of each. Specific examples of this process are given in later sections of this chapter, but for now let us continue with other methods of problem recognition.

14.4 AUDITING THE ORGANIZATION

While there is merit in initially gaining a better understanding of the problems facing the organization as perceived by various organizational personnel, a second approach, which can also be done in parallel, is to conduct a systematic management audit of the current operations.

A management audit finds answers to the following questions:

1. Is the current organization "under proper control?"
2. If not, what parts of the organization require improvement?
3. How can the costs of producing the current products or services be reduced?
4. What future change in the products or services should be made?
   a. Should other products or different services be added?
   b. Should some products or services currently produced be dropped?
5. What change in the organization will be required in implementing these changes?

A systematic method of answering these questions is now described.

14.4.1 Is the Current Organization under Proper Control?[3]

This question of control may be interpreted as, "Is the current organization meeting its current objectives satisfactorily in accordance with the current plan and expectations of management?"

To answer this question, we should obtain the most current plan of the organization and convert this into the operational flow model of Figure

[3]See Case 2 (Chapter 8) for further discussion of management control systems.

14.1. If such a formal plan including the various performance standards does not exist (which also gives an indication of the effectiveness of the management control system), the planner must then construct the operational flow model from observations of the various operations and from data gathered as required.

Each of the systems shown can also be modeled in greater detail as a series-parallel network of activities, and the output of each activity described by the same set of generic descriptors. The operational flow diagram of the entire main-line system, including the various standards set by the organization for each activity, can be used as the basis for controlling the entire operation producing the products or services. Thus the planner should see whether a suitable management control system, such as that described in Chapter 8, exists. This type of management control system can be used by the various managers to determine how well each activity is meeting its set of standards, how current performance compares with past performance, and where deficiencies exist. These serve as prime indicators for recognizing and diagnosing problems.

If such a management control system does not exist, the planner should point this out to the management and initiate the development of such a control system as soon as possible. This requires that each organizational activity set appropriate standards. If such standards are not all available, a data collection system should be initiated to gather historical data to be used as the standard.

Two definitions of standards can be made. The first is the set of customer expectations, and these, of course, may vary somewhat with each customer. The second definition is the set of company standards which are based on the particular market segment for which the product and its selling price has been designed. It is important to monitor the customers' perception of quality, to see to what extent it may differ from the manufacturing facility's standards of quality. It is particularly important to monitor the amount of service required during the time the product is under warranty because these repairs cost the organization money. An analysis of the causes of these deficiencies may lead to changes in product design to reduce future in-warranty costs, possible product recalls by the government, as well as increasing future sales.

14.5 FUNCTIONAL ANALYSIS OF THE ORGANIZATION

Even if the management control system indicates that the current products are being manufactured and produced in accordance with available standards, improvements in these processes may still be possible. Remember that the audit basically compares current performance with existing standards. However, these standards are generally based on past performance or some past analysis of how an activity should be performed. Since conditions may have changed since that time (or will change in the future), look for ways of improving the various activities in the organization. One systematic way of looking for such improvements is to reexamine the planning process generally used to plan for an organization which manufactures and distributes products.[4] Figure 14.2 is a simplified model of the product planning process showing the various interrelated activities involved.

The process is best explained by starting with the analysis of user needs, considering:

1. Who are the potential customers and users of the product?[5]
2. What are the functions which the product will perform?
3. What are the deficiencies of the products currently available?
4. What resources do potential users have which could be used to obtain such a product or the service it provides?

Alternative sets of desired performance characteristics are then converted into specific product designs, including:

1. An operational concept for the use of this product.
2. The parts comprising the product.

Both output performance and the cost characteristics of the product are determined by the design of the manufacturing system used,[6] as determined by such factors as:

1. How well can the parts be fabricated?
2. How easily and quickly can the parts be fabricated?

For each product design a set of financial conditions of sale is postulated, somewhat based on the cost of manufacture and distribution and the

[4]The same process is also used for organizations providing services.

[5]The user is defined as that organization or those individuals who ultimately use or consume the product. The customer is that organization or individual who buys the product. Examples of customers who do not actually use a product include wholesalers, retailers, and manufacturers of parts bought for production.

[6]The distribution system influences product cost.

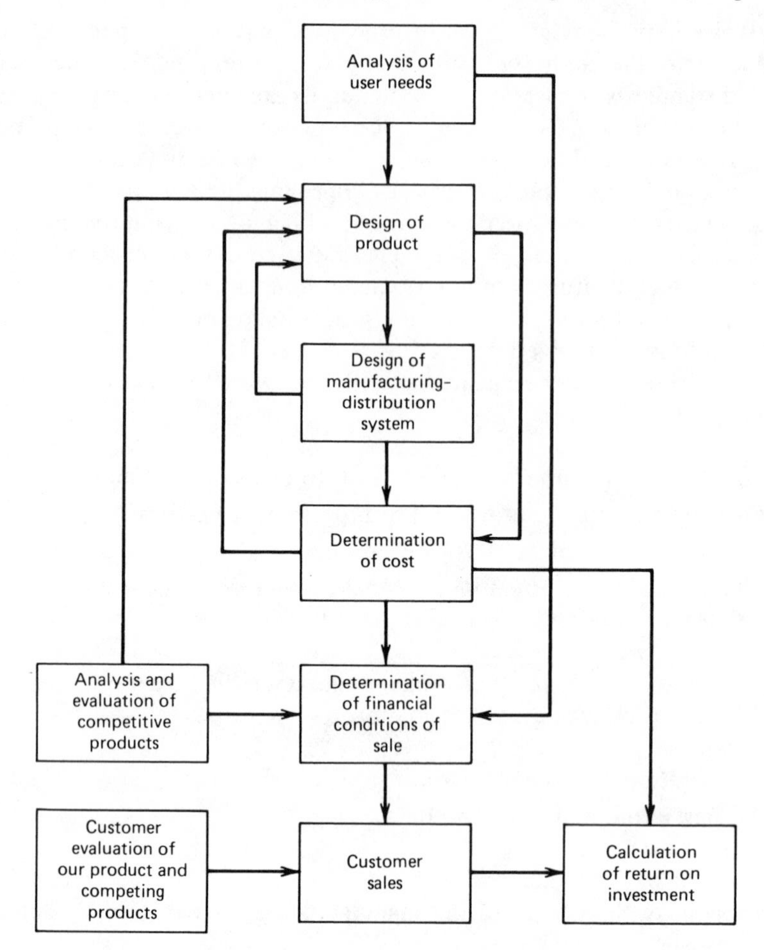

Figure 14.2. Product planning process.

financial conditions of sale of comparable competing products. This leads to an estimate of sales volume and sales revenue obtainable.

Based on this model of the product planning process, it is possible to partition the total improvement problem in such a way as to identify each of the degrees of freedom or key variables available to the product planner. This is modeled in decision tree form in Figure 14.3. These degrees of freedom (and constraints) can then be clustered into various combinations, resulting in the following scenarios:

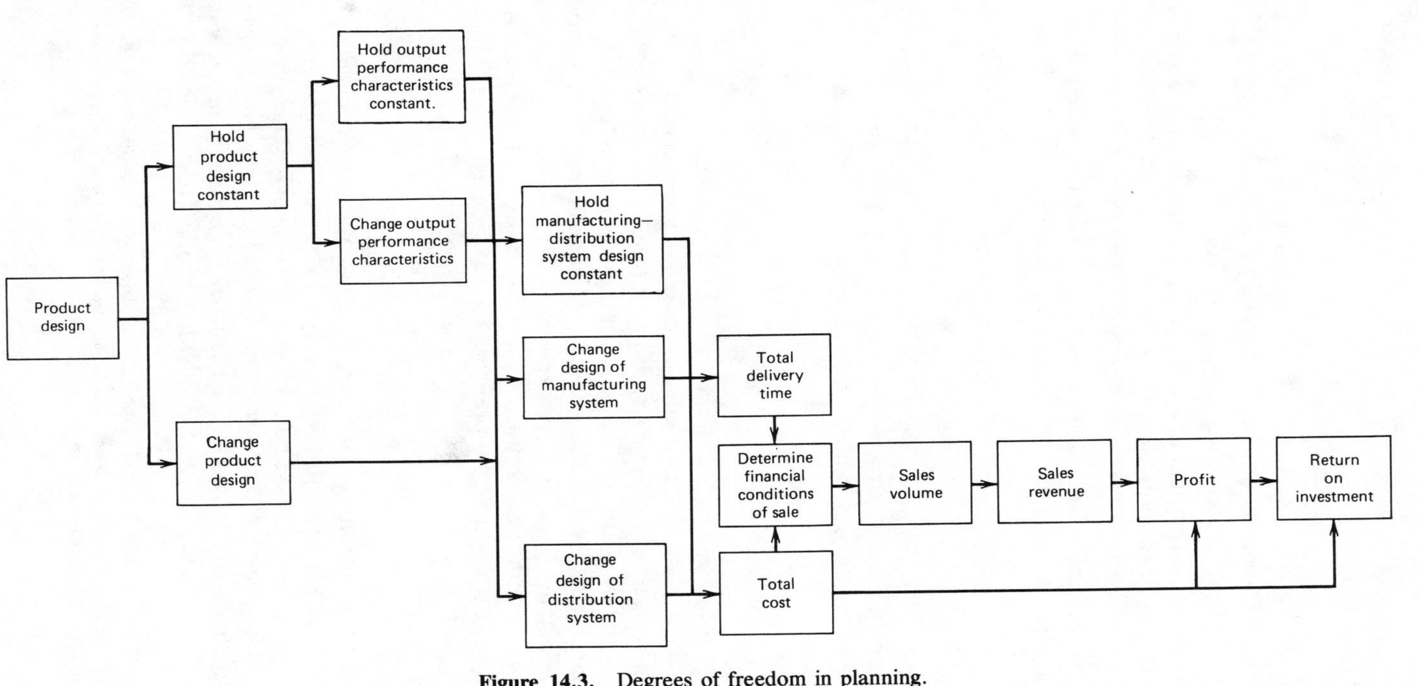

Figure 14.3. Degrees of freedom in planning.

SCENARIO 1

1. Keep all characteristics of the current products (i.e., user functions or product features, output performance, parts design) fixed.
2. Keep the present manufacturing equipment.
3. Analyze the current manufacturing systems to uncover ways that will reduce costs and increase profits.[7]

This is the traditional operations research approach in which only the operational procedures are changed.

SCENARIO 2

Scenario 2 is the same as Scenario 1, except that you also consider replacement of manufacturing equipment. This is the same approach as in Case 1, where new equipment alternatives were considered.

SCENARIO 3

Scenario 3 is the same as Scenarios 1 or 2, except that you also consider changing the design of the product to make it less expensive to manufacture. However, do not change any of the output performance characteristics of any of the existing products (i.e., product features, reliability, etc.). This is the case of engineering redesign, using methods of value analysis and value engineering. Finally, expected sales revenue and cost determine the profitability and return on investment for each product alternative being analyzed, and the preferred product design selected.

SCENARIO 4

This scenario concentrates on changing the output performance characteristics of the product (providing either improved performance at the same or higher cost, or less performance at less cost). This scenario also includes adding new products or dropping products which no longer have sales attractiveness. Now the higher-level objective is no longer cost reduction, but increasing the profitability of the organization.

[7]It is possible to reduce costs in ways that also decrease profits, as by drastically reducing marketing efforts, resulting in a drastically reduced sales volume and sales revenue; hence both criteria must be considered simultaneously in this scenario.

SCENARIO 5

This scenario attempts to increase the profitability of the organization by concentrating on the characteristics of the distribution system. Efforts are made to decrease distribution costs or to make it easier or more pleasant for the customer to obtain the product, thereby increasing sales volume and revenue.

Having defined the design constraints and variables comprising each scenario, each of the parts of the total manufacturing-distribution system will now be analyzed in turn, focusing on the key characteristics describing each activity of:

- Quality.
- Time.
- Cost.
- Financial conditions of sale.

Each is examined to see how it can be changed under the scenario to improve the situation being analyzed. Factors that should be considered in "optimizing" the system design will be identified and discussed. Scenarios 1 and 2 are primarily concerned with product cost reduction, and will be discussed in this chapter. Scenarios 3 and 4 are primarily concerned with product redesign and will be discussed in Chapter 15, as will Scenario 5.

14.6 ANALYSIS OF SCENARIO 1: REDUCING COST OF CURRENT OPERATIONS

Here the primary objective is to reduce the cost to manufacture current products to customers, keeping the product characteristics constant.

14.6.1 Reducing Production Costs

Each activity of any production process may be structured in the form of the operational flow diagram of Figure 14.4, and containing the following elements:

- An input store where input to the production activity is kept ready so

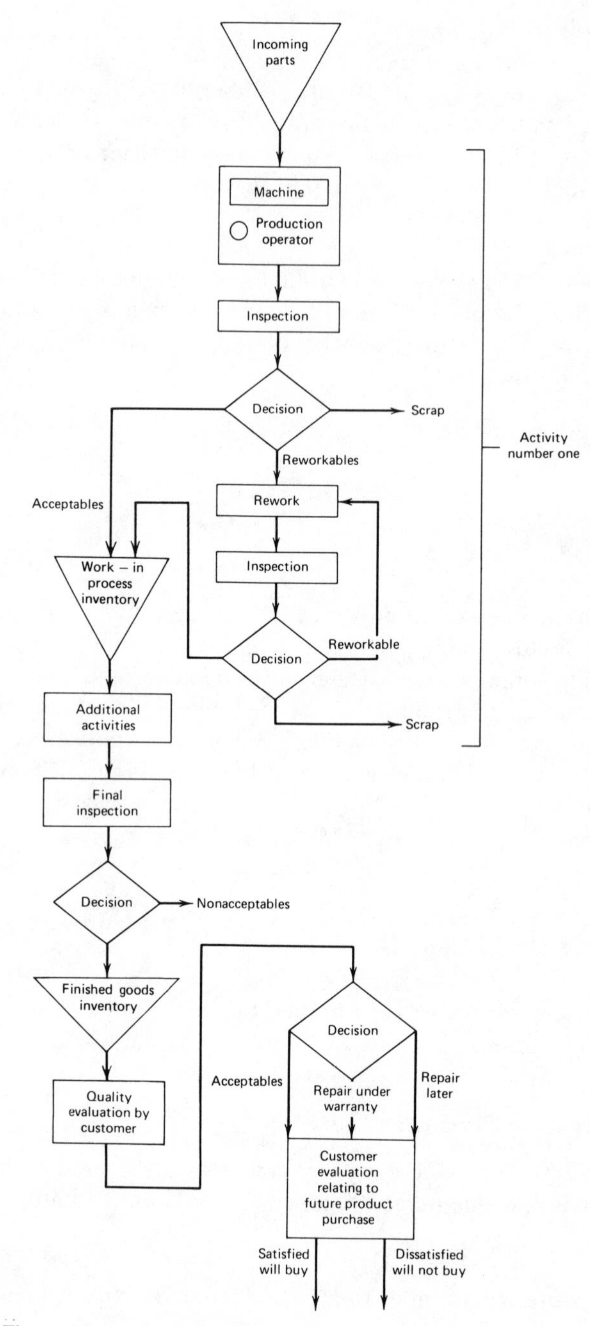

Figure 14.4. Operational flow diagram for production process.

that there will be no time lost when the worker has finished working on one unit and is ready to begin the next unit.

- A man-machine production activity.
- An inspection of the completed output is made (generally by the worker—sometimes by an independent inspector), to determine whether the output meets the acceptance standards. This inspection results in the unit being classified as acceptable, scrap, unacceptable but reworkable (meaning that it can be transformed into an acceptable unit by using an amount of resources less than the incremental cost of producing a new unit).
- Following the entire process, all acceptable units are placed in an output store for delivery to the next activity or, if this is the last activity, for delivery to the customer. In the former case, the store is called work-in-process inventory. In the latter case, the store is called finished-goods inventory.
- All scrap is placed in some store for later disposal.
- Occasionally, a subsequent inspection reveals that a unit originally classified as acceptable is actually not, and must be reworked or scrapped. One example of this is an automobile that the buyer later finds to contain a manufacturing defect and returns to the dealer for repair under warranty. Sometimes the manufacturer issues a recall of a particular model when failure statistics indicate a safety problem may exist.

Based on this structure, each of the individual activities of our organization can then be analyzed for possible cost improvements. For each case considered, we shall identify the key factors and show how to find the proper design that will achieve the lowest total cost for the organization.[8]

14.6.1.1 Focus on Time Losses

One of an analyst's important tasks is to observe sequential activities to see how the total time is being used. Here are examples of possible time reductions that would result in fewer man-hours being required for a unit of production.

Operator Setup and Teardown Time. During setup and teardown the machine is not engaged in active production. Hence any reduction in these times would reduce production costs. Besides seeking ways of reducing the times to do each setup or teardown, the planner should also

[8]Sometimes there may be other constraints which require the manager to pay more than lowest total cost. However, he should be aware of how much extra cost this constraint requires so that he may judge if the constraint is really worth this extra cost.

try to reduce the number of times these functions need to be done. Batch production can be accomplished by collecting similar jobs that might ordinarily be done over a period of one week or one month and doing them at one time, maintaining a stock of work-in-process inventory that can be drawn on as need arises. Consider the following example:

A major electrical manufacturer was producing medium-sized electrical transformers. One of the operations consisted of producing the transformer cores. This involved winding heavy paper around a mandrel until the desired thickness of the core was obtained. The edges were then trimmed to the desired width. Since each transformer was a custom design for one customer each core production operation was considered unique and the worker was paid for one setup (mounting the proper-sized mandrel), one operation (winding and trimming edges), and one teardown (removing the mandrel). What management did not know was that the worker had his own information system. He found out from the production scheduling unit that next week he would be scheduled to make a core using the same mandrel. Hence, instead of winding one core, he wound two, one after the other (the width could still differ) and kept his own work-in-process inventory until the next week. Thus he was paid for two setups but he actually did the job with one.

Analysis of the flow of parts through the production system over time would have revealed these similarities and resulted in a redesign of the process to reduce the number of setups and teardowns.[9] Obviously, such savings are obtained by spending some money for work-in-process inventory. As most books on operations management[10] indicate, the analyst can apply the economic production quantity formula to this problem to calculate the number of parts that should be produced from one setup. Further details of this approach will be described in Section 15.3.4.

Obtaining Work Inputs to the Activity. It is essential that input materials always be available to the worker for his activity. If not, the worker will have to wait for these materials and thus lose productive time. When the production line consists of a moving conveyor line, where the work moves from one activity to the next, as in automobile manufacturing, a major production system design problem is to make certain that there is enough time at each station, otherwise product quality will suffer. In an unbalanced line some workers have too much time available, an implicit additional cost.

[9]Providing suitable employee incentives might also have provided the organization with this cost reduction information.

[10]As examples see E. S. Buffa, *Modern Production Management,* Wiley, 1969, and J. L. Riggs, *Production Systems, Planning, Analysis, and Control,* Wiley, New York, 1970.

When work stations are stationary, activities should be decoupled from one another by using a work-in-process inventory to make certain that there is no waiting for work from the previous station. Of course, having a set of inventories also requires inventory control, including questions of where to store and safeguard the inventory—at the input to the work stations, or away from the work area. If the latter case is used, material handlers have to move the work in process in and out of the work area. Thus the analyst must also be concerned with equipment layout and the distance over which the materials move, since this will require time and incur movement costs.

Worker Idle Time While Machine Is Operating. Sometimes a worker is required to start and stop the machine but needs to monitor it only occasionally during its operation. Under these conditions an operator can operate more than one machine.[11] However, the machines must be located within close proximity of the worker if he is to monitor their operations and minimize the time spent in walking between the various machines to which he is assigned. This may require changes in the layout of the machines.

If worker utilization is low, try to find other jobs within that worker's job rating that must be done.[12] Having a worker who is "cross-trained" to handle several jobs is also advantageous for these situations.

Worker Idle Time (Taking Breaks). Occasionally, some workers abuse the traditional time off for coffee breaks and such abuse may spread to other workers. However, rigid enforcement of work times, use of time clocks, and so on may affect employee morale and may also reduce worker productivity. (It may even produce sabotage.) Perhaps the best approach is to establish reasonable work quotas, informing the workers of what is expected of them. In this way each worker can regulate his own production rate. As long as he meets his quota of acceptable units each day he is allowed to set his own work schedule.[13] This also helps to increase production quality. Incentive pay such as profit sharing often provides financial rewards for exceptional production. However, this did not work for one organization that employed working mothers. A better incentive system for these workers was to allow them to leave early when they finished their daily quota, so they could be home when their children

[11] As described in the fleet center operation of Case 3 (Chapter 10).

[12] Recall in Case 3 that the transmitter and receiver stations at Italy required two workers per shift for safety reasons, but there was not enough work for two radio operators. This utilization problem was solved by using an electronic technician as one of the workers; he could do required maintenance work during the shift and still be available to handle peak transmitter or receiver operations as they arose.

[13] Case 2 describes a management control system which incorporates this approach.

finished school. In this case time off was of more value to these workers than extra money. "Flexitime" (flexible hours on the job) is another fringe benefit workers seem to prefer.

Worker Idle Time During Equipment Malfunctions. This will be discussed under maintenance considerations, Section 14.5.1.6.

Other Worker Idle Time. Certain jobs may not require the assignment of a full-time worker. Having a worker who is "cross-trained" to handle several jobs is advantageous in these situations, since this reduces potential idle time.

Higher Than Normal Employee Absenteeism and Turnover. This is an extra cost since less experienced substitute workers generally operate at lower production rate and yield. Also, the extra cost of hiring, training, and severance adds to the total cost to the organization. Here, we need to balance these additional costs against the total wage and benefit package offered. Sometimes a benefit that employees like, such as flexitime or the other benefits mentioned previously, does not cost anything and increases employee morale. Worker attitude surveys, which may uncover some incentives that make the job more satisfying, may help here.

14.6.1.2 Production Quality Considerations

The previous discussion focused on the characteristic of time. The analyst can next focus on the quality of the products. There are a number of factors that should be examined and simultaneously traded off with respect to product quality to reduce the average cost of an acceptable unit. The operational flow diagram of Figure 14.4 helps in identifying these factors. Basically there are two major criteria in designing the quality aspects of the production system:

1. What is the average cost to the organization to produce one acceptable product unit? Here we must include not only the total cost of originally producing all units but also the cost of inspection, rework, loss of unacceptable units, scrap disposal, and product repair cost under warranty or under product recall by the company.
2. How many of the original customers are sufficiently satisfied with the product that they will purchase another one at a later time? The customer's evaluation of the product influences future sales (by himself and his friends) and, hence, future profits.

Here are the factors that must be considered in designing this part of the system:

Designing the Product Inspection and Quality Control (QC) Jobs. Proper quality control and production inspection methods provide the benefits of (1) motivating the production workers to improve their level of product acceptability and (2) decreasing the number of unacceptable products reaching the customer. However, as the effectiveness of the quality control program increases, so does its cost. Thus the planner must relate these two factors to find the preferred design of the QC system.

Basically, design of the QC program involves the following steps:

1. First determine alternative sets of quality standards that could be used by the company's inspection team in accepting or rejecting a product after it has been manufactured.
2. For each set of product acceptability standards, determine the method to be used to verify that each standard has been met (i.e., visually, by using measurement instruments, etc.) and at what stage in the production process this should take place.
3. Find the accuracy of the method being used to determine product acceptability (i.e., how well it identifies unacceptable units). This may be determined experimentally.
4. Determine the effect of the measurement on the production process (i.e., what increase in acceptability rate is obtained using the measurement as compared with not having it).
5. How important is the inaccuracy in measurement (i.e., the proportion of unacceptable units recognized as such by the customer). How many are returned for warranty repair, and at what cost?
6. What is the cost of performing the quality inspection as defined?

Using data describing the frequency and cost of corrective actions taken both during production as well as through the customer warranty period, this approach can be used to relate the cost of additional inspections to the gains obtained from the inspecting. The gains take two forms: (1) the decreased number of customers who encounter the unacceptable characteristic and (2) the reduced cost of defects (reductions in both production costs and warranty repairs). Obviously, all inspections in which total costs are reduced are worthwhile.

Finally, as indicated previously, it is also important to choose the proper place for the inspection (between the first point that the defect can occur and the final inspection). Consider the two extremes with respect to a defect that would result in scrapping the product. If the inspection occurs at the end of production, all the labor and material costs subsequent to the activity where the defect occurred are wasted, whereas if it occurs immediately after the activity, only the costs of labor and materials through that point are wasted. Thus the earlier the inspection, the less the waste. However, each inspection costs money. Also, several inspections may be combined at some "economy of scale." This is another problem whose solution is amenable to analysis: What inspections should take place and at what points during the production process?

Sometimes it may be the deliberate policy of a company to do only a minimum amount of inspection, but to provide the customer with a liberal warranty period within which he can bring the unit back for exchange or service. For example, vacuum tube producers generally perform only lot sample tests and depend on the customer to do final testing at the retail store (or in his TV or radio receiver), and individual unacceptable tubes are detected in that way. Exchanges are then readily provided. The automobile industry may also employ this strategy to a certain extent in servicing manufacturing defects during the warranty period. Such a policy may be advantageous in reducing costs. Actually, the only way to confirm this is to find the *total* costs of unacceptables, including warranty costs, and devise an inspection policy as previously described that considers such total costs.

However, this minimum inspection policy may cause customer ill-will because the customer's time and effort are required to return the product for service. In fact some customers may not even bother to return the product for repair. In both cases future sales may be affected. Furthermore, some product deficiencies may be catastrophic, such as for the individual electrocuted by a defective electric drill. Government recall of all units sold may also be involved if the condition involves a potential safety hazard. For these reasons it may be worthwhile to spend even more for quality and inspections than the monetary gain immediately available.

14.6.1.3 Measuring Customer Satisfaction

One necessary input in changing the design of a product (Scenario 4) is measurement of the customer's evaluation of the current product. Although many customers return a product for refund or repair under warranty, some who do not may be dissatisfied enough not to buy again or to influence acquaintances not to buy. Hence it is important that the company keep track of customer satisfaction through both market research and analysis of repairs done under warranty. The latter information is especially helpful in determining whether the production inspectors are doing their jobs. The former information is important in determining whether customers' expectations exceed production standards and hence if these standards are really irrelevant.

14.6.1.4 Determining Skills Required for a Job

Many times it is possible to subdivide the various labor tasks involved in a particular job so that some tasks can be done by lower-cost personnel. Several examples of this are:

- Using material handlers to move the input and output between operations instead of using skilled operators.
- Using hospital orderlies to move patients from their room to other locations (X-ray department, operating room, etc.) rather than higher-paid nurses.
- Having a hospital technician set up a large number of X-ray photos onto a special machine designed for stacking a large number of such photos. This permits the radiologist to concentrate on his primary function of diagnosis.

There is one pitfall the planner must avoid in analyzing such proposals. Is there a sufficiently large, justifiable work load that can fully utilize both of these workers? For example, if the radiologist were paid by the hour, or if only one radiologist is needed rather than two, the potential cost savings may justify the added expense. Any savings in the time of a more highly paid individual must be justified in terms of how the saved time will be used. Thus if the use of the lower-skilled individual only results in the radiologist having more free time, the extra expenditure would not be warranted (unless it can be shown that this extra time would result in a higher quality job being done).

14.6.1.5 Disposal of Scrap

Another cost that should be examined is the cost of disposing of all scrap connected with production. What should be done with unacceptable units? Can they be reworked, and, if so, at what cost? Can they be sold as "seconds"? How do we dispose of scrap materials? Sawdust, ash, old auto tires have long been a problem. With new environmental controls, disposal of materials such as iron-ore tailings and coal smoke will be an even greater problem than before. Can such scrap materials be used for some other purpose (e.g., converting sawdust into fuel pellets)? Here the challenge is to look for alternative uses of such materials, as well as lower-cost waste disposal methods.

14.6.1.6 Reducing Corrective Maintenance Costs

The basic objective of the maintenance function is to keep all the equipment in good operating condition at lowest total cost to the organization. Here total costs include not only those costs directly attributable to the maintenance function (such as maintenance test equipment, spare parts, maintenance labor), but also the indirect costs of all downtime associated with equipment failures. The various elements to be considered in the process of planning for the maintenance function can best be seen by first analyzing the time sequence of events involved in the repair of equipment (Figure 14.5).

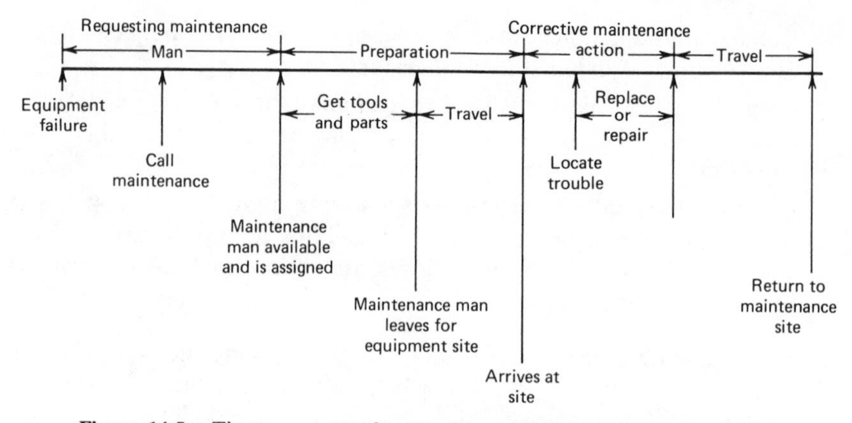

Figure 14.5. Time sequence of events in a corrective maintenance cycle.

Two basic times may be derived from these events:

1. *The total time required by maintenance personnel.* Any method of reducing this time is beneficial, since it tends to reduce maintenance manpower costs. Such methods include
   - Improving fault location equipment (automated type).
   - Improving communications to the maintenance man, so that he will know the location of the next maintenance job.
   - Adding more maintenance men, so as to increase their availability.
2. *Equipment downtime.* During the time the equipment is not operative the machine operator may be idle unless there is a spare machine or some other work that the operator may do while waiting. For this reason it generally pays either to have a spare machine available or to have the maintenance man replace the defective subassembly with a spare unit to minimize operator downtime. The defective subassembly may then either be thrown away or repaired back at the maintenance shop (sometimes while waiting for the next call). Such alternatives can be analyzed to determine which results in reduced total cost, taking into account lost operator time and the overtime required to return production to the appropriate schedule.

Since part of operator downtime is spent waiting for a maintenance man to be available to respond to the call, the planner must determine how many maintenance men are required. At a minimum there should be a large enough number available to perform all corrective maintenance actions.[14] This is the average manpower number. However, in addition there must also be an additional number to cope with peak loads caused by "bunching" of failures or a rash of higher than average number of parts to repair or replace.[15] Analysis must determine the minimum cost system, considering the cost of the idle time of operators waiting for the maintenance men and the cost of the idle time of maintenance men waiting for a service call. The ideal solution is to have work (such as preventive maintenance) that the maintenance men can do while waiting for a corrective maintenance call.

14.6.1.7 Preventive Maintenance to Reduce Corrective Maintenance Actions

Preventive or schedule maintenance (PM) consists of those maintenance actions periodically taken to prevent catastrophic failure or to extend the life of a piece of equipment. For example, one rule of thumb is that 90% of all tire flats and blow-outs take place over the last 10% of tire tread. Hence, to prevent such catastrophic failure, we replace tires when they reach a certain degree of wear. Because aircraft malfunctions can cause high losses in lives and property, the Federal Aviation Administration (FAA) requires many PM actions whose objective is to reduce the number of catastrophic failures. The design of nuclear power stations is another example in which conservative design techniques and PM policies are imposed as constraints to reduce the chances of a catastrophic failure.[16]

In addition, there are a number of PM steps recommended by equipment manufacturers that are directed toward increasing the useful life of equipment. We are all familiar with the various periodic PM actions involving automobiles: lubrication, changing of engine and transmission oil, tune-ups. The major objective here is to reduce the *total cost* of obtaining a given amount of equipment life (whether measured in total years available for duty or operating time). Here the total cost is the total maintenance cost (both CM and PM) including the cost of equipment overhaul and replacement, as well as the cost of downtime when the equipment is not available.

The objective of preventing catastrophic failures, which is primarily concerned with damage to human beings, is difficult to analyze. Here we are dealing with the trade-off of risk to life versus cost required to reduce this risk. Generally there is quite a bit of uncertainty in the data and

[14]See Case 3, Chapter 9, for details of this calculation.

[15]See Buffa, op. cit., pp. 60–67, 757–773, and Riggs, op. cit., pp. 333–342, for discussion of the use of waiting line theory in these types of problems.

[16]See W. D. Rowe, *An Anatomy of Risk*, John Wiley and Sons, New York, 1977.

opinions describing these relationships.[17] The objective of reducing total system life cost (not including injury or mortality) is common to all systems and is easier to treat systematically. Here the planner is concerned with this question of whether the proper amount of PM is being done or whether too much or too little is being done. One way of dealing with this is by comparing the costs of maintaining similar equipment at similar locations, as was done in Case 3. Recall that Norfolk's PM costs were uniformly much higher than at all other sites. Yet their CM costs were also much higher, and their availability (proportion of uptime) was no greater than at the other sites. This leads to the conclusion that too much PM was being done.

A second type of analysis that should be made is to relate the different types of failures requiring CM to the PM that is currently performed, and to investigate the result of further PM. Here the planner tries to determine whether the cost of additional PM actions is justified in terms of reducing CM expenses.[18] Buffa considers alternative PM policies in the following way[19]:

Simulation of Alternate Practices

When the maintenance is being performed anyway, it is a fairly common practice to replace parts that have not yet failed in order to prevent a future breakdown. The incremental cost of replacing these parts is often small since the machine is already partially disassembled. For example, if an automobile engine is disassembled to replace piston rings, other parts, such as the connecting rod bearings, also can be replaced for little more than the cost of the parts. If these parts are not replaced and fail later, the cost to replace them will be high because the engine must be disassembled again. Whether such practices are economical or not for individual cases depends on the distribution of part lives and the relative magnitudes of maintenance labor, part costs, and down-time costs. Because of the complexity of interacting probable lives of parts, simulation is often a practical way of evaluating alternative practices.

A Preventive Maintenance Example

Let us take, for example, the case of a company that maintained a bank of machines which were exposed to severe service, causing bearing failure to be a common maintenance problem. There were three bearings in the machines that caused trouble. The general practice had been to replace bearings at the time that

[17]See Rowe, op. cit. for an in-depth treatment of this topic.

[18]The Department of Defense also uses this type of data to identify those parts that fail most frequently. These data are used to determine which of these parts should be redesigned to reduce the number of such failures.

[19]Buffa, op. cit., pp. 607–611. Reproduced by permission. Also see 5th edition, 1977.

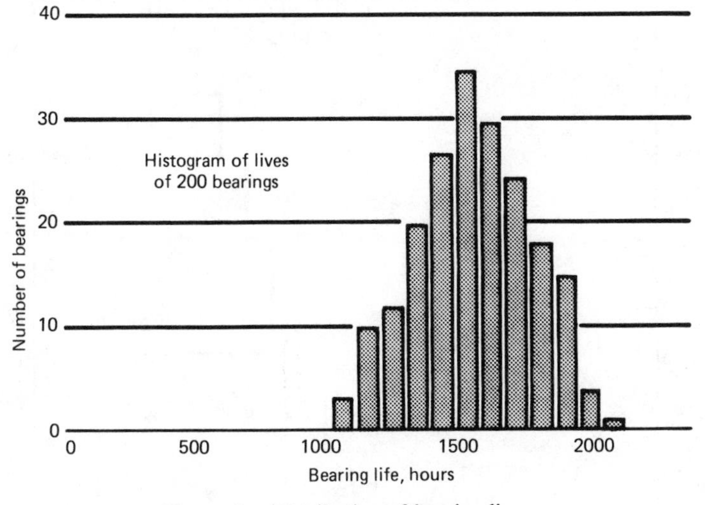

Figure 1. Distribution of bearing lives.

they failed. However, excessive down-time costs raised the question of whether or not a preventive policy was worthwhile. The company wished to evaluate three alternate possible practices.

1. The current practice of replacing bearings that fail.
2. When a bearing fails, replace all three.
3. When a bearing fails, replace that bearing plus other bearings that have been in use 1700 hours or more.

To simulate operation under the three alternate policies, data on bearing lives were needed, together with cost data. Figure 1 shows the distribution of bearing lives; Table I summarizes pertinent time and cost data.

To simulate the alternative maintenance operations, we convert the distribution to a cumulative distribution, establishing a percentage or probability scale, as in Figure 2. We can now use Figure 2 to select bearing lives at random from the

Table I. *Maintenance Time and Cost Data for Bearing Replacement*

Maintenance mechanic's time:	
Replace one bearing	5 hours
Replace two bearings	6 hours
Replace three bearings	7 hours
Maintenance mechanic's wage rate	\$3/hour
Bearing cost	\$5 each
Down-time costs	\$2/hour

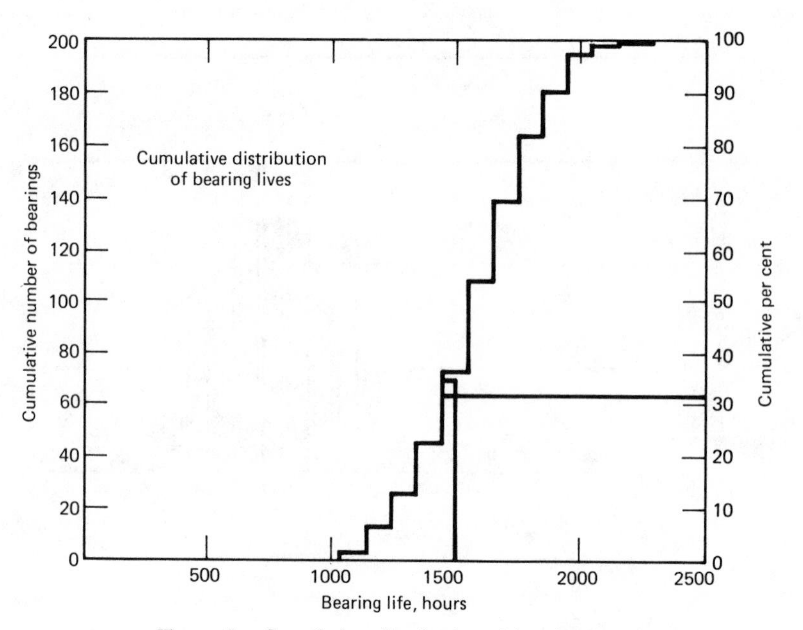

Figure 2. Cumulative distribution of bearing lives.

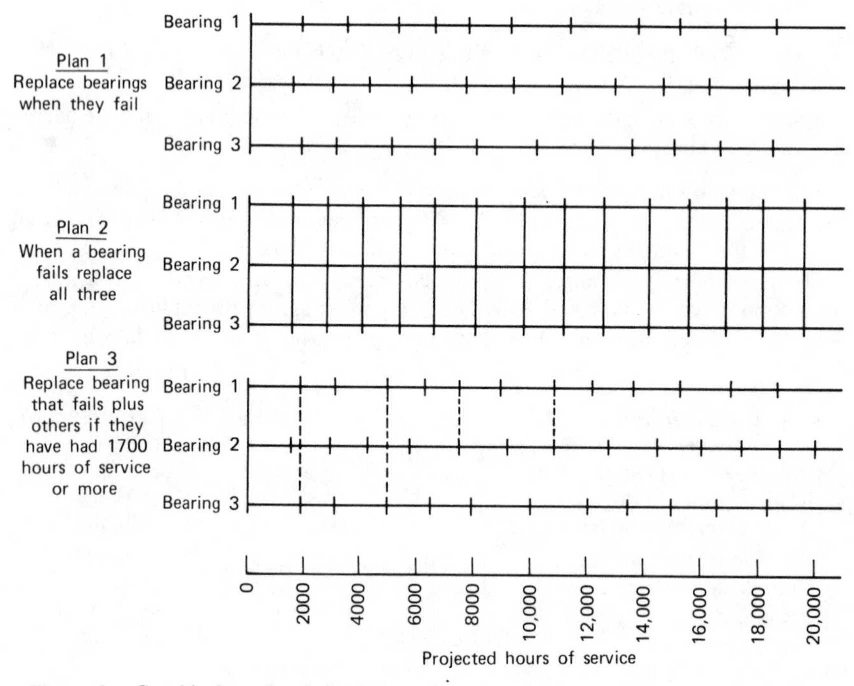

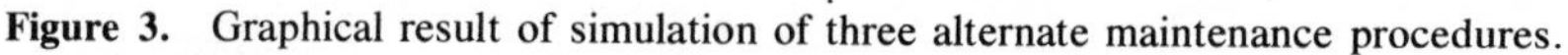

Figure 3. Graphical result of simulation of three alternate maintenance procedures.

Table II. *Results of 20,000 Hours of Simulated Operation under Three Alternate Maintenance Plans*

	Plan 1—Replace Bearings When They Fail	Plan 2—When a Bearing Fails, Replace All Three	Plan 3—Replace Bearing that Fails plus Others If They Have Had 1700 Hours of Service or More
Number of single replacements	34	—	28
Number of double replacements	—	—	4
Number of triple replacements	—	14	—
Total number of bearings replaced	34	42	36
Down-time, hours	170	98	164
Costs:			
Maintenance mechanics	$ 510	$294	$ 492
Cost of bearings	170	215	180
Down-time	340	196	328
Total	$1020	$705	$1000

distribution. By using a random number table, or some other system for selecting numbers at random between 0 and 100, we can proceed to select bearing lives and simulate operation. For example, the random number 32 selects a bearing with a life of 1500 hours, as shown in Figure 2. To simulate the first alternate practice, bearings are selected serially for the three positions and costs can be calculated according to the maintenance time, bearing cost, and down time that results. To simulate the second practice, three bearing lives are drawn at random, the shortest of the three determining when they will be replaced, etc. Figure 3 shows the resulting graphical representation of 20,000 hours of simulated operation for the three plans.

Table II summarizes the comparative results. Plan 2 is somewhat cheaper than the other two. Although bearing cost is higher, the time demand for maintenance mechanics is lower and, therefore, down-time costs are lower. Plan 3 is next best. Other plans between 2 and 3 could be tested as well, shortening the permitted running life of bearings below 1700 hours, etc. We should note that a change in the structure of costs could easily change results. For example, if the value of the new bearing is $100 instead of $5, Plan 1 is cheapest, the new totals being, respectively, $4250, $4690, and $4420.

Analysts should also consider a third situation in which the PM policies have been prescribed and it is desired to implement them at reduced total cost. Since the exact time for performing PM is not precisely fixed, it should be done when most convenient to reduce costs. This means that spare equipment should be available, if possible, as a means not only of minimizing both operator downtime and premium overtime to meet the required production schedule. If spare equipment is not available, PM should be done during operational off hours or so that operator downtime is minimized.

Since the loss caused by tying up an expensive jet airliner during normal flying hours is high, the airlines have worked out systematic procedures for reducing total cost, based on the following considerations. United Airlines, for example, has specialized its maintenance activities. Chicago and San Francisco are prime stations for the extensive maintenance activities that may require the aircraft to be down for an entire week. However, Washington, D.C., is the prime station for all PM done on all 727 and 737 aircraft. A PM schedule is generated listing each PM action to be taken as a function of flight hours and landing cycles (number of landings made). A computer keeps track of these occurrences for each aircraft and compares them against the PM schedule. As the time for a group of PM actions approaches, the routing department routes the last daily flight of the aircraft to Washington. Thus the aircraft may arrive in Washington about 9:00 P.M. It is then readied for work by the midnight maintenance shift (11:00 P.M. to 7:00 A.M.). A sufficiently large number of people are assigned to the shift so that the scheduled work plus any pilot complaints noted can be completed in time for the aircraft to be ready for operations by 7:00 A.M., the normal morning departure time.

14.7 ANALYSIS OF SCENARIO 2: INTRODUCING NEW EQUIPMENT

The objective of Scenario 2 is also to reduce the cost of manufacturing the same set of products. However, there is now an additional variable: changing any of the equipment used in the production system. Thus this part of the problem recognition phase becomes: "Is there any equipment that should be considered in order to reduce the total cost of producing the existing products?"

Many successful organizations assign responsibility for advanced planning to at least one individual in each functional area. This responsibility must include keeping up to date on the progress of new technology, new tools and devices that can improve operations in that area. The application of computers to each area is a good example of this type of planning.

Once new manufacturing equipment alternatives that appear to offer cost improvements are recognized, these alternatives may be evaluated using the same principles and approach used in Case 1. Hence no further discussion of this scenario will be given.

15

Other Ways of Improving an Organization

The major emphasis of Chapter 14 was to organize the planning effort and to recognize and generate various ways of changing the manufacturing-system so as to reduce costs. Chapter 15 describes three other types of improvements:

- Redesigning the product for the same performance but at a lower cost for manufacturing and distribution.
- Redesigning the product for a higher level of performance and a higher selling price, or a lower level of performance and lower selling price.
- Improving the distribution system.

The chapter then concludes by describing how the same planning principles apply to other types of organizations, such as service and nonprofit organizations.

15.1 ANALYSIS OF SCENARIO 3: REDESIGNING THE PRODUCT FOR THE SAME PERFORMANCE BUT FOR LOWER COST

The key principles of value analysis and value engineering apply to this objective. Basically, the analytical approach used in value analysis approximates that of systems design:

1. Examine the product's component parts and subassemblies.
2. Identify the *function* each part or cluster of parts performs.
3. Identify any alternative part or parts that can also perform the same function, but at lower cost.

The automobile industry has practiced this approach for years. Saving pennies on a part used in each of millions of autos can offer attractive increases in profits. One example is substitution of materials. When copper prices rose, many companies began to explore how many parts made of copper could be made from lower-cost materials such as aluminum or plastic.

Another approach that should be examined is the possibility of redesigning the product parts so that they may be purchased or fabricated and assembled at lower manufacturing costs. Using standard, readily available parts instead of fabricating special parts should also be considered.

Sometimes the objective is not reduced cost, but improving some other characteristic. The current trend in the automotive industry is toward reduction of the weight of automobiles to aid in increased mileage, a constraint placed upon the industry by the federal government. Autos are redesigned to make them smaller on the outside while keeping the same inside space and lighter-weight materials are used, as mentioned previously. All these techniques can also lead to cost reductions.

15.1.1 Product Structuring

The previous discussions have concentrated on methods of reducing the cost of producing an individual product. By enlarging the scope of the problem to include the entire product line, more opportunity for cost savings may present themselves. For example, the planner can redesign the entire product line so as to maximize the commonality of parts. An analytical technique useful in doing this is called *product structuring*. It entails constructing a matrix-type structure of the various parts and subassemblies comprising each product in the product line. In this way the planner not only can see how many parts and subassemblies are already common to different products but can actively plan to maximize the commonality of parts when designing new products or redesigning the product line. For years General Motors has done this with their entire line of automobiles. Thus although the fenders, radiator grills, and so on, may be different for Chevrolets, Pontiacs, Oldsmobiles, Buicks, and Cadillacs, parts that are not as noticeable (chassis, engine parts, etc.) may be the same. This produces a higher volume of parts production, which not only reduces the setups and teardowns required, but also provides many other economies. These include a smaller number of tools and dies to be designed and built, a decrease in training and unit labor time required as the volume of production increases. Greater commonality of parts also simplifies the entire logistics problem for the company, since there are fewer parts to keep track of and total inventory stocks can be reduced for the same level of demand. Greater commonality also aids the production planning process. Even though the demand for individual products may fluctuate widely over time, the demand for individual parts over time may "average out" and show fewer fluctuations.

The planner should also be alert for other opportunities for production standardization. In the transformer production example described in Chapter 14, the analysts found that over the entire product line there were four different ways used to insulate the transformer wires. These included wrapping insulating material around the wire or dipping the wire in varnish. Standardizing this production process led to cost reductions. Obviously, the different customers must be shown that the resulting products still meet their *functional* specifications. Some of these cost reductions may then be passed on to the customers to aid in this transition and may be useful in increasing sales.

15.2 ANALYSIS OF SCENARIO 4: CHANGING PRODUCT OUTPUT PERFORMANCE CHARACTERISTICS

The previous changes in the product design (Scenario 3) were directed toward reducing the time and cost of producing the product. Presumably the user of the product would not be able to detect these changes since they did not affect the performance of the product as the user saw it. We shall now consider ways of systematically recognizing changes that can be made to the product characteristics from the users' point of view that may result in greater profits to the organization.

15.2.1 Performing User Analysis

One of the first steps that should be taken in product planning is for the analyst to identify all he can about (1) who the end users of the product are (or will be); (2) how these users currently use (or will use) the product; and (3) who are the customers of the product.[1] This section will focus on analyzing the needs of the end user.

An important principle is to consider that the product is used in the operation of the user's "production system." This is literally true when machine tools or consumables are purchased by a manufacturing organization. It is equally true when a commercial laundry buys a washing

[1]As was mentioned in Chapter 14, a customer such as a home builder or distributor may not be the end user of the product, yet he may be important to the sales process, as will be described later.

machine. Similarly when a consumer buys a product such as a washing machine for home use, he or she operates the machine as part of a "home production facility" in a similar fashion to the commercial laundry.

Hence, for each of these cases, insights into the use of the product can be obtained by constructing an operational flow model showing the various activities needed to do the user's entire job. This is helpful in identifying

1. The time required for each activity and the manpower required by the user for doing the entire job for which the product is being used (obtained from the time required).
2. All other resources required, such as energy and materials.
3. The quality obtained from each activity in doing the total job.

Although the primary operational flow model constructed depicts the complete operation of the activity using the product, it is also helpful to extend it to include activities that relate to the primary activity, but occur before and after the primary activity. It may be desirable to redesign the product to include some of these activities, thereby increasing the product's value. Another operational flow model can be constructed showing all the activities that take place from the initial purchase of the product through its entire operational life of service until disposal. These activities would include (1) distribution, sales, and delivery; (2) maintenance, overhaul, and other support activities; and (3) final disposal procedures. Such analysis may generate ideas for reducing these resources currently required by the user during the total product life cycle.

Identification of users may be obtained by considering (1) geographical areas of users; (2) where they live (cities, suburbs, country); (3) income levels; (4) age. To aid in obtaining this information the analyst could trace the time sequence of events in one's life that impact on the product (i.e., birth, childhood, marriage, parenthood, retirement, death).

Now consider the following example of this type of analysis as it was applied by a large electrical appliance manufacturer for their home laundry products, specifically in their analysis of washing machines.

First, the analysts divided their market segments as follows:

1. Product environment or where the users live (e.g., apartments versus homes). In the case of the former, the space for a washer is generally quite limited. In addition, many apartments have coin-operated units in the basement for tenant use. Furthermore, apartments are generally in high-density areas that often have coin-operated or service laundromats nearby.

2. Product timing or when did the user purchase the product. Following the life cycle of a user from birth to death, the laundry load goes up after marriage (washing for two) and increases even more when children arrive. Washer sales go up at this time.
3. What is the user's income. Income is also related to product sales. Low-income people who live in apartments may not have the space for a washer. High-income people may use a professional laundry, but many also have a washer at home, used by a maid or housekeeper. Middle-income home occupants make greatest use of a washer.

Construction of the operational flow model depicts the various activities associated with the total washing process, including

1. Gathering of clothes.
2. Sorting according to color and fabric.
3. Delivery to washing machine.
4. Washing clothes, which includes subtasks of
   a. Entering the proper amount of water corresponding to the size of load, at the proper temperature for the selected bath.
   b. Adding proper amount of soap.
   c. Adding proper amount of bleach at the proper time.
   d. Water removal through spin cycle, including water entry for preliminary rinse.
   e. Water entry for final rinse cycle, including agitation.
   f. Adding fabric softener.
   g. Final spin-dry cycle.
5. Removing clothes.
6. Removing certain clothes for air drying.
7. Inserting remaining clothes in clothes dryer (or air-drying these also).
8. Removing dry clothes from dryer.
9. Ironing clothes requiring ironing.
10. Folding all other clothes.
11. Delivering clothes to storage facilities.

15.2.2 Improving the Functional Activities

Analysis of these functions can indicate certain measures describing the quality, time, and cost of each of these activities, as well as opportunities for improvement; for example:

1. Quality improvement could be investigated. Here the designer can compare the quality (i.e., degree of cleanliness) obtained from the three washing methods conventionally used:
   a. "Side-side" agitator type, which GE, Maytag, and most washers use.
   b. The "tumble" type used by Bendix.
   c. The "up-and-down" jet agitator type used by Whirlpool and Sears.
2. The water required for the load could be minimized. One advantage of the tumble type is the small amount of water required. Each of the other types can have an adjustment to match water to load size. GE also has a "mini-basket" that can sit on top of the agitator and be used for very small loads.
3. The temperature of the water could be automatically set to conform to the type of fabric in the load.
4. The time of each cycle could be adjusted for best cleaning or it could be treated as a variable adjustment.
5. Bleach and fabric softener could be stored internally and automatically added in the proper amount at the proper time.

Since the entire home laundry process was modeled, it is possible to detect opportunities for selling other products related to washing machines. For example, washers and dryers are generally designed as a set to encourage joint sales of the two (from a styling viewpoint). The GE combination washer-dryer was an attempt to package the two products as one unit to reduce floor space requirements (for apartment dwellers) and to minimize the handling required between the two functions.

Finally, analyzing the entire process may present other opportunities for improvement. For example, even with wash and wear fabrics, some ironing or "forming" may be needed. Load size probably meets the span of requirements. Reliability and maintainability always constitute a challenge for improvement (although the sealed transmission unit is guaranteed for a life of 5 years). Space is also a consideration for many people. To aim toward this market, Westinghouse once built a washer that could have a dryer mounted on top to minimize floor space.

The key principle here is that a functional analysis dissects the problem into key activities and their related characteristics. From this, the product planner can explore various ways of implementing each function to determine the performance obtained and cost required. Sometimes the appearance of higher product quality may be an illusion. Over the years, Chrysler Corporation always stressed its reputation for automotive engineering and quality, but seemed to lag in styling during the 1950s. Its sales increased substantially during the early 1960s when it added the 5-year, 50,000-mile warranty on its main drive trains. There is some question whether Chrysler autos were actually better. However, many people believed the auto was better, and this gave Chrysler a competitive edge.

This analytical approach started with the user's problems and jobs; another approach to creating new improvements is to start with solutions and look for a way to apply them to problems. For example, computers and calculators are rapidly decreasing in price. A product designer might ask, "Where can I use a calculating device in my product?" An electronics designer might also start with different locations where there might be a need for his product. For example, he might focus on each room of a house and see what might be used in each.

15.2.3 Would the Product Change be Worthwhile?

The next step in the planning process is to estimate how much it will cost the manufacturer, over the estimated sales life, to make these product changes. Such costs will include all development, investment, and operations and maintenance costs for manufacturing and distribution of the product. In addition, the financial conditions of sale should be set, as well as the expected resulting sales volume.[2]

Three types of changes in the product can be identified and may constitute improvements from a business point of view:

1. Increase the quality or timeliness of performance of the product (i.e., allowing it to perform its functions better or more quickly, or to perform more functions, at the same or lower financial conditions of sale). Since the product improvements would increase the benefits to the user at the same or lower selling price, product sales should tend to increase.
2. Increase the quality or timeliness of the performance of the product at higher financial conditions of sale (generally because the higher quality results in higher cost). This is like the previous example except the price has been increased. Obviously, the question to be answered is whether the new sales volume and sales revenue minus the new costs will result in higher profitability to the organization. An investment analysis similar to Case 1 (Chapters 4–7) can be performed to answer this question.

[2]Different combinations of "price" and volume may be examined.

3. Reduce the quality or timeliness of performance of the product at lower financial conditions of sale. This is the opposite of example 2. Now we may be appealing to a broader "mass market" with a lower priced product having a lower quality than before. Again the investment analysis must be performed to see if the product change is a profitable one.

In addition, for examples 2 and 3, the product planner must determine if the new product should replace the original product or merely be an addition to the product line. The latter approach, particularly when product structuring is used, may enable the organization to serve a much broader market at relatively small additional cost.

15.2.4 Hierarchy of Customers

The previous description of a user analysis focused on user needs and how the product could provide a service to the user. However, there are occasions when the ultimate user is not the original customer or buyer of the product. For example, home builders buy many products such as kitchen appliances on behalf of future users. This may present additional opportunities or new markets to a manufacturer. Consider this example. Carpet manufacturers have long sought to increase their market by selling wall-to-wall carpeting to new home buyers. However, some of these end users were reluctant to make this purchase for several reasons. First, they might not want to cover up new oak floors completely. Second, having just made the downpayment on the new home, the customer might be reluctant to spend more at that time. With the rising costs of new homes, builders were ready for various cost reductions. Thus the concept of providing builders with good carpeting that could be installed on top of plywood flooring was a way both of saving the cost of expensive oak flooring and of providing the home owner with a good-looking floor.

15.2.5 Synergistic Effects

Another technique is to extend your business into other areas that are related to yours and thereby to create new products. Texas Instruments and Fairchild Electronics have been leaders in the production of microelectronic circuits. Then Texas Instruments integrated forward and offered electronic calculators based on these components. Similarly, Fairchild manufactured digital watches using their components. Another example is provided by Bulova, long a leader in the production of accurate mechanical watches. Perceiving a growing market for even more accurate digital watches, Bulova added them to its product line. The company obviously felt it could not ignore these markets, and since the Bulova name is associated with a quality watch, it should be helpful in selling this type of product.

15.3 ANALYSIS OF SCENARIO 5: IMPROVING THE DISTRIBUTION SYSTEMS

In this scenario the incoming as well as the outgoing distribution systems are analyzed in the same way as was done for the manufacturing system in the previous scenarios; that is, reducing costs or improving their characteristics to increase profitability.

15.3.1 Analyzing the Incoming Distribution System

The incoming distribution system is sometimes called purchasing or procurement. Since it is analogous to the product distribution system, it is defined here as the incoming distribution system. The objective of the incoming distribution system is to provide the required number of different parts, supplies, and materials to the production and other systems at the time they are needed by the systems, and at lowest total cost to the organization. Thus timeliness and cost are the major characteristics involved.[3]

The key activities of the incoming distribution system are

- Ordering.
- Transport from vendor to the organization.
- Storage at the organization until needed for use.

Storage is required since materials are used at a rate determined by the production schedule, yet orders for these materials must be placed in larger lots or batches to minimize transportation and other costs.

In implementing the distribution function, the following questions must be considered:

1. How many of each type of part and material are needed over time?
2. What number should be ordered?
3. When should orders be placed?
4. What vendors are preferred?

[3]This assumes that the product designers have already selected quality specifications of each part or material.

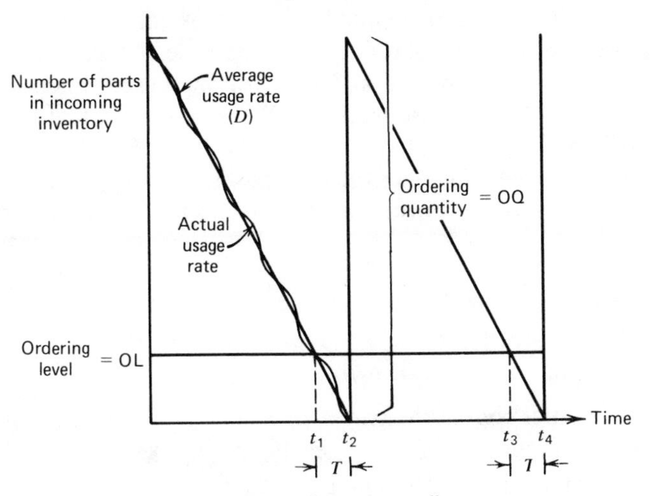

Figure 15.1. Inventory policy.

To begin the analysis of this problem, consider the following simplified scenario involving a demand rate for a certain part as illustrated in Figure 15.1. Notice that the number of units remaining in the incoming inventory is constantly reduced. At some point (t_1) the process of reordering these parts begins. Assume that it takes a total time of T for the entire ordering and delivery process and that this time is always constant.[4] If t_2 corresponds to the lowest level (zero units in inventory), the minimum order time (t_1) must occur when the inventory is at a level (OL) corresponding to the usage that will occur during time T, where $OL = (D)(T)$, and D equals the average demand rate for the part.

15.3.1.1 Determining Economic Ordering Quantity (EOQ)

The proper amount to order at one time (Q) involves a consideration of the following factors:

Holding Cost. The larger the size of each order, the larger the investment cost required (resulting in a reduced return on investment). In addition, this higher investment in parts may result in higher taxes, a requirement for more storage space, spoilage over long storage time, or obsolescence if the product is discontinued. Thus this holding cost will be at least as high as the standard discount rate used by the organization for investment analyses.

[4]Thus T includes the time for all activities, from placing the order to vendor filling order, shipment, incoming inspection, and storage at the inventory bins.

Ordering Cost. For a given number of units required over one year, if the ordering quantity (Q) is reduced to decrease the holding cost, the number of orders to be placed and followed increases. Each additional order requires additional resources and these resources (called the "ordering cost") must be established.

The value of Q that minimizes the total cost for ordering and holding the part is called the economic ordering quantity (EOQ), which may be found as follows:

- Number of orders per year $= D/Q$.
- Annual cost of ordering $= OD/Q$.
- Annual holding cost $= (H + iP)Q/2$.

  Where P = unit cost of the part and assuming that the average inventory is $Q/2$ (assuming $OL = 0$; see Figure 15.1).
- Thus the total annual ordering and handling cost is $OD/Q + (H + iP)$ $Q/2$.

The value of Q that minimizes the total annual cost may be found using calculus, by taking the derivative of this expression with respect to Q, setting this equal to zero, and finding the value of Q (defined as EOQ).

Thus

$$\text{EOQ} = \sqrt{\frac{2OD}{H + iP}}$$

Quantity Discounts. Many times price discounts are offered if the organization buys larger quantities at one time. In addition, shipping and handling costs are usually less as the ordering quantity increases. This factor may be taken into account as follows. First, calculate the EOQ without a discount and the total annual cost including the purchase price. Then determine the next higher quantity at which a price discount occurs and repeat the total annual cost calculation, comparing this against the previous cost calculations. This is repeated until the total annual cost begins to increase, indicating that the minimum total annual cost has been passed. The savings for the higher ordering level are then compared against the risk factor of ordering large quantities. This risk is a function of the stability of the market demand, other uses for the part, and so on.[5]

Other Considerations. There are several other factors that the analyst should consider. First, he may wish to achieve "economies of scale" through "centralized procurements." New York State, for example, an-

[5]See J. L. Riggs, *Production Systems, Planning, Analysis and Control*, John Wiley and Sons, New York, 1970, pp. 372–374 for further details.

nually conducts sealed bid procurements for all state needs in many areas, such as office supplies, food, and so on. In this way one contractor supplies all the state's needs for the item involved. In fact, the state merely indicates the number of units it intends to buy during the year. These are then delivered as needed, which saves on holding costs. In the same way, some companies buy centrally for all of their divisions. However, although the selling price may be lower, many times the service will not be as good as dealing with a local supplier. In either case, some competition should be maintained to minimize risks of not receiving materials on time. In the case of specialized subcontractors, some companies maintain competition by setting up two sources, giving the second-place competitor an order of 10 to 25% of the total requirement, with the winner getting the remainder. Additional procurement competitions are held periodically (say annually or as new orders are obtained by the prime contractor).

15.3.1.2 Determining the Proper Stocking Level Point

The initial analysis done to determine the proper level for reordering a part (OL) ignored several important factors:

- The actual demand may sometimes exceed the average demand rate.
- The actual lead time may sometimes be longer than the average lead time.

For this reason an additional safety stock level (SSL) must also be established to compensate for these uncertainties that may lead to running out of inventory. By gathering statistical data of these characteristics and using them to develop probability distributions, it is possible to determine for different values of OL how often the inventory would be out of stock.[6] This risk as well as the cost of being out of stock basically determines the proper total stocking level.[7] As an example, in Case 2, the Apex Company used asbestos fibers as the primary substance for making brake linings. Its primary supplier was a Canadian firm. One year a severe snowstorm prevented railroad deliveries, and the supply of asbestos was reduced to a very low level. Apex was then forced to dispatch its own truck to an alternative supplier to obtain the materials, preventing a possible production shutdown. They never forgot this occurrence and subsequently kept

[6]By having more accurate data on events affecting supplies, such as is obtained from modern information systems, safety stock inventories can be reduced without major risk of being out of stock.

[7]See Riggs, op. cit., pp. 378–382, for an example illustrating these principles.

their asbestos inventory level exceedingly high. However, the proper safety stock level should always be chosen based on the consideration of alternative supply sources and the additional costs required to use these alternative sources if needed.

15.3.2 Analyzing Outgoing Distribution System

The objective of the outgoing distribution system is to provide the number of produced units to the organization's customers at the proper time required and at lowest total cost to the organization. As such, it contains many of the same elements as the incoming distribution system, namely:

1. A finished-goods inventory that stores sufficient quantities of the final products so that they can be delivered to the customer in a timely fashion once an order is received.
2. A transportation system to move the product to the customers when ordered.
3. A control system that receives orders (and may also anticipate them), so that deliveries to the customer can be made in a timely fashion.

Several comments can be made about outgoing distribution systems:

1. A system of this kind is basically a storage and delivery system designed to deliver products to customers in an undamaged condition, within certain timeliness standards.
2. The distribution system may be designed in different hierarchical fashions. From the factory inventory (1) directly to the customer, (2) to a retail store that then stocks its own inventory for a customer, or (3) through a wholesaler or distributor who supplies the retailer.

Each of these designs has its own advantages and disadvantages. Among the latter is the extra cost involved as the number of distribution layers increases. Advances in data processing should aid in reducing the number of these distribution layers needed.

Many operations management and management science books deal with the distribution problem in terms of the amount of inventory required at each distribution level, type of transportation, and speed of response to provide a given level of service at lowest cost (or to increase the level of service at a given cost).[8]

[8]Also see B. H. Rudwick, *Systems Analysis for Effective Planning*, John Wiley and Sons, New York, 1969, Chap. 13, for dealing with an uncertain demand.

15.3.3 Production Planning

Having described the major elements of a manufacturing-distribution system and having considered various ways of improving the total flow of parts into a product to the customer, now consider the production planning and control function. The objective of this system is to control the entire production process to meet the delivery schedule at lowest total cost, taking into account risks and uncertainties. The following paragraphs describe how the "systems approach" can aid in implementing this objective.

The starting point is the systems demand function. Because of the time required for manufacturing and distribution, a forecast must be made of the number of units expected to be sold over time.[9] However, sales may vary so radically over time that it would be very inefficient for the production plan to follow the sales forecast. For example, this might require costly changes in manufacturing personnel during the year. Frequent large-scale hirings and firings could not be tolerated. Buffa describes alternative production policies for meeting the demand function and how to evaluate them.[10] Three alternative production plans have been designed to meet the forecasted sales demand for the year:

For the example of the production program, several alternate schedules for meeting sales requirements could be calculated, as shown in Figure 1. Each of the three alternate plans shown involves a hypothesis about the most effective way to meet requirements, using normal and overtime capacity, subcontracting, and seasonal inventories. The effects on incremental costs of the three plans is rather startling as is shown in Table I. Plan 1, level production, has no extra labor turnover cost, overtime premium, or subcontracting cost; however, the cost of carrying seasonal inventory to meet the high peak of sales in the summer is very large. Plan 2 involves some fluctuation in production level, but not so much that plant capacity with overtime work is exceeded. The result is a somewhat smaller seasonal inventory cost than for Plan 1 and a total incremental cost which is $20,000 per year less than Plan 1. Plan 3 tends to follow the sales requirement curve more closely and this involves considerable hiring and laying off of labor. Also, because of limitations of plant capacity, even with the use of overtime capacity, it is necessary to resort to subcontracting to meet the peak requirements. Nevertheless, the tremendous reduction in inventory cost makes Plan 3 the cheapest of the three plans. There are obviously other possible programs for which the total incremental costs could have been computed. For example, using a computer search programming model, we could determine a program that

[9]See Riggs, op. cit., Chap. 3, for forecasting techniques.
[10]See E. S. Buffa, *Modern Production Management*, 3rd ed. John Wiley and Sons, New York, 1969, pp. 51–53. Reproduced by permission. Also see 5th ed., 1977.

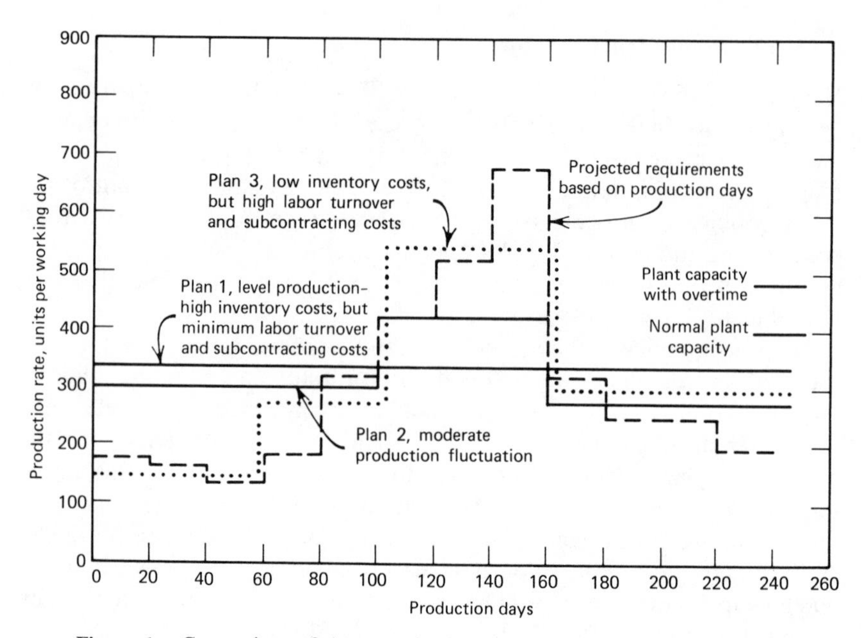

Figure 1. Comparison of three production programs that meet requirements.

Table I. *Comparative Incremental Costs for Three Production Programs*

	Plan 1	Plan 2	Plan 3
Inventory requirements, units	8000	6885	3691
Peak capacity required:			
(Plan 1 = 100)	100	126	165
Incremental costs:			
Seasonal stock cost*	$318,000	$239,000	$ 47,300
Labor turnover cost†	0	48,000	104,000
Overtime premium‡	0	11,000	44,000
Subcontracting cost**	0	0	57,750
	$318,000	$298,000	$253,050

*Inventory carrying cost computed at $60 per unit per year.
†A change in production rate at 20 units per day requires the employment or separation of 40 men at a cost of hiring and training an employee of $200.
‡Units produced at overtime labor rates cost $10 per unit extra.
**Units produced by subcontractors cost $15 per unit extra.

would approximate the minimum possible or optimum cost for the items of cost considered.

Weigh and decide what course of action to take based on a balancing of quantitative analysis and the nonquantitative factors in the situation. It might seem on the surface that the model should make this step unnecessary. After all, the measure of effectiveness is supposed to measure the effect of alternate courses of action. But does it do this completely? Usually, measures of effectiveness are not capable of reflecting all aspects of performance. There will be nonquantitative factors which must be given weight. In the production programming example, one factor which might be given heavy weight in today's economy would be the amount of hiring and laying off to be tolerated. The organization has a social responsibility and also feels the pressure of labor, public, and community reactions. The decision maker must balance off the $65,000 per year advantage of Plan 3 over Plan 1 against the disadvantage of the labor turnover induced by Plan 3. Compared to Plan 2, which involves more moderate swings in production level, Plan 3 has an economic advantage of $45,000 per year. In addition, Plan 3 involves considerable subcontracting to meet peak sales requirements. Can quality levels be held if we work through a subcontractor? Is it worth the difference to use either Plan 1 or 2 in order to maintain closer in-house control over quality?

The generation and evaluation of alternatives are of extreme importance in the overall process because they give flexibility to the decision maker. Quite often, alternate solutions may have differences in effectiveness which are not large, but one alternative may deal with a nonquantitative factor in a more satisfactory way. One great trap in quantitative analysis is the siren song of the "optimal solution." People are often drawn to mathematically provable optimal solutions, but we must remember that such solutions face the often severe limitations of the definition of the measure of effectiveness. If E (effectiveness) is really all inclusive for a given situation, then judgment may not need to enter, but this is seldom the case. In the really large-scale systems with which we may deal, it is often impractical to think in terms of optimality because the system is so complex. In these situations we must think in terms of improving existing conditions rather than in terms of optimizing the overall system.

15.3.4 Economic Production Quantity

All the production situations considered thus far have involved continuous production of parts in a product. However, there are also situations in which a machine has a higher production rate than the effective production rate required. Hence this machine may be used only part of the time for manufacturing a particular product, and thus is available for manufacturing other products. Invariably there is some setup cost involved between production runs, which leads to the production planning question of how many units should be produced for each production lot. This involves determining the balance between setup costs and holding costs of the inventory that must be stored between production runs. Notice the similarity between the previous problem of determining the economic ordering quantity for inventory control and this problem of determining the economic production quantity (EPQ).

Riggs derives the formula for EPQ and uses this formula in the following problem.[11]

Economic Production Quantity

The conditions for instantaneous replenishment of supplies are modified slightly when the supplies are manufactured on order rather than shipped from a stockpile of already manufactured items. The difference is the supplies are shipped instantaneously *as* they are manufactured. This means items are used during the replenishment period as represented by the sloping lines rising from each reorder point in Figure 1.

The principal expense of procurement is the set-up cost when a firm produces its own supplies. The inventory pattern in Figure 1 shows production beginning the moment supplies on hand are exhausted. In practice, the reorder point would be set at some inventory level above zero to notify production that supplies soon would be needed. This lead time should allow sufficient leeway for scheduling the set-up procedures.

The replenishment period, t', is the length of time required to produce the economic production quantity, *EPQ:*

$$t' = \frac{Q}{M} = \frac{\text{quantity ordered}}{\text{production output per day}}$$

When D and M are stated in daily rates, the inventory level increases each day during the replenishment period by the amount $M - D$. The stock on hand reaches its peak at the end of the replenishment period where:

$$\text{maximum inventory level} = (M - D)t' = (M - D)\frac{Q}{M}$$

$$= \left(1 - \frac{D}{M}\right) Q$$

Then,

$$\text{average inventory level} = \left(1 - \frac{D}{M}\right) \frac{Q}{2}$$

[11]See Riggs, op. cit., pp. 374–375. Reproduced by permission.

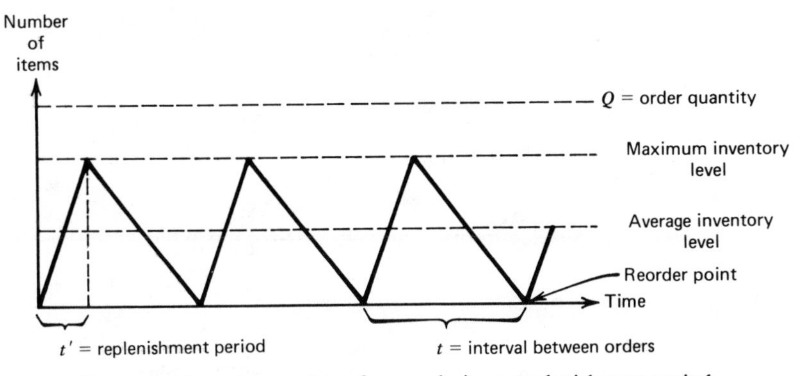

Figure 1. Inventory pattern for use during a replenishment period.

which makes

$$\text{total annual } EPQ \text{ cost} = \frac{OD}{Q} + \frac{(H + iP)(1 - D/M)Q}{2}$$

where O includes set-up costs and P is the production cost, and leads to

$$Q = \sqrt{\frac{2OD}{(H + iP)(1 - D/M)}}$$

Calculation of an economic production quantity

One product produced by Moore-Funn Novelties is a voodoo doll. It has a fairly constant demand of 40,000 per year. The soft plastic body is the same for all the dolls, but the clothing is changed periodically to conform to fad hysterics. Production runs for different products require changing the molds and settings of plastic-forming machines, new patterns for the cutters and sewers, and some adjustments in the assembly area. The production rate of previous runs has averaged 2000 dolls per day. Set-up costs are estimated at \$350 per production run.

A doll that sells for \$2.50 at a retail outlet is valued at \$0.90 when it comes off the production line. Complete carrying costs for production items are set at 20% of the production cost and are based on the average inventory level. From these cost figures, the economic production quantity is calculated as

$$Q = \sqrt{\frac{2OD}{iP(1 - D/M)}} = \sqrt{\frac{2 \times \$350 \times 40{,}000}{(0.20 \times \$0.90)(1 - 40{,}000/400{,}000)}}$$

$$= \sqrt{172{,}840{,}000} = 13{,}146$$

where

$$M = \frac{2000 \text{ dolls}}{\text{day}} \times \frac{200 \text{ days}}{\text{year}} = \frac{400{,}000 \text{ dolls}}{\text{year}}$$

Using the calculated Q value, production can anticipate:

$$\text{number of production runs per year} = \frac{D}{Q} = \frac{40{,}000}{13{,}146} = 3$$

$$\text{length of production run, } t' = \frac{Q}{M} = \frac{13{,}146}{2000} = 6.6 \text{ days}$$

and warehousing can expect:

$$\text{maximum inventory level} = \left(1 - \frac{D}{M}\right) Q = (1 - 0.1)(13{,}146) = 11{,}831$$

15.3.5 Improving the Output Distribution System

The output distribution system can also be modeled as a series of activities required to transport the product to the first customer. Several distribution layers from the factory warehouse to wholesalers or distributors to retailers may be involved. Each of these layers involves the functions of inventory storage, transportation, as well as control of the distribution process and its resources of product inventory, manpower, and equipment to implement, maintain, and support these functions. Hence the output distribution system can also be viewed as a production system, and the same approach for reducing costs as previously described may also be applied to the analysis of this distribution system. That is, given the existing set of products, focus one at a time on the quality, timeliness, and cost of the existing distribution system, and determine how each of these characteristics can be improved. Here the criterion is to determine the least costly way of meeting a given level of demand within acceptable time limits.

15.3.5.1 Improving Quality and Timeliness of Distribution

Quality of distribution may be interpreted in several ways. First, a product should arrive at the final destination in an undamaged state. The analyst should examine the records of how much damage in shipment or spoilage in storage has occurred in the past, find out where the responsibility lies, and determine what remedial action is necessary.

Quality may also be interpreted as the ease and pleasure with which a customer can obtain the product. For example, some Middle Eastern stores serve tea when the customer first enters the store. Some American stores serve coffee. Some women's fashion stores use live models to display their fashions. Some restaurants also have models displaying fashions. Some stores, such as Bloomingdale's and Neiman-Marcus, have built reputations as "fun places" to shop and as stores that provide unusual merchandise. The "old-fashioned" hardware store carries a wide variety of stock and provides the technical advice needed to "do it yourself." Some stores deliver promptly to your house. Some auto service centers provide customers with a ride to work when they bring in their cars for repair; some provide free taxi service. Some auto service centers actually send out a service truck (as the appliance maintenance organizations do) to tune the engine, lubricate the car, and change the oil right at the customer's home or place of business. Many service centers provide coffee and T.V. while the customer waits for his car to be repaired.

"Shopping at home" is another improvement in distribution. Sears' catalog was a great step forward, particularly for those not living near a retail store. Sears has gained greatly from good product descriptions and a reputation for honesty and customer satisfaction. Ordering by phone or by the use of closed-circuit T.V. to view the "specials of the day" may be the way shopping of the future will be accomplished.

Several years ago, General Electric was faced with competition in its small pole transformer business. It would announce a price and then find its smaller competitors offering a price reduction of 5 to 10%. One counterstrategy the company found useful was to improve its distribution system instead of engaging in a price war. It set up stocking centers at various parts of the country and provided them with an improved distribution control system. This enabled the company to deliver orders of this type of equipment to electric power utilities within 24 hours after the order was placed. This enabled the utilities to reduce their inventory size drastically and still depend on rapidly getting the equipment they needed. The utilities felt that the improved service was worth the additional price.

15.4 OTHER TYPES OF IMPROVEMENTS

The planner should explore various ways of increasing sales revenue leading to increased profits. Here are a few of them.

15.4.1 Increasing Sales Volume

Many times sales volume may be increased through increased marketing efforts, sometimes accompanied by price reductions. Rexall's annual "two-for-one" sale on drugs is an example of such a promotional activity.

To analyze the effectiveness of such a strategy, the planner must consider these factors:

1. How much will sales volume increase?
2. Considering the reduced price, how much will sales revenue increase?
3. How much will marketing costs increase?
4. What are the incremental costs for the increased volume of the products? Within certain limits, as the volume of production increases, the incremental cost, and hence the average cost of the product, tends to decrease. This "economy of scale" is due to "learning" efficiencies on the part of workers, quantity discounts for materials, the use of high-production automated equipment and so on.
5. How much will sales volume increase after the sale is over? There is always some carry-over following a sale as new customers become acquainted with a store.

Sometimes this "loss leader" strategy is accomplished more subtly, by reducing the initial product investment price, but then making much larger profits on operations and maintenance sales. It was not too long ago that IBM customers could not buy punched cards from any manufacturer other than IBM. Only recently has film for Polaroid cameras been available from anyone but Polaroid. Following an automobile accident, replacement parts such as fenders, in general, must be obtained from the manufacturer. Many people obtain their auto and appliance service from franchised dealers who sell factory supplied parts to this "captive market" at profit margins which may be higher than that of the original product.

Sometimes the financial conditions of sale may be improved by changing other characteristics than selling price. More liberal credit terms such as low down-payments and extended payment lengths may be offered (often at a higher risk). Trade-ins on cars and even homes are offered as a sales inducement. Equipment leasing is a way of increasing sales, since it permits companies to reduce the total investment that would normally be required, thereby increasing their return on investment.

Thus the analyst must calculate the outcomes which may be expected from alternative pricing strategies and estimate which is thought to be the most beneficial one in terms of profit and other appropriate factors.

15.4.2 Obtain Additional Income from Idle Resources

Another opportunity for improving the profitability of the organization is by using idle resources to bring in additional revenue. To do this requires knowledge of the following characteristics:

1. What is the current utilization of all resources within the organization? As described in Chapter 14, this is one of the first analyses a planner should make, without which proper allocation of personnel and equipment cannot be made.
2. What can be done with the idle capacity? Several opportunities are available.
   a. Try to find some productive use for this idle capacity. Auto manufacturers maintain full production and then hold end of the year sales, giving rebates to the dealer and the customer to reduce end of model year inventories.
   b. Use the resource for some unrelated production. Selling idle computer time to other companies in need of it is such an example.

The obvious last resort consists of furloughs or termination of people and the sale of idle equipment.

15.4.3 Using Government Research to Aid Commercial Product Development

Another synergistic effect is to apply to commercial products the knowledge gained from development programs funded by the government. Although this is a fairly obvious strategy, not many companies have been successful in employing it. One reason may be because the government and commercial segments of the company are generally kept apart in most companies. IBM, however, is said to have been successful in sharing such information within its computer business. Planned effort is essential for success in this strategy.

15.5 APPLICATION OF THESE PRINCIPLES TO OTHER TYPES OF ORGANIZATIONS

This section describes how the principles of problem recognition and subsystem design previously developed for a manufacturing-distribution system (MDS) can be applied in the same way to other types of organizations. This can be done since, in general, every organization can be shown to contain the same generic types of subsystems and activities found in a MDS. Thus the major role of the planner is to relate each function from his own system to the corresponding function in the MDS and to apply the same principles of systems planning as described for that function of the MDS. Obviously, the planner must also consider any special constraints appropriate to the particular problem.

15.5.1 Defining Types of Organizations

For purposes of our analysis all organizations may be classified by the following major characteristics:

1. Is the organization's output a product or a service?
2. Does the organization collect revenues as a direct charge for its products or services?
3. If so, is the organization's objective to
   a. Make a profit?
   b. Break even (revenues minus costs)?
   c. Minimize its losses (revenues minus costs)?
4. If not, is the organization's objective to minimize its costs of providing a given level of services?

Thus the first step that a planner should take in applying the planning principles described in this chapter is to identify the type of organization he is dealing with in terms of these characteristics.

We next illustrate how to do this and then state a series of principles that will help show how to apply the previous principles of planning.

15.5.2 Profit-Making Organizations that Manufacture and/or Distribute Products

Profit-making organizations that manufacture and distribute products include large manufacturing-distribution firms such as General Electric. It also includes firms that only manufacture as well as those that only distribute, such as wholesalers or retailers. Planning for these organizations was described in Chapters 14 and 15.

15.5.3 Profit-Making Organizations that Offer Services

These are defined to include such organizations as the airlines, hotels, travel agencies, laundries, and profit-making nursing homes. In these cases the operational flow model contains all the activities required by this system to produce a service and provide (i.e., distribute) it to the customer. This service can be characterized in the same way as the product was using the four characteristics of

1. Quality.
2. Timeliness.

3. Cost.
4. Financial conditions of sale.

The objectives of the four scenarios previously described still hold as illustrated by the following examples applicable to airline companies:

1. The reduction of the costs of producing and distributing the service by redesigning the characteristics of the production-distribution facilities, keeping the characteristics of the service as seen by the user constant.
2. The introduction of cocktails, gourmet food, and movies in-flight to increase sales.
3. Increase in sales volume by offering "no frills service" at lower cost.
4. Selective price reductions, such as reduced fares for wives accompanying husbands and for children, and stand-by service for military personnel, students, or senior citizens, with a guarantee of a seat on the next flight if a stand-by seat is not available on the first flight. In this way the normal price to business people is not lowered; hence the airlines can concentrate on reducing the number of empty seats. Since the incremental revenue can far exceed the incremental cost of filling the seat, profits will rise.
5. Conversely, operation of the supersonic Concorde providing much faster service at a higher price.

15.5.4 Not-for-Profit Organizations that Offer Services

Two types of organizations are included in this category:

1. Those that collect revenue as a direct charge for the service.
2. Those that do not.

15.5.4.1 Revenue-Collecting Organizations

Not-for-profit organizations that collect revenue for their services include mutual insurance companies, nonprofit hospitals and nursing homes, the U.S. Postal Service, the Tennessee Valley Authority, and private universities. Each of these cases can be analyzed in the same way as the profit-making service organizations. These organizations have the same objective of reducing their cost of providing the service. Where these organizations differ from their profit-making counterparts is that the nonprofit organization is required to assign a sales price such that the sales revenue does not exceed the costs. Thus in the case of mutual insurance companies dividends are assigned so that no profits result. Sometimes the revenues collected do not equal the costs, as in the case of most universities and the U.S. Postal Service. In such cases a subsidy is required (alumni contributions or a U.S. government subsidy) to make up the difference. Thus the same approach to planning for improvements as that described in Chapters 14 and 15 also holds here:

1. Finding ways of reducing the costs of providing the service, including the use of automated machinery to reduce labor costs.
2. Changing the characteristics of the service offered versus the price obtainable so that the sales revenue will be sufficiently close to the total costs that any subsidy is within acceptable bounds. An example of this is the consideration given by the Postal Service to eliminating Saturday deliveries, rather than raising the postage cost, thereby reducing the subsidy required. Improving the service, such as guaranteed Express Mail Service, at a higher price, is another example.

15.5.5 Not-for-Profit Organizations that Manufacture and Distribute Products

An example of this is the charitable rehabilitation center that trains handicapped people in manufacturing skills and sells the resulting products. The same comments as those described for the not-for-profit service organization collecting revenue also apply here.

15.5.6 Not-for-Profit Organization Providing Services Without Direct Charge

Primary examples of the not-for-profit organization that provides services without direct charge are the various governmental agencies offering various services paid for through tax revenues rather than a direct "sale" of the service. Thus these organizations should be considered cost centers rather than profit centers. Hence a primary planning objective is reducing costs to provide the set of services. A more difficult planning objective is to determine whether a particular service is worth the cost required, since generally this is decided in an intuitive fashion comparing the increase in "quality of service" provided to the incremental cost required. The objectives of government agencies differ. Some, like the police, fire, and health departments, provide direct protective services to the people in their area. Some provide the management functions of planning and regulation of certain areas operated by the private sector. These include the Securities and Exchange Commission, Federal Trade Commission, Food and Drug Administration, Department of Commerce, and Department of Transportation. In these cases the planning for services offered by these organizations should include the benefits offered to the citizens served, including lives saved, injuries reduced, dollars saved, and so on. Unfortunately, it is difficult to determine what these benefits may be, using only analytical methods. Trying some of these proposed improvements on a trial or limited test basis can help in evaluating the benefits and costs involved. Such a test and evaluation phase is needed before full scale implementation is begun.

It should be noted that there are a number of governmental agencies that manufacture products for use by other agencies. These include:

1. The U.S. Mint, which manufactures coins, paper money, and federal bonds.
2. Government arsenals, which produce weapons.

Since these organizations do not collect revenue directly for their products, they are more related to this category than the previous one of a not-for-profit organization manufacturing a product.

The planning approach for this type of organization may be defined as follows. Initially,

1. Determine the set of product characteristics obtainable as a function of costs required.
2. Decide what set of product characteristics are acceptable and justifiable on the basis of
   a. Benefits obtainable as compared with costs required, if possible.
   b. Minimum acceptable product standards.

As periodic planning efforts occur, efforts should be made to:

1. Reduce the costs of providing the same type of product output characteristics.
2. Obtain a "more favorable" relationship between product output characteristics and cost.

16

Conclusions

16.1 SUMMARY OF KEY PRINCIPLES OF SYSTEMS PLANNING

Table 7.1, Section 7.7 offers a detailed listing of the various steps involved in systems planning. This can be utilized as a checklist to aid in the inclusion of the key factors to be considered during the planning process.

The most difficult problem a planner can encounter involves problem recognition, the subject of Chapters 14 and 15. How does a planner enter an organization, conduct an audit to see if the operations of the organization are under proper control, and determine what problems exist, so that replanning of the pertinent operations to produce improvements may occur? To implement these tasks, the following steps may be followed.

16.1.1 Obtaining Initial Perceptions of Problems within the Organization

There are several approaches that can be taken to obtain an initial insight into the problems of the organizations. The most direct way is by eliciting from various members of the organization their perceptions of problems facing the organization, either currently or within the future planning horizon under consideration. These qualitative perceptions should be obtained from

- The "patron" who has brought the planner into the organization to conduct the analysis,
- The general manager or the highest executive who is interested in the project and who is willing to meet with the planner to provide his perceptions of problems faced by the organization, and
- Individuals at lower levels of the organization who can provide such perceptions.

The process of obtaining ideas from these individuals may be helped if the planner subdivides his request into categories such as those mentioned in Chapter 3:

- Perceived deficiencies in the current products or services or the ways of producing them.
- Ideas for better ways of producing current products or services.
- Ideas for new products or services.
- In what ways are our competitors doing things better than we are?
- How would you spend a larger budget?

16.1.2 Forming the Planning Team

Another, more permanent source of planning information is obtained by forming a planning task force consisting of representatives of each of the functional areas of the organization being analyzed. Such an arrangement provides a greater continuity of the effort and, at the least, a "contact point" for obtaining information as required during the project. This type of organization can also provide active participants supplying good ideas to the effort. For the generalized case in analyzing a total organization the planning task force should include representatives of

- The primary "production" system.
- The incoming distribution system.
- The outgoing distribution system.
- Other support systems (as appropriate).
- The "new product planning and development" system.

16.1.3 Evaluating the Current Operations

One of the next questions to be examined is, "Is the current organization 'under proper control'?" Referring back to the Apex case of Part III and the discussion of management control systems, this question may be interpreted as, "Are the current operations being performed in accordance with an existing overall plan, which defines the objectives of each of the activities of the organization and provides the quantitative standards by which its outputs may be measured?" Specifically, such standards should include

- Quality of product or service.
- Timeliness or delivery schedule.

- Costs or resources required (manpower, materials, etc.).
- Financial conditions of sale.

The management control system should be detailed enough that it not only recognizes exceptions to the standards but also provides diagnostic information indicating why the exceptions are occurring, so that the proper corrective action can be taken. This, in itself, is a form of problem recognition. If the current management control system is not adequate, steps must be taken to improve it sufficiently so that the appropriate control data are obtainable. The Apex case can serve as a good example of a control system useful for evaluating product quality and timeliness, but also the labor and other resources, required for such production.

16.1.4 Setting Priorities

Once the initial deficiencies and opportunities have been identified they are clustered in terms of related problems, so that related "symptoms" can be considered together. These clusters are then identified as separate planning projects and ranked in terms of

1. Magnitude of the impact on the organization if the problem is solved.
2. Time that the benefits are apt to be obtained.
3. Planning resources required to solve the problem.

Each project is then initiated in order, depending on the time and analytical resources available. The planning process described in the previous chapters is then followed to implement each project.

16.1.5 Subsystem Analysis, Taking into Account Intrasystem Relationships

Another method for systematically recognizing problems (or opportunities for improvement) is to focus on each subsystem, determine what the generic subsystem, design containing all the functions involved, would consist of, compare the current design with this, and determine where improvements might be made. A number of examples of this type of analysis were presented. In each example the analysis was performed as follows:

1. Partition the entire system into a number of subsystems and define the subsystem being examined in turn.

2. Partition the entire planning activity into a number of key "planning scenarios," each corresponding to different degrees of freedom for making changes in the current and programmed system. These key planning scenarios are as follows:

   Scenario 1. Find ways of reducing the cost of producing the same products (or services), keeping fixed the product (or service) specifications and the equipment used in manufacturing and distributing the products (or services). Thus only the procedures used can be changed.

   Scenario 2. Same as Scenario 1 (i.e., reducing cost) but with the additional degree of freedom of changing any of the current manufacturing entities.

   Scenario 3. Same as Scenario 2 but with the additional degree of freedom of changing the design of the product to make it less costly to manufacture. However, the output performance characteristics of the product remain fixed as in Scenario 1.

   Scenario 4. Same degrees of freedom as that in Scenario 3 but with the additional element of changing performance characteristics of the product and its selling price. The new product may be added to the existing product line, or old products deemed obsolescent may be dropped. Now the higher-level objective is no longer cost reduction, but increasing the profitability of the organization.

   Scenario 5. Same as Scenarios 1–4, except change the characteristics of the distribution systems to increase the sales revenue, leading to increased profitability of the organization.

16.1.6 Subsystem Synthesis and Finding Preferred Alternatives

Each of the subsystems should be analyzed in turn, always trying to design the preferred system that would accomplish the objective of the subsystem at lowest total cost (or highest profit) to the total system. Such total system costs take into account not only the total cost of the elements of the particular subsystem being analyzed but also any costs to the remainder of the total system being affected by the subsystem design. An example of this is the cost of operational downtime caused by the repair time required by the maintenance subsystem.

Identification of such subsystem improvements is aided by focusing in turn on the characteristics of quality, timeliness, and cost in seeking to reduce the resources required to meet the objective. The characteristics of this ideal subsystem are then compared with the current subsystem to determine how many of the improvements can be made, considering the environment in which the actual organization is operating. In certain cases management chooses to operate under certain constraints (e.g., no high labor fluctuations, and a small finished-goods inventory, to minimize the risk of loss resulting from product obsolescence). In these cases the analyst might calculate how much each of these constraints is costing the organization as compared to what the preferred system, not containing the constraint, would cost. In this way the decision maker can judge whether the constraint is really worth its cost, or if some middle ground might not also be acceptable at lower cost.

Thus the concept of the ideal system and its formulation not only is an aid in identifying problems but is a design tool to indicate the types of improvements that are theoretically possible.

16.1.7 Coping with Risks and Uncertainties

Finally, the various risks and uncertainties associated with each alternative must be considered, as described in Chapter 7.

CONCLUDING REMARKS

The various chapters in this book have been devoted to describing the key principles of planning and to illustrating how they may be applied to a wide variety of problems. A number of generic problem types have been identified so that when the planner approaches a new problem, he may relate it back to the generic type whose approach to solution he is familiar with. This does not say that a "canned solution" will result, since it is the approach that has been standardized, not the solution. In fact, one of the first types of information the planner should obtain is the set of constraints under which the organization is forced to operate, since these are the primary determinants of what alternatives will be acceptable in the given environment.

At the very least, the experienced planner can use these principles as a checklist to supplement his own planning approach in an actual problem-solving situation. The beginning planner can use these principles as a guide in developing and evolving his own approach. The world has acute need for better planners and problem solvers. Hopefully this book has contributed to this objective.

Index

DATE DUE

JUL 1 '81			
JAN 9 '81			
MAR 1 '83			
APR 5 '83			
APR 26 '85			
APR 27 1991			
DEC 19 1992			
MAR 03 1995			